DIRECTORY OF SCIENTIFIC DIRECTORIES

REFERENCE BOOKS AVAILABLE FROM LONGMAN GROUP LIMITED

Longman Reference on Research Series

Agricultural Research Centres
Directory of Scientific Directories
Earth and Astronomical Sciences Research Centres
Electronics and Computer Research Centres
Engineering Research Centres
European Research Centres
European Sources of Scientific and Technical Information
Industrial Research in the United Kingdom
International Medical Who's Who
International Who's Who in Energy and Nuclear Sciences
Materials Research Centres
Medical Research Centres
Who's Who in Science in Europe
Who's Who in World Agriculture
World Energy Directory
World Nuclear Directory

Longman Guide to World Science and Technology

Science and Technology in the Middle East
   by Ziauddin Sardar
Science and Technology in Latin America
   by Latin American Newsletters Limited
Science and Technology in China
   by Tong B. Tang
Science and Technology in Japan
   by Alun M. Anderson

In preparation
Science and Technology in the United Kingdom
   by Anthony P. Harvey
Science and Technology in South-East Asia
   by Ziauddin Sardar
Science and Technology in Australasia, Antarctica and the Pacific
   by Jarlath Ronayne
Science and Technology in the United States of America
   by Albert H. Teich

DIRECTORY OF SCIENTIFIC DIRECTORIES

a world bibliographic guide to medical, agricultural, industrial, and natural science directories

Q145
A12
H336d
1986

Longman

DIRECTORY OF SCIENTIFIC DIRECTORIES: a world bibliographic guide to medical, agricultural, industrial, and natural science directories

First edition 1969
Second edition 1972
Third edition 1979
Fourth edition 1986

Published by Longman Group Limited, Professional Reference and Information Publishing Division, Longman House, Burnt Mill, Harlow, Essex CM20 2JE, UK
Telephone: Harlow (0279) 442601

Distributed exclusively in the USA and Canada by Gale Research Company, Book House, Detroit, MI 48226, USA

British Library Cataloguing in Publication Data

Directory of scientific directories: a world
   bibliographic guide to medical, agricultural,
   industrial and natural science directories. – 4th ed.
   1. Science – Directories – Bibliography
   I. Harvey, Anthony P.
   016.502'5   Z7405.D55

ISBN 0-582-90151-0

© Longman Group Limited 1986. All rights reserved; no part of this publication may be reproduced, stored in a retrieval system, or transmitted in any form or by any means, electronic, mechanical, photocopying, recording, or otherwise, without the prior written permission of the Copyright Holder.

Typeset by Stibo Datagrafik
Printed and bound in Great Britain by
Robert Hartnoll (1985) Ltd., Bodmin, Cornwall

PUBLISHER'S INTRODUCTION

The first edition of the **Directory of Scientific Directories** was published in 1969. Since then this publication has been the natural first point of call for enquirers about national, scientific and technical directories. This edition is the fourth and contains double the number of directories as the first edition. This edition contains about 2000 entries.

The aim has been to list directories of scientific organizations, individuals, classifications, collections, museums, libraries, guides to technical facilities, products and manufacturers, and year books to technical trades where the technical content and employment of scientifically trained personnel are high, such as the mining, water, and electricity generating industries. Some business and commercial directories have been included but only where they contain substantial listings of technical centres or individuals. Some extensive listings in journals have been included even where they are not separately available.

Overall national and international coverage has been an important criteria for our editors. Directories published prior to 1976 have been excluded, unless the editor felt that they were of overwhelming importance. Some out of print items are included where it is felt that these are important and have not been superseded, or where details of a forthcoming edition have been received.

This publication will appeal to market researchers, laboratory equipment manufacturers, scientific journalists, patent agents, information scientists, marketing specialists, and libraries holding large scientific and technical collections.

Directory of Scientific Directories is divided into seven territorial chapters. Individual directories are allocated to the territorial chapter on which their main subject matters refers. Directories which provide information which should fall into several chapters, are found in the 'International' directories chapter.

Within each chapter directories have been placed into the following subject areas:

General science, incorporating all general science directories, including those of societies, institutions, organizations, documentation centres, libraries, higher education institutes, meetings and museums which cover more than one of the subject groupings given below.

Agriculture and food science, incorporating horticulture, veterinary medicine, forestry, fisheries and aquaculture, plant and animal production.

Chemical and materials sciences, incorporating pure chemistry, industrial chemistry and chemical process engineering, hydrocarbon processing, air and water pollution, mineral beneficiation studies, refining technology, metallurgy, synthetic materials and fibres, composite materials, fine chemicals.

Earth and space sciences, incorporating geology, cartography and surveying, ocean studies, meteorology and climatology, planetary galactic observations, geochemistry, mineralogy and petrology, mining studies, earthquake control, geophysics.

Electronics and computer science, incorporating computer theory, control and information theory, navigation, guidance and control, telecommunications, radar, cybernetics, semiconductor physics, laser technology and optics, ultrasonics, electronic engineering.

Energy sciences, incorporating coal conversion technology, solar collectors, wind power, wave power, biomass conversion, magnetohydrodynamics, electrical generation and distribution, energy conservation, geothermal, oil and gas.

Engineering and transportation, incorporating aerodynamics and structural aeronautics, industrial and mechanical engineering, transportation, including rail, road and ship technology, ocean engineering, civil engineering, mechanical aspects of robotics, fluid mechanics.

Industry and manufacturing, incorporating trade and industrial directories, equipment and product directories not specifically linked to scientific research.

Medical and biological sciences, incorporating the medical specialties, health and safety, dentistry, human nutrition, pharmacology, biomedical technology, biochemistry, botany, zoology, entomology, anthropology.

Physics, mathematics and nuclear sciences, incorporating high and low energy physics, plasma physics, fusion technology, nuclear reactor design, radiation methods and devices, radioactive waste disposal, radiation health and safety.

Directory of Scientific Directories is completed by three indexes. The Author, editor and compiler index lists personal names of individuals associated with the particular directory in alphabetical order. The Directory titles index lists alphabetically titles, English translations of titles (printed in italics), revolved titles and English translations (also printed in italics). The Publisher index lists the directories issued by a particular publisher.

In all the indexes the reader is referred back to the entry by chapter and publication number with the chapter. The three-letter chapter abbreviations are shown on the contents page.

Thus the reader can find a particular directory by title or revolved title; the reader can look up the relevant chapter and browse to find the directory or selection of directories he wants; the reader can see which directories are produced by each publisher.

The entry for each directory includes the following, where available: full title (as on title page); title in English, where appropriate; name of author(s), editor(s), or compilers(s); name of publisher; address of publisher; year of publication of most recent edition; number of pages of the most recent edition (xx + 748 indicates 20 prelim pages and 748 text pages); price in 1985 of most recent edition (OP means out of print); ISBN followed by a number indicates the International Standard Book Number; ISSN followed by a number indicates the International Standard Serials Number; the name of any series to which the directory belongs; geographical area shows the area of coverage of the material in the directory; the content indicates the number of entries, the arrangement and content of entries, the titles of indexes, and will give information if the material is available on-line; the month and year of publication of the next edition and its approximate price as gleaned in summer 1985.

The information in this directory has been obtained from questionnaires, from publisher's catalogues, from direct enquires, and in some cases from British Library records. Every reasonable effort has been made to ensure that the bibliographic details are correct. Nevertheless we invite comments from readers and users for suggestions regarding omissions or inaccuracies. These are important to us as we intend to publish future editions at about two early intervals.

Colin P. Taylor   September 1985
Publisher

CONTENTS AND INDEX ABBREVIATIONS

page

1	1 International directories	int
60	2 Africa	afr
66	3 Asia	asa
81	4 Australia and the Pacific	aus
87	5 Europe including USSR	eur
149	6 Latin America and the Caribbean	lat
154	7 North America	usa

197 Author, editor and compiler index

202 Directory titles index

226 Publisher index

Note: In the indexes, entries are referred to by chapter and entry number. To assist the reader country codes of publisher are inserted. These abbreviations are: arg, Argentina; aus, Australia; aut, Austria; bel, Belgium; can, Canada; chi, Chile; col, Colombia; den, Denmark; egy, Egypt; fin, Finland; fra, France; gdr, German Democratic Republic; gfr, German Federal Republic; gha, Ghana; hng, Hong Kong; ice, Iceland; ind, India; ino, Indonesia; ire, Ireland; isr, Israel; ita, Italy; jap, Japan; leb, Lebanon; lux, Luxembourg; may, Malaysia; mor, Morocco; net, The Netherlands; nor, Norway; nze, New Zealand; pak, Pakistan; pol, Poland; saf, South Africa; sin, Singapore; swe, Sweden; swi, Switzerland; tai, Taiwan; tha, Thailand; uni, United Kingdom; uru, Uruguay; usa, United States of America; zim, Zimbabwe.

1 INTERNATIONAL DIRECTORIES

GENERAL SCIENCES

Abstracting and indexing services directory, first edition   1

Editor: John Schmittroth
Publisher: Gale Research Company
Address: Book Tower, Detroit, MI 48226, USA
Year: 1982-83
Pages: 583 (in 3 vols)
Price: $170.00 (for the set)
ISBN: 0-8103-1649-8
Geographical area: World.
Content: The directory lists and describes more than 2000 current and continuing publications, including abstracts, indexes, digests, bibliographies, catalogues, and similar works in all fields. Also covered are abstracts and bibliographic compilations that are regular features of standard journals. In addition to full names and addresses, entries include editor's name, year first published, description and scope, subject coverage, arrangement, subscription information, computer access, and other details.

Appropriate technology directory, volume 2*   2

Editor: N. Jequier, G. Blanc
Publisher: Organization for Economic Cooperation and Development
Address: Publications Office, 2 rue André-Pascal, 75775 Paris Cedex 16, France
Year: 1985
Price: £12.00
ISBN: 9264126430
Geographical area: World.
Content: Entries for 316 new organizations not included in volume one, which was published in 1979.

Appropriate technology in situations: a directory*   3

Editor: Angela Sinclair
Publisher: Intermediate Technology Publications Limited
Address: 9 King Street, Covent Garden, London WC2E 8HW, UK
Year: 1984
Pages: iii + 124
Price: £4.95
Series: Sourcebooks
Geographical area: World.
Content: Institutions are listed by country.

Arab and Islamic International Organization Directory 1984-85   4

Editor: Union of International Associations
Publisher: Saur Verlag, K.G.
Address: Postfach 711009, D-8000 München 71, German FR
Year: 1985
Pages: 484
Price: DM248.00
ISBN: 3-598-21651-3
Series: Guide to international organizations
Geographical area: Arab and Islamic countries.
Content: The directory consists of three main parts: descriptions of organizations; international organization secretariats in each country; international organizations with Arab or Islamic membership. Indexes to each part giving names and keywords in English, French, and other working languages, old names, names of key officers, name initials or abbreviations, sponsoring organizations. International secretariat/principal contact addresses are given, in whatever country in the world they may be, as well as regional secretariat/secondary contact addresses.

Awards for Commonwealth university academic staff 1984-86 5

Publisher: Association of Commonwealth Universities
Address: John Foster House, 36 Gordon Square, London WC1H 0PF, UK
Year: 1984
Pages: 212
Price: £8.40
ISBN: 0-85143-085-6
Content: The handbook is published every two years and lists 670 separate schemes of fellowships, visiting professorships and lectureships, travel grants etc open to university academic staff in a Commonwealth country who wish to undertake research, make study visits, or teach for a while in another Commonwealth country. It contains four appendices and an index to awards listed.
Next edition: October 1986.

Brief guide to centres of international lending and photography 6

Editor: Richard J. Bennett
Publisher: IFLA Office for International Lending
Address: c/o British Library Lending Division, Boston Spa, Wetherby, West Yorkshire LS23 7BQ, UK
Year: 1984
Year: 153
Price: £10.00
ISBN: 0-7123-2020-2
Geographical area: World.
Content: Information on international lending and photocopying arrangements of institutions in 90 countries - arranged alphabetically by country indexes to countries in English, French, German, Russian; also contains *International Lending: principles and guidelines for procedure*.
Not available on-line.

Commonwealth universities yearbook 1985 7

Editor: A. Christodoulou, T. Craig
Publisher: Association of Commonwealth Universities
Address: John Foster House, 36 Gordon Square, London WC1H 0PF, UK
Year: 1985
Pages: ix + 2800 (in 4 vols)
Price: £62.00
ISBN: 0-85143-094-5
Content: A chapter for each of 370 university institutions of good standing in 30 Commonwealth countries. Most chapters contain: a list of principal officers; a list of all teaching staff arranged by subject departments and showing degrees held; a list of senior administrative staff; general information including history, location, constitution, income, faculties and deans, libraries, laboratories, entry requirements, first degrees, higher degrees, diplomas, certificates, study leave, fees, academic year, student residences and other facilities, publications, university bookstore and press; statistics of student enrolment, and of degrees and other qualifications awarded; a list of constituent, affiliated or associated institutions.
There are twelve major national introductory articles; and six appendices which include a description of the Commonwealth Scholarship and Fellowship Plan; statistics of students from abroad in six Commonwealth countries; and articles on admission to universities in eight countries.
There is a major bibliography; and index to all 165 000 personal names in the book; Commonwealth-wide indexes to institutions; topics and subjects of study indexes; lists of Commonwealth university degree and diploma titles; 23 indexes to facilities for university study in fifteen countries; thirteen maps showing location of universities.
Next edition: May 1986.

Current Asian and Australasian directories: a guide to directories published in or relating to all countries in Asia, Australia, and Oceania 8

Editor: I.G. Anderson
Publisher: CBD Research Limited
Address: 154 High Street, Beckenham, Kent BR3 1EA, UK
Year: 1978
Pages: xii + 264
Price: £27.00
ISBN: 0-900246-25-1
Content: Name of publisher, address, telephone, price, frequency, contents, etc of publications; subject indexes for major countries.
Not available on-line.
Next edition: October 1986; £40.00.

Current bibliographic sources in education* 9

Publisher: United Nations Education, Scientific, and Cultural Organization
Address: Palais des Nations, CH-1211, Genève 10, Switzerland
Year: 1984
Geographical area: World.
Content: Entries, arranged alphabetically by country with separate section for regional and international organizations, give: content of publication, name and address of source, contact for further information.

Dictionary of scientific biography 10

Editor: Charles Coulston Gillispie
Publisher: Charles Scribner's Sons
Address: 115 Fifth Avenue, New York, NY 10003, USA
Year: 1970-81
Pages: 16 vols, bound as 8
Price: $750.00
ISBN: 0-684-16962-2
Geographical area: World.
Content: Over 5000 biographies of mathematicians and physical and natural scientists from antiquity to the present. Originally published in sixteen volumes from 1970 to 1981, the work is now available in eight double volumes. Volume 15 was Supplement I and Volume 16 was a comprehensive index volume. A second supplement is scheduled for publication in Autumn 1986.
Next edition: October 1986; $75.00.

Directory information service 11

Editor: James M. Ethridge
Publisher: Gale Research Company
Address: Book Tower, Detroit, MI 48226, USA
Year: 1985
Price: $100.00
ISBN: 0-8103-0271-3
Geographical area: World.
Content: The Directory Information Service is a series of periodical supplements to update the Directory of Directories (see separate entry). The DIS covers well over 1000 directories between editions of DOD. It contains cumulative subject and title indexes.
Next edition: March 1987.

Directory of directories 1985, third edition 12

Editor: James M. Ethridge
Publisher: Gale Research Company
Address: Book Tower, Detroit, MI 48226, USA
Year: 1984
Pages: 1325
Price: $145.00
ISBN: 0-8103-0274-8
Geographical area: World.
Content: The fully revised and updated third edition thoroughly describes and indexes over 7800 directories of all kinds. To facilitate browsing, the DOD is arranged in sixteen broad subject categories. The detailed subject index, which employs more than 2100 subject headings and many cross references, provides access to entries on a specific subject. Including numerous foreign directories, DOD covers business and industrial directories, professional and scientific rosters, and other lists and guides of all kinds.
The Directory of Directories is updated periodically by the Directory Information Service (see separate entry).
Next edition: July 1986; $160.00

Directory of engineering societies and related organizations, eleventh edition, 1934 13

Publisher: American Association of Engineering Societies
Address: 415 Second Avenue NE, Washington, DC 20002, USA
Pages: xviii + 286
Price: $100.00
ISBN: 0-87615-003-2
Geographical area: World, concentration USA.
Content: Entries for 485 engineering societies and organizations of engineering-related activities arranged alphabetically by title; geographical, keyword, and acronymic indices.
Not available on-line.
Next edition: Summer 1986; $125.00.

Directory of institutions and individuals active in the field of research in information science, librarianship and archival records management 14

Publisher: Fédération Internationale de Documentation (FID)
Address: PO Box 90402, 2509 LK 's-Gravenhague, Netherlands
Year: 1980
Pages: iv + 138
Price: Dfl50.00
Geographical area: World.
Content: Listing of institutions and individuals active in the field of research in information science, librarianship, and archival records management.

Directory of institutions of higher education in Asia and the Pacific engaged in distance education 15

Publisher: Unesco Regional Office for Education in Asia and the Pacific
Address: PO Box 1425 GPO, Bangkok 10500, Thailand
Year: 1982
Pages: i + 55
Geographical area: Asia-Pacific.
Content: Entries for 50 institutions, arranged alphabetically by country and alphabetically by institution within a country.
Not available on-line.

Directory of international non-governmental organizations in consultative status with UNIDO 16

Publisher: United Nations Industrial Development Organization
Address: POB 300, A-1400 Wien, Austria

Year: 1984
Pages: 102
Geographical area: World.
Content: The directory provides factual profiles on the structure and activities of 75 international and regional non-governmental organizations presently in consultative status with UNIDO. It also includes as annexes information on the procedure for granting consultative status to international non-governmental organizations concerned with the promotion of industrial development and lists of the organizations alphabetically, by the country in which headquarters are located, by region, and by sector. The fields of cooperation include investment promotion, transfer of technology, provision of expertise, international sub-contracting, development research, marketing management, and training.

Directory of museums, second edition*   17

Editor: Kenneth Hudson, Ann Nicholls
Publisher: Macmillan Press
Address: Houndsmill, Basingstoke, Hampshire, UK
Year: 1981
Price: £60.00
ISBN: 0-333-24515-6
Geographical area: World.
Content: Listed alphabetically by country, entries give details of museums with indication of specialities and important exhibits.
Next edition: November 1985.

Directory of online databases   18

Publisher: Cuadra Associates Incorporated
Address: 2001 Wilshire Boulevard, Suite 305, Santa Monica, CA 90403, USA
Year: 1985
Pages: xii + 440
Price: $95.00 USA; $110.00 Canada and Mexico; $120.00 elsewhere; (for one year's subscription)
ISSN: 0193-6840
Geographical area: World.
Content: Over 2700 entries arranged alphabetically, with details of more than 2700 publicly available on-line data bases. For each data base, provides name, type (bibliographical, referral, numerical, textual-numerical, full-text, software), description of content, subject, language, geographic coverage, time span, frequency of updating and, if applicable conditions of access. Also provides names and addresses of data base producers and on-line services.
Indexes to name, subject, data base producer, on-line service, and telecommunications network, as well as master index.
Next edition: September 1985 (update supplement).

Directory of scientific directories   19

Publisher: Longman Group Limited
Address: Westgate House, The High, Harlow, Essex CM20 1NE, UK
Year: 1985
Pages: 350
Price: £80.00
ISBN: 0-582-90151-0
Series: Reference on research
Geographical area: World.
Content: There are 1800 entries, divided into seven geographical chapters: international; Asia (including Japan and the Middle East); Europe (East and West); North America; Africa; Australasia and Pacific; Latin America and Caribbean, and sub-divided into ten subject headings: general science; agriculture and food science; chemical and materials science; earth and space science; electronics and avionic sciences; energy sciences; engineering and transport; industry and manufacturing; medical and biological sciences; physics, mathematics, and nuclear sciences. The directory is indexed by title of directory, publisher, and editor/compiler.
Not available on-line.
Next edition: 1988.

Directory of training institutions and resources in Asia and the Pacific*   20

Publisher: Asian and Pacific Skill Development Programme
Address: International Labour Office, PO Box 1423, Islamabad, Pakistan
Year: 1983
Pages: 928
Geographical area: Bangladesh, Fiji, Hong Kong, India, Indonesia, Japan, Korea, Malaysia, New Zealand, Nepal, Pakistan, Papua New Guinea, Philippines, Singapore, Sri Lanka, and Thailand.
Content: The directory presents information on the training institution, its name, address, contact person and status in relation to the public education system; information on the trades offered, their duration and periodicity, language and technique requirements, average enrolment capacity, deadline for admission, fees, average seating capacity, training facilities, living arrangements, source of financing, fellowship or traineeship from sponsors and so on; and information on trainers or resource persons. The index gives an alphabetical list of trade offerings.

Directory of United Nations databases and information systems and services   21

Publisher: United Nations Organization
Address: Palais des Nations, CH-1211 Genève 10, Switzerland
Year: 1984

Price: $35.00
ISSN: 0255-920X
Content: The directory is a guide to over 600 selected databases, informatikon systems, and information services which are operated by 38 organizations of the United Nations system or in association with non-UN organizations. The entries in the directory represent many types of information systems. These include libraries, documentation centres, referral centres/clearinghouses, research centres and many others (eg the World Weather Watch) which are difficult to categorize but which function as indispensable information sources to the clientele which they serve.
Information given includes title, address, telephone, telex, contact person, type of system/service, year established, availability, working languages, subject coverage, geographical coverage, services provided, etc.

Educational media yearbook, tenth edition 1984*

Editor: James W. Brown
Publisher: Libraries Unlimited
Address: PO Box 263, Littleton, CO 80160, USA
Price: £51.50
ISBN: 0-87287-423-0
Content: Includes listing of media-related groups, programmes in library and information science, sources for funds and grants, directory of publishers and distributors of media-related products.

Encyclopedia of associations: volume 4, international organizations

Editor: Katherine Gruber
Publisher: Gale Research Company
Address: Book Tower, Detroit, MI 48226 USA
Year: 1984
Pages: 508
Price: $170.00
ISBN: 0-8103-0128-8
Geographical area: World.
Content: This publication provides details on over 2000 non-profitmaking groups which are international in scope and interest, headquartered outside the USA. All varieties of associations are covered, from commerce and industry to public and cultural affairs. Two softcovered supplements update the publication between issues ($125.00 for both).
See separate entry for Encyclopedia of Associations: Volume 1, National Organizations of the US.
Next edition: October 1985.

Encyclopedia of information systems and services: volume 1 - international volume*

Editor: John Schmittroth
Publisher: Gale Research Company
Address: Book Tower, Detroit, MI 48226, USA
Year: 1984
Pages: 669
Price: $175.00
ISBN: 0-8103-1538-6
Geographical area: World.
Content: This volume covers more than 1100 international and national information organizations, systems, and services located in some 65 countries excluding the United States.
Falling within the scope of the encyclopedia are the following categories of information systems: information providers, including publishers, professional associations, libraries, commercial firms, government agencies, educational institutions, and others who produce electronically accessible information. This edition notes an increasing number of databases devoted to high technology as well as a proliferation of full-text information on-line due to reduced computer and communication costs; information access services, including on-line host services, time-sharing companies, videotex/teletext information services, demographic and marketing data companies, and others; information sources on the information industry; and support services.
Entries give: full name, address, and telephone number; year founded; head; staff; related organizations; description of system or service; scope and/or subject matter; input sources; holdings and storage media; publications; microform products and services; computer-based products and services; other services; clientele/availability; projected publications and services; remarks/addenda; contact person.
Entries are arranged alphabetically within each volume, and each contains the following indexes: master index; data bases index; publications index; software index; function/service classifications index; personal name index; geographical index by country; subject index.
See separate entry for: Encyclopedia of Information Systems and Services 1985-86, Volume - United States
The two volumes are also available as a set, price $325.00, ISBN 0-8103-1537-8.
Next edition: September 1986

Europa year book 1985: a world survey

Publisher: Europa Publications
Address: 18 Bedford Square, London WC1B 3JN, UK
Pages: xxv + 2949 (in 2 vols)
Price: £106.00
ISBN: Volume 1 - 0-905118-79-0; volume 2 - 0-905118-92-8

Geographical area: World.
Content: There are approximately 50 000 entries. Volume One covers 1650 international organizations and countries of Europe, also the rest of the World Afghanistan to Burundi. Volume Two covers countries from Cameroon to Zimbabwe.
Not available on-line.
Next edition: Spring 1986; £110.00.

Exploring science: a guide to contemporary science and technology museums 26

Editor: Carol Bannerman
Publisher: Association of Science-Technology Centers
Address: 1413 K Street NW, Tenth Floor, Washington, DC 20005, USA
Year: 1980
Pages: 72
Price: $9.75
Geographical area: World.
Content: A guide to more than 100 ASTC-member museums in North America and abroad. Individual narratives describe each museum. Includes charts detailing facilities, programmes, and statistics on attendance, operating budgets, and square footage on exhibit space. Designed for use by the general public as well as museum staff.
Next edition: 1988.

Far East and Australasia 1984-85* 27

Publisher: Europa Publications
Address: 18 Bedford Square, London WC1B 3JN, UK
Year: 1984
Price: £54.00
Content: Survey and reference book containing a general review of the region followed by a chapter on each of the countries giving directories of government, diplomatic corps, political parties, communications, finance, and industry.
Next edition: November 1985.

Financial aid for first degree study at Commonwealth universities 1984-86 28

Publisher: Association of Commonwealth Universities
Address: John Foster House, 36 Gordon Square, London WC1H 0PF, UK
Year: 1984
Pages: 39
Price: £2.25
ISBN: 0-85143-084-8
Content: Listing of more than 100 separate schemes of scholarships, grants etc for Commonwealth students who wish to study for a first degree at a Commonwealth university outside their own country. Most of the awards listed are for students from developing countries. There is an index to organizations and award titles. The handbook is published every two years.
Next edition: October 1986.

Forthcoming international scientific and technical conferences 29

Publisher: Aslib
Address: Information House, 26-27 Boswell Street, London WC1N 3JZ, UK
Year: 1985
Price: £45.00 per annum
ISSN: 0046-4686
Geographical area: World.
Content: FISTC is issued quarterly. The main issue of each year is published in February and this supersedes and updates any previous issues: this is followed by a supplement in May and cumulative supplements in August and November.
Conferences are listed in chronological order and each entry gives the date, title, and location of the conference and a contact address for enquiries. The entries are indexed by subject, location, and organization.
Not available on-line.
Next edition: August 1985; £45.00 per annum.

Grants for study visits by university administrators and librarians 1985-87 30

Publisher: Association of Commonwealth Universities
Address: John Foster House, 36 Gordon Square, London WC1H 0PF, UK
Year: 1984
Pages: 28
Price: £2.75
ISBN: 0-85143-091-0
Geographical area: Commonwealth.
Content: Describes 40 sources of financial aid for university administrators and university librarians who want to undertake study visits to another country. Part A lists grants tenable in one Commonwealth country by staff from another. Part B lists grants for movement (either way) between Commonwealth and non-Commonwealth countries. An appendix lists courses and conferences for administrators. Index to organizations and awards.
Next edition: October 1987.

Guide to reference books, ninth edition* 31

Editor: Eugene P. Sheehy
Publisher: American Library Association
Address: c/o Eurospan, 3 Henrietta Street, London WC2E 8LU, UK
Year: 1982

Price: £18.00
ISBN: 0-8389-0361-4
Content: List of more than 2100 items mostly published between 1978 and 1980; emphasis on scholarly works.

Guide to reference books for small and medium-sized libraries 1970-82*   32

Publisher: Libraries Unlimited
Address: PO Box 263, Littleton, CO 80160, USA
Year: 1984
Price: $34.00
ISBN: 0-87287-403-6
Content: Emphasis is on standard works in all subjects; 1195 titles included.

IFLA directory 1984-85   33

Publisher: International Federation of Library Associations and Institutions
Address: POB 95312, 2509 CH s-'Gravenhague, Netherlands
Year: 1984
Pages: 245
Price: Dfl35.00
Geographical area: World.
Next edition: April 1986; Dfl50.00.

Information sources in science and technology, second edition   34

Editor: C.C. Parker, R.V. Turley
Publisher: Butterworth and Company Limited
Address: Borough Green, Sevenoaks, Kent TN15 8PH, UK
Year: 1985
Pages: 234
Price: £16.00
ISBN: 0-408-01467-9
Series: Butterworths guides to information sources

Intergovernmental organization directory 1984-85   35

Editor: Union of International Associations
Publisher: Saur Verlag, K.G.
Address: Postfach 711009, D-8000 München 71, German FR
Year: 1985
Pages: 680
Price: DM248.00
ISBN: 3-598-21653-X
Series: Guide to international organizations
Geographical area: World.
Content: Descriptive entries on over 700 organizations; index of names, initials, and keywords in English and other working languages; classified lists (with headquarters address) by country of secretariat; statistics.

International congress calendar, 25th edition 1985   36

Editor: Union of International Associations
Publisher: Saur Verlag K.G.
Address: Postfach 711009, D-8000 München 71, German FR
Pages: 280 per issue (4 issues per year)
Price: DM280.00 (for the 4 issues)
ISSN: 0538-6349
Geographical area: World.
Content: Each quarterly issue contains three sections: geographical-listing of events by where they take place, alphabetically by country and by city within the countries, giving date and theme of event, address of organizing body, estimated number of participants, and concurrent exhibitions; chronological - listing of events under dates also giving place, name of organizer, type and subject of event, etc; subject/organizations index, also giving place and date.

International directory of higher education research institutes 1982*   37

Publisher: United Nations Educational, Scientific and Cultural Organization
Address: Palais des Nations, CH-1211 Genève 10, Switzerland
Year: 1982
Price: £3.60
ISBN: 9230019283
Geographical area: World.
Content: Published in English, French, Spanish, and Russian.

International foundation directory   38

Publisher: Europa Publications Limited
Address: 18 Bedford Square, London WC1B 3JN, UK
Year: 1983
Pages: 450
Price: £24.00
ISBN: 0-905118-90-1
Geographical area: World.
Content: Each of the 1100 entries contains aims, address, finance and officers of foundations. There are indexes of foundations and of main activities.
Not available on-line.
Next edition: September 1986; £30.00.

International guide to library archival and information science associations, second edition   39

Editor: Josephine R. Fang, Alice H. Songe
Publisher: Bowker Publishing Company
Address: 58/62 High Street, Epping, Essex CM16 4BU, UK

Year: 1981
Pages: 448
Price: £38.50
ISBN: 0-8352-1285-8
Geographical area: World.
Content: Information on libraries and information science associations located in over 100 countries, including organizations concerned with the processing and dissemination of information and information networks. The main text is divided into two separate sections: an alphabetical listing of the international associations and a geographical listing of those within the United States. Entry information for each includes: official name, acronym (if any), address; names and titles of major officers, and type and number of staff. Also noted are major fields of interest, languages used, historical data, structure and goals of the organization, financial status and the latest budget and membership details.

International handbook of universities and other institutions of higher education   40

Editor: International Association of Universities
Publisher: Macmillan Press
Address: 4 Little Essex Street, London WC2R 3LF, UK
Year: 1983
Pages: 1144
Price: £60.00
ISBN: 0-333-27446-6
Geographical area: World, except USA and Commonwealth.
Content: Guide to higher education facilities in 112 countries excluding the USA and the Commonwealth, and provides the following information: name, address and telephone number; names of senior administrators; complete list of faculties, departments, schools and institutes in university; historical background in brief; notes on administration and governing bodies; dates of academic year; admission requirements; details of fees; degrees and diplomas offered; length of study required; student enrolment figures; numbers and breakdown of academic staff; language(s) of instruction; library facilities; publications by the institution; arrangements for cooperation with other universities.

International index on training on conservation of cultural property   41

Editor: Cynthia Rockwell
Publisher: ICCROM (International Centre for Conservation)
Address: Via di San Michele 13, Roma RM, Italy
Year: 1982
Pages: xv + 141
Price: $5.00
Geographical area: World.
Content: Entries for 304 institutions, giving the following information: name and address of institution; subject or material taught; titles of courses; nature of training; duration of training courses; personnel in charge; sources of financial aid for students; admission requirements; special conditions; language of instruction; certificate or diploma granted; date programme founded; number of students who have completed the programme.
Indexes: subject, institutions, country codes, courses.
Not available on-line, at present.
Next edition: End of 1986.

International organization abbreviations and addresses 1984-85   42

Editor: Union of International Associations
Publisher: Saur Verlag, K.G.
Address: Postfach 711009, D-8000 München 71, German FR
Year: 1985
Pages: 536
Price: DM248.00
ISBN: 3-598-21652-1
Series: Guides to international organizations
Geographical area: World.
Content: List of over 10 000 initials/abbreviations with corresponding name of international organization, whether in English, French, Spanish, or other languages; list of names of over 10 000 international organizations, whether in English, French, Spanish, or other languages, together with their corresponding initials/abbreviations.

International organizations: a dictionary and directory   43

Editor: Giuseppe Schiavone
Publisher: Macmillan Press
Address: 4 Little Essex Street, London WC2R 3LF, UK
Year: 1983
Pages: 328
Price: £25.00 (hardback); £8.95 (paperback)
ISBN: 0-333-32423-4 (hardback); 0-333-40502-1 (paperback)
Content: Details of intergovernmental institutions serving political, economic, social and cultural purposes, including alphabetically arranged descriptions of a large number of organizations - from the United Nations to specialized agencies and regional institutions in Europe, the Americas, the Middle East, Africa, Asia and the Pacific.
For each agency, after a brief historical sketch, there is an analysis of objectives, functions and powers, classes and conditions of membership, acts, institutional structure and external relations. Next comes an account of the work of the organization and an assessment of its role and perspectives in its peculiar field of action. A list

of basic references and a list of publications is appended to each entry. There is also a directory section in the book listing addresses and principal officers for a further 160 international organizations. A detailed index provides a key to readers.

International research centers directory, second edition 44

Editor: Anthony T. Kruzas, Kay Gill
Publisher: Gale Research Company
Address: Book Tower, Detroit, MI 48226, USA
Year: 1984
Pages: 739
Price: $295.00 (set)
ISBN: 0-8103-0467-8
Content: Entries are arranged by country for productive browsing. Entries contain detailed information, including: full name, address, and telephone number; date established; name and title of director; parent agency and organizational affiliation; size and description of staff; type of activity; fields of research; library holdings; publications and seminar programmes.
Following the main entries is a name and keyword index, making it easy to locate the entry for a particular research unit. Completing the directory is a country index, with entries cumulated and arranged alphabetically, by name, under appropriate country headings. Two softbound supplements provide full entries describing newly formed and newly discovered research centers located around the world (ISBN 0468-6) 1984-85.
Next edition: February 1986; $350.00.

International who's who 1984-85 45

Publisher: Europa Publications
Address: 18 Bedford Square, London WC1B 3JN, UK
Year: 1984
Pages: xix + 1583
Price: £50.00
ISBN: 0-905118-97-9
Geographical area: World.
Content: There are 17 000 entries listed alphabetically. Not available on-line.
Next edition: July 1985; £54.00.

International who's who of the Arab world* 46

Publisher: International Who's Who of the Arab World Limited
Address: c/o Graham and Trotman Limited, Sterling House, 66 Wilton Road, London SW1V 1DE, UK
Year: 1984
Pages: 608
Price: £120.00; $192.00
ISBN: 0-9506122-1-9
Content: This book concentrates on the biographies of eminent Arabs. The alphabetical listing of 3000 names includes businessmen, lawyers, oil officials, scientists, religious leaders, academics, writers, and artists. Individual details include full name, nationality, profession/position, education, career history, membership of professional associations, languages, personal interests and leisure activities, full contact address with telephone and telex numbers where available.
This title is not available from Graham and Trotman in the USA, Saudi Arabia or the Gulf States.

International yearbook and statesmen's who's who 47

Publisher: Thomas Skinner Directories
Address: Windsor Court, East Grinstead House, East Grinstead, Sussex RH19 1XB, UK
Year: 1985
Pages: 1250
Price: £65.00
Content: Information on political leaders, industrial development, transport and communication systems, import and export figures.
Next edition: May 1986; £70.00.

International yearbook of educational and instructional technology 1984-85 48

Editor: C.W. Osborne
Publisher: Kogan Page
Address: 120 Pentonville Road, London N1 9JN, UK
Year: 1984
Pages: 656
Price: £20.00
ISBN: 0-85038-784-1
Content: Includes directory of centres of activity, list of producers and distributors of programs and audio-visual software, and a guide to currently available audio-visual hardware.
Next edition: 1986.

Internationales ISBN-verlagsverzeichnis einschliesslich verlagsadressen aus ländern ohne ISBN-system 49

[International ISBN publishers' directory including publishers' addresses from some countries outside the ISBN system]
Editor: International ISBN Agency, Staatsbibliothek Preussischer Kulturbesitz, Berlin
Publisher: Buchhändler-Vereinigung
Address: Postfach 2404, Grosser Hirschgraben 17/21, D-Frankfurt am Main 1, German FR
Year: 1985
Pages: 2121
Price: DM318.00

ISBN: 3-88053-022-X
Content: Nearly 140 000 publishers' addresses from 55 countries, arranged as follows: Part 1 - alphabetical directory; Part 2 - numerical ISBN directory; Part 3 - alphabetical publishers' index (classification by countries).
Next edition: April 1986; DM318.00.

Inventory of data sources in science and technology*   50

Editor: Committee on Data for Science and Technology of the International Council of Scientific Unions
Publisher: United Nations Educational, Scientific and Cultural Organization
Address: 7 Place de Fontenoy, F-75700 Paris, France
Year: 1982
Pages: 235
Price: F50.00
Content: Entries for over 650 institutions and other organizations which offer data in renewable energy sources, fertilizers, hydrological sciences and water resources, nutrition, pesticides, and soil science; includes more than 90 countries.

List of company directories and summary of their contents, second edition*   51

Publisher: United Nations Organization
Address: Palais des Nations, CH-1211 Genève 10, Switzerland
Year: 1983
Price: £14.40
ISBN: 0-11-907919-4
Geographical area: World.

Major libraries of the world: a selective guide   52

Editor: Colin R. Steele
Publisher: Bowker Publishing Company
Address: 58/62 High Street, Epping, Essex CM16 4BU, UK
Year: 1976
Pages: 479
Availability: OP
ISBN: 0-85935-012-6
Content: Entries cover 300 major libraries throughout the world. National, university, special, and public libraries in 79 countries are arranged geographically. Data includes a description of each library, comments on the most important books and manuscripts, and other practical information.

Marine environmental centres: Mediterranean   53

Publisher: Food and Agriculture Organization
Address: Via delle Terme di Caracalla, 00100 Roma, Italy
Year: 1985
Pages: xiii + 302
Series: UNEP regional seas directories and bibliographies
Content: Compiled jointly by the United Nations Environmental Programme and the Food and Agriculture Organization, the directory lists 147 research centres in 17 countries.

McGraw-Hill modern scientists and engineers   54

Publisher: McGraw-Hill Book Company (UK) Limited
Address: Shoppenhangers Road, Maidenhead, Berkshire SL6 2QL, UK
Year: 1980
Pages: 1420 (3 vols)
Price: £140.00
ISBN: 0-07-045266-0
Content: The directory contains some 1250 entries.

Middle East and North Africa 1984-85*   55

Publisher: Europa Publications
Address: 18 Bedford Square, London WC1B 3JN, UK
Year: 1984
Price: £44.00
Content: Facts and figures on all countries of the Middle East and North Africa, including directories of government, diplomatic corps, media, banks, finance, transport, and tourism.
Next edition: October 1985; £48.00.

Museums of the world, third edition   56

Publisher: Saur Verlag, K.G.
Address: Postfach 711009, D-8000 München 71, German FR
Year: 1981
Pages: viii + 623
Price: DM320.00
ISBN: 3-598-10118-X
Series: Handbook of international documentation and information
Content: Entries are arranged alphabetically by country; a list of subject headings with cross references precedes the main index.
Next edition: 1986.

INTERNATIONAL DIRECTORIES

Online bibliographic databases 57

Editor: James L. Hall, Marjorie J. Brown
Publisher: Aslib
Address: 26-27 Boswell Street, London WC1, UK
Year: 1983
Pages: 383
Price: $95.00
ISBN: 0-8103-0530-5
Geographical area: World.
Content: The work provides essential details on nearly 200 on-line bibliographic data bases accessible through about 40 on-line service suppliers. Bibliographic databases are those which furnish access to information contained in other sources. Many bibliographic data bases are indexes and often include abstracts of the material indexed.
The directory section furnishes nearly 200 detailed entries arranged alphabetically by the name of the data base, from ABI/INFORM to WTI. Each entry shows: name and/or acronym; supplier of the data base, with full name and address; printed versions, if any; subject field description; on-line file details, including time span, total citations, update frequency, and whether or not the file contains abstracts or keywords; on-line service supplier; typical access charge per hour and per off-line citation; documentation available.
Next edition: 1987.

Population and related organizations: international address list 58

Editor: Ruth Sandor, Jane Vanderlin
Publisher: Association for Population/Family Planning Libraries and Information Centers - International (APLIC)
Address: c/o Carolina Population Center Library, University of North Carolina, University Square - East 300A, Chapel Hill, NC 27514, USA
Year: 1984
Pages: 87
Price: $45.00
ISBN: 0-933438-09-5
Series: APLIC special publication, 5
Geographical area: World.
Content: Approximately 900 entries (cross-referenced by language (into English if original is in a foreign language); arrangement of main list is alphabetical by institutional name. Also indexed by acronym and by country and state. Appendices are listed by statistical offices and family planning agencies.
Not available on-line.

Research opportunities in Commonwealth developing countries 59

Publisher: Association of Commonwealth Universities
Address: John Foster House, 36 Gordon Square, London WC1H 0PF, UK
Year: 1984
Pages: 235
Price: £12.00
ISBN: 0-85143-092-9
Content: There are 1770 entries for areas of research strength at 122 universities and descriptions of the kind of research undertaken by each of 300 non-university institutions. Three appendices list, inter alia, centres of advanced study and departments of special assistance at Indian universities. Index to subjects.
Next edition: November 1986.

Scholarships guide for Commonwealth postgraduate students 1985-87 60

Publisher: Association of Commonwealth Universities
Address: John Foster House, 36 Gordon Square, London WC1H 0PF, UK
Year: 1984
Pages: 356
Price: £9.70
ISBN: 0-85143-090-2
Content: The handbook is published every two years and lists nearly 1300 separate schemes of scholarships, grants, assistantships, and other forms of financial assistance open to graduates of Commonwealth universities who wish to undertake postgraduate (including postdoctoral) study or research at a Commonwealth university outside their own country. There are appendices on awards tenable at non-university institutions and on awards tenable in the UK by UK graduates; and an index to organizations and award titles.
Next edition: October 1987.

Sourcebook of global statistics 61

Editor: George Kurian
Publisher: Longman Group Limited
Address: Sixth Floor, Westgate House, The High, Harlow, Essex CM20 1NE, UK
Year: 1985
Pages: 400
Price: £60.00
ISBN: 0-582-90266-5
Content: A comprehensive checklist and descriptive guide to over 2000 statistical publications and organizations dealing with global and regional statistics. Among the unique features of the book are the inclusion of time series global data available in data banks and the extensive coverage of library resources and private data banks, such as Business International and Euromonitor. The directory is indexed by subject and by title and there is also a geographical index.

State economic agencies: a world directory 62

Editor: Alan J. Day
Publisher: Longman Group Limited
Address: Sixth Floor, Westgate House, The High, Harlow, Essex CM20 1NE, UK
Year: 1985
Pages: 546
Price: £48.00
ISBN: 0-582-90253-3
Geographical area: World.
Content: This reference guide provides detailed information on some 2000 state agencies and organizations (including government departments and commercial companies) currently active in economic spheres such as national, regional and sectoral planning, industrial development, agrarian reform, export promotion, etc. All countries of the world are covered, and the individual entries not only give the name, address and top officials of each agency but also present concise information on history, aims, current activities and financial structure. Each section of the book is introduced by a brief description of the prevailing economic conditions of the particular country. The volume also contains both subject and names indexes.

Statistics America: sources for social, economic and market research (North, Central and South America), second edition 63

Editor: Joan M. Harvey
Publisher: CBD Research Limited
Address: 154 High Street, Beckenham, Kent BR3 1EA, UK
Year: 1980
Pages: xiv + 385
Price: £43.50
ISBN: 0-900246-33-2
Content: Bibliography of sources of statistical information for each country, showing publisher, frequency, price, contents, etc; central statistical offices (address, telephone, scope of activities); other organizations collecting statistics (s.i.); alphabetical index of organizations; alphabetical index of titles; subject index.
Next edition: December 1986; £55.00.

Statistics Asia and Australasia: sources for social, economic and market research 64

Editor: Joan M. Harvey
Publisher: CBD Research Limited
Address: 154 High Street, Beckenham, Kent BR3 1EA, UK
Year: 1983
Pages: xiv + 440

Price: £48.00
ISBN: 0-900246-41-3
Content: Bibliography of sources of statistical information for each country, showing publisher, frequency, price, contents, etc; central statistical offices (address, telephone, scope of activities); other organizations collecting statistics; alphabetical index of organizations; alphabetical index of titles; subject index.
Next edition: May 1987.

Top 3000 directories and annuals 1985-86* 65

Editor: Allan Wood
Publisher: Armstrong, Alan, and Associates Limited
Address: 76 Park Road, London NW1 4SH, UK
Year: 1985
Price: £29.00
ISBN: 0-946291-10-1
Geographical area: World.
Content: Listing of 3000 directories currently in print, as well as reports on out-of-print or changed titles, giving: publisher, price, content, ISBN, date of publication, etc. Lists of publishers and their publications, publishers and their addresses, directories and annuals by month and year of issue; subject index.
Next edition: June 1986.

Union internationale des laboratoires independants: register of members 66

[International union of independent laboratories register of members 1983]
Publisher: International Union of Independent Laboratories
Address: The Laboratories, Fortune Lane, Elstree, Hertfordshire WD6 3HQ, UK
ISBN: 92-9011-007-4
Content: Member laboratories listed alphabetically by country giving name, address, telephone, key personnel, and specialization in native language and in English. Subject index in English, with cross reference in French, German, and Spanish.

Who's who in technology today* 67

Editor: Jan W. Chunchwell, LouAnn Chaudier
Publisher: Dick, J., Publishing
Address: 801 Green Bay Road, Lake Bluff, IL 60044, USA
Year: 1984
Pages: 4000 (in 5 vols)
Price: $425.00 (for set); biographical volumes $70.00 each; keyword index volume $145.00
Content: Entries for 32 000 engineers, scientists, inventors, and researchers in the fields of electrical, mechanical, chemical, civil, and biomedical engineering, and in the earth sciences, physics, and other technological

fields. Entries include: name, title, affiliation, address; personal, educational, and career data; publications, inventions, consulting experience, expert witness experience; technical field of activity; area of expertise. Arrangement: by major field, then by discipline. Indexes: personal name, expertise keyword, discipline.
Next edition: 1986.

Who's who in the world 68

Publisher: Marquis Who's Who Incorporated
Address: 200 East Ohio Street, Chicago, IL 60611, USA
Year: 1984
Pages: xv + 1149
Price: $125.00
ISBN: 0-8379-1107-9
Geographical area: World.
Content: More than 22 000 entries, alphabetically arranged, giving name, occupation, vital statistics, parents, marriage, children, education, certification, career, writings and creative works, civic and political activities, military record, awards, professional and association memberships, political affiliation, religion, clubs, lodges, home address, office address.
Next edition: October 1985; $197.00.

World directory of environmental 69 organizations, second edition

Editor: T.C. Trzyna, E.V. Coan
Publisher: Public Affairs Clearing House
Address: PO Box 30, Claremont, CA 91711, USA
Year: 1976
Pages: xxx + 258
Availability: OP
ISBN: 0-912102-20-9
Series: Sierra Club special publications, international series 1
Content: Four parts. Users' guide. International and national organizations for thirty six major areas (eg grasslands, polar regions). Intergovernmental organizations. International non-governmental organizations - alphabetical listing. National organizations - arranged by country (basic information only, name, address). Organization and acronym indexes.

World directory of national science 70 and technology policy making bodies

Editor: M. Abtahi
Publisher: UNESCO
Address: 7 place de Fontenoy, 75700 Paris, France
Year: 1984
Pages: xix + 99
Price: F22.00
ISBN: 92-3-002236-5
Series: Science policy studies and documents
Content: There are 163 entries arranged by countries (101) in alphabetical order, which give: address; telephone number; name of head; year of establishment; references of legislative texts governing the organization; aims and responsibilities; functions; linkages; publications; data bases, etc.
On-line availability planned.
Next edition: 1987.

World directory of peace research 71 institutions

Publisher: UNESCO
Address: 7 place de Fontenoy, 75700 Paris, France
Year: 1984
Pages: viii + 228
Price: F26.00
ISBN: 92-3-102261-X
Series: Reports and papers in the social sciences
Content: There are 282 entries arranged alphabetically by country of headquarters, then alphabetical by name. Entries include: synonymous names; abbreviations/acronyms; address of headquarters; date of creation; name of head; type of organization (eg private); type of activity (research, publishing, etc); main disciplines; geographical areas covered; other information (aims, research orientation, etc); form of data processing; types of publication; titles of publications; current research topics; number of researchers; total staff; names of main researchers; annual budget; sources of revenue.
Alphabetical list of institutions (names and abbreviations); list of heads of institutions; subject index.
Not available on-line.
Next edition: 1985; F30.000.

World directory of research projects, 72 studies and courses in science and technology policy

Editor: M. Abtahi
Publisher: UNESCO
Address: 7 place de Fontenoy, 75700 Paris, France
Year: 1981
Pages: xxviii + 424
Price: F58.00
ISBN: 92-3-001979-8
Series: Science policy studies and documents
Content: There are 1117 entries, arranged by countries in alphabetical order, giving: address, telephnone number; name of head; facilities and activities; title of the research projects/studies with their major foci; courses of the unit with their major foci; material published by the unit; name of publishers having published material produced by the unit.
On-line availability planned.
Next edition: 1986-87.

World environmental directory, volumes 1 and 2, third edition   73

Editor: B.E. Gough
Publisher: Business Publishers Incorporated
Address: 951 Pershing Drive, Silver Spring, MD 20770, USA
Year: 1978
Pages: Volume 1 - 1104; volume 2 - 544
Price: Volume 1 - £26.25; volume 2 - £22.70
ISBN: Volume 1 - 0471-99662-9; volume 2 - 0471-99686-6
Content: Universal in scope, the two volumes together cover 150 countries. Volume one includes listings of personnel, companies, government agencies, educational institutions, and organizations from the United States and Canada. Volume two covers listings from Africa, Asia, Australia, New Zealand, Europe, Mexico, and South America. The combined listings of persons, companies, agencies, institutions and organizations exceed 3000, spread over some 1500 pages.
Not available on-line.

World guide to abbreviations of organizations, seventh edition   74

Editor: F.A. Buttress
Publisher: Blackie Publishing Group
Address: Wester Cleddens Road, Bishopbriggs, Glasgow G64 2NZ, UK
Year: 1984
Pages: x + 731
Price: £50.00
ISBN: 0-249-44167-5
Geographical area: World.
Content: Over 43 000 entries, including every type of organization - international, national, governmental, individual, large, and small - covering the fields of commerce, industry, administration, and education.
Not available on-line.

World guide to libraries/ internationales bibliothekshandbuch, sixth edition   75

Editor: Helga Lengenfelder
Publisher: Saur Verlag, K.G.
Address: Postfach 711009, D-8000 München 71, German FR
Year: 1983
Pages: xlviii + 1186
Price: DM380.00
ISBN: 3-598-20523-6
Series: Handbook of international documentation and information
Content: The directory lists 42 7000 libraries in 167 countries, including national, federal, and state libraries, and college, public, university, and other academic libraries with holdings of over 30 000 volumes, as well as special, parliamentary, government, religious, and business libraries with holdings of at least 3000 volumes. Alphabetical index.
Next edition: November 1985; DM380.00.

World guide to scientific associations and learned societies/internationales verzeichnis wissenschaftlicher verbände und gesellschaften   76

Editor: Helmut Opitz
Publisher: Saur Verlag, K.G.
Address: Postfach 711009, D-8000 München 71, German FR
Year: 1984
Pages: xix + 947
Price: DM268.00
ISBN: 3-598-20522-8
Series: Handbook of international documentation and information
Content: Entries for approximately 22 000 associations from 150 countries listed alphabetically by name within a country alphabet giving name (supplementary or foreign language name as well as English name), address, telephone, telex, names of key personnel, year of foundation, number of members, area of specialization; index of official abbreviations of organization names; subject index.
Next edition: 1986.

World guide to special libraries/ internationales handbuch der spezialbibliotheken   77

Editor: Helga Lengenfelder
Publisher: Saur Verlag, K.G.
Address: Postfach 711009, D-8000 München 71, German FR
Year: 1983
Pages: xxx + 990
Price: DM298.00
ISBN: 3-598-20528-7
Series: Handbook of international documentation and information
Content: The guide is divided into five main subject categories with 300 entries in the general section, 9000 in the field of humanities, 8000 in social science, 4500 in medicine and life sciences, and 10 500 in science and technology. Each entry gives address, telegraphic address, telephone and telex, year of foundation, name of director, special holdings and collections, connection to electronic information systems, etc. Subject index.

World guide to trade associations/ internationales verzeichnis der wirtschaftsverbände 78

Editor: Helmut Opitz
Publisher: Saur Verlag, K.G.
Address: Postfach 711009, D-8000 München 71, German FR
Year: 1985
Pages: 1259
Price: DM480.00
ISBN: 3-598-20527-9
Series: Handbook of international documentation and information
Next edition: 1987.

World list of universities 79

Editor: International Association of Universities
Publisher: Macmillan Press
Address: 4 Little Essex Street, London WC2R 3LF, UK
Year: 1985
Pages: 632
Price: £30.00
ISBN: 0-333-35388-2
Content: Data on 8500 universities and other institutions of higher education in 154 countries.
Part 1 lists establishments alphabetically by country. Entries give: name of institution; postal address; faculties; departments; schools; institutes; title of official to whom general correspondence should be addressed. Also included in Part 1 are other institutes of higher education; teacher training establishments; principal academic, student and government bodies; and National Commissions of Unesco.
Part 2 is a guide to the main international and regional organizations in higher education and student affairs.

World meetings: outside United States and Canada 80

Publisher: Macmillan Publishing Company
Address: 200D Brown Street, Riverside, NJ 08370, USA
Price: $160.00 (for four quarterly issues)
ISSN: 0-02-695280-7
Content: Full data on all forthcoming major scientific, technical and medical meetings and conferences to be held in countries outside the United States and Canada, giving the deadline dates for submitting papers for each meeting; the names and addresses to whom the papers should be sent; dates on which abstracts, papers, reprints, preprints, and conference proceedings will be published, their source and price.
Indexes as follows: keyword index; location index; date index; deadline index; sponsor directory and index.

World meetings: social and behavioral sciences, human services and management* 81

Publisher: Macmillan Publishing Company
Address: 200D Brown Street, Riverside, NJ 08370, USA
Price: $145.00 (for four quarterly issues)
ISSN: 0-02-695290-4
Content: Information on meetings and conferences, in the US and throughout the world, to be held within the next two years in the fields of the social and behavioural sciences, human services, and management. Each quarterly issue covers all of the social and behavioural sciences, as well as the applied areas of social welfare, population, public health and rehabilitation, psychiatry and mental health, law and criminology, education, management and public administration, marketing, statistics and operations research, urban affairs, social aspects of environment and energy, information science and communications, and many other areas.
Indexes: keyword index; date index; deadline index; sponsor directory and index.
Next edition: 1985

World museums publications* 82

Publisher: Bowker Publishing Company
Address: PO Box 5, Epping, Essex CM16 4BU, UK
Year: 1982
Price: £147.50
Content: Details of more than 10 000 museums and major art galleries and the 20 000 publications available from them, listed by country; museum publications and audio-visual materials index.

World of learning 1984-85, 35th edition 83

Publisher: Europa Publications Limited
Address: 18 Bedford Square, London WC1B 3JN, UK
Year: 1985
Pages: xv + 1852
Price: £65.00
ISBN: 0-946653-01-1
Geographical area: World.
Content: Addresses and other details of more than 25 000 universities, colleges, schools of art and music, libraries and archives, learned societies, research institutes, museums and art galleries - all arranged alphabetically by country; names of over 150 000 professors, librarians, academicians, university chancellors and vice-chancellors, deans, bursars, presidents, rectors, curators and other officials; details of over 400 international cultural, scientific and educational organizations. Each entry gives the name, address, telephone and telex numbers, chief executives, aims and publications of the organization concerned.

A hundred-page index lists all the institutions in the book.
Not available on-line.
Next edition: January 1986; £69.00.

World problems and human potential, second edition 1985-86   84

Editor: Union of International Associations
Publisher: Saur Verlag, K.G.
Address: Postfach 711009, D-8000 München 71, German FR
Pages: 1500
Price: DM428.00
ISBN: 3-598-21864-8
Series: Yearbook of international organizations
Geographical area: World.
Content: Details of over 4500 world problems listed in fourteen categories, interlinked by 50 000 cross references; problem atlas; index to title and keywords.

World yearbook of education 1985: research, policy and practice   85

Editor: John Nisbet, Stanley Nisbet
Publisher: Kogan Page
Address: 120 Pentonville Road, London N1 9JN, UK
Year: 1985
Pages: 336
Price: £17.95
ISBN: 0-85038-907-0
Content: Reviews of educational research and development in 14 different countries; information on organizations responsible for educational r&d.

Yearbook of international organizations 1984-85, 21st edition   86

Editor: Union of International Associations
Publisher: Saur Verlag, K.G.
Address: Postfach 711009, D-8000 München 71, German FR
Year: 1985
Pages: Volume 1 - 1260; volume 2 - 1456; volume 3 - 1072
Price: Volume 1 - DM428.00; volume 2 - DM428.00; volume 3 - DM248.00; 3 volume set - DM980.00
ISBN: Volume 1 - 3-598-21860-1; volume 2 - 3-598-21861-3; volume 3 - 3-598-21862-X; 3 volume set - 3-598-21863-X
Geographical area: World.
Content: Volume one gives descriptions of 22 456 organizations including: name in all applicable languages; address, telephone and telex; main activities and programmes; technical and regional commissions; name of principal executive officer; history, goals, and structure; consultative relationships; inter-organizational links; membership by country. It has indexes of keywords in English and French, personal names of executive officers, name initials or abbreviations, and organizations cited as sponsoring bodies.
Volume two covers the involvement of any given country in the international community of organizations.
Volume three lists: 8000 organizations by subject; organizations by regional activity in 30 categories; contacts between organizations (indexed under English name); publications indexed by title in single alphabetical sequence.

Yearbook of the International Council of Scientific Unions   87

Editor: F.W.G. Baker
Publisher: International Council of Scientific Unions
Address: 51 boulevard de Montmorency, Paris 75016, France
Year: 1985
Pages: viii + 239
ISSN: 0074-4387
Geographical area: World.
Next edition: March 1986

AGRICULTURE AND FOOD SCIENCE

Acarologists of the world*   88

Editor: Donald E. Johnston
Publisher: Acarology Laboratory, Ohio State University
Address: 484 W. 12th Avenue, Columbus, OH 43210, USA
Year: 1984
Content: Entries for over 1400 persons concerned with the study of mites and ticks. Entries include: name, address, interests. Arrangement: alphabetical. Indexes: geographical, subject.

Agricultural research centres: a world directory of organizations and programmes, seventh edition   89

Editor: Nigel Harvey
Publisher: Longman Group Limited
Address: Sixth Floor, Westgate House, The High, Harlow, Essex CM20 1NE, UK
Year: 1983
Pages: 1276
Price: £175.00
ISBN: 0-582-90014-X
Series: Reference of research
Content: In two volumes the directory lists over 10 000

INTERNATIONAL DIRECTORIES

laboratories and establishments which conduct or finance research in agriculture, fisheries, food, forestry, horticulture and veterinary science. Each entry includes, where appropriate, the address, type of organization and affiliation, senior staff, a summary of research activities, and publications.
There is a title of establishments index and a subject index.

Bee world - directory of the world's beekeeping museums issue*   90

Editor: Eva Crane
Publisher: International Bee Research Association
Address: Hill House, Gerrards Cross, Buckinghamshire SL9 0NR, UK
Year: 1979 (supplement 1980)
Pages: 20
Price: $1.90
Content: Entries for about 90 beekeeping museums. Entries include: name, address, year established, name of contact, hours open, accessibility to the public, description of holdings, catalogues and other publications available.

Commonwealth forestry handbook 1981, tenth edition   91

Publisher: Commonwealth Forestry Association
Address: c/o Commonwealth Forestry Institute, South Parks Road, Oxford OX1 3RB, UK
Pages: xii + 250
Price: £6.00
Geographical area: World, Commonwealth in particular.
Content: Details of senior personnel of Commonwealth Forest Services research and educational institutes, organizations, and societies; list of forestry publications; technical information section; list of world timbers with their standard botanical names and countries of origin.
Next edition: 1988; £10.00.

Directory of specialist workers on phytophthora palmivora with special reference to cacao   92

Editor: P.H. Gregory
Publisher: Cocoa, Chocolate and Confectionery Alliance
Address: 11 Green Street, London W1Y 3RF, UK
Year: 1978
Pages: ii + 37
Availability: OP
Geographical area: World.
Content: There are 30 entries arranged alphabetically by surname, with a geographical index.

Guide to agricultural information sources in Asia and Oceania   93

Editor: G.R.T. Levick
Publisher: Fédération Internationale de Documentation (FID)
Address: PO Box 90402, 2509 LK 's-Gravenhague, Netherlands
Year: 1980
Pages: vi + 72
Price: Dfl40.00
ISBN: 92-66-00592-4
Content: Brief descriptions of secondary sources, bibliographies and directories, of relevance to agricultural topics, with special reference to the Asia and Oceania region.

Horticultural research international: directory of horticultural research institutes and their activities in 61 countries   94

Publisher: PUDOC, Centre for Agricultural Publishing and Documentation
Address: PO Box 4, 6700 AA Wageningen, Netherlands
Year: 1981
Pages: 698
Price: Dfl144.00
ISBN: 90-220-0765-0
Geographical area: World.
Content: About 1400 horticultural research institutes with their addresses and the names of 14 000 research workers mentioning their field of interest: the following information is supplied by country: a country map showing the site of institutes; a brief survey of horticultural research in the country; names and addresses of the research institutes; names of principal workers in each institute; main activities of each institute and its personnel.
There are indexes to names of places and names of research workers.

Information sources in agriculture and food science   95

Editor: G.P. Lilley
Publisher: Butterworth and Company Limited
Address: Borough Green, Sevenoaks, Kent TN15 9PH, UK
Year: 1981
Pages: 216
Price: £35.00
ISBN: 0-408-10612-3
Series: Butterworths guides to information sources

Information sources on bioconservation of agricultural wastes* 96

Publisher: United Nations Industrial Development Organization
Address: Vienna International Centre, PO Box 300, A-1400 Wien, Austria
Price: $4.00
Series: UNIDO guides to information sources
Geographical area: World.

Information sources on essential oils* 97

Publisher: United Nations Industrial Development Organization
Address: Vienna International Centre, PO Box 300, A-1400 Wien, Austria
Price: $4.00
Series: UNIDO guides to information sources
Geographical area: World.

Information sources on grain processing and storage 98

Publisher: United Nations Industrial Development Organization
Address: Vienna International Centre, PO Box 300, A-1400 Wien, Austria
Year: 1982
Pages: xii + 97
Price: $4.00
Series: UNIDO guides to information sources
Geographical area: World.
Content: There are 840 entries, arranged as follows: professional, trade and research organizations, learned societies, and special information services; directories; sources of statistical and marketing data; basic books and monographs; proceedings, papers, and reports; periodicals; abstracting, indexing, and bibliographical material; dictionaries; other potential sources of information.

Information sources on the animal feed industry* 99

Publisher: United Nations Industrial Development Organization
Address: Vienna International Centre, PO Box 300, A-1400 Wien, Austria
Price: $4.00
Series: UNIDO guides to information sources
Geographical area: World.

Information sources on the beer and wine industry* 100

Publisher: United Nations Industrial Development Organization
Address: Vienna International Centre, PO Box 300, A-1400 Wien, Austria
Price: $4.00
Series: UNIDO guides to information sources
Geographical area: World.
Content: List of professional, trade, and research organizations.

Information sources on the coffee, cocoa, tea and spices industry* 101

Publisher: United Nations Industrial Development Organization
Address: Vienna International Centre, PO Box 300, A-1400 Wien, Austria
Price: $4.00
Series: UNIDO guides to information sources
Geographical area: World.

Information sources on the dairy product manufacturing industry* 102

Publisher: United Nations Industrial Development Organization
Address: Vienna International Centre, PO Box 300, A-1400 Wien, Austria
Price: $4.00
Series: UNIDO guides to information sources
Geographical area: World.

Information sources on the fertilizer industry* 103

Publisher: United Nations Industrial Development Organization
Address: Vienna International Centre, PO Box 300, A-1400 Wien, Austria
Price: $4.00
Series: UNIDO guides to information sources
Geographical area: World.

Information sources on the flour milling and bakery products industries* 104

Publisher: United Nations Industrial Development Organization
Address: Vienna International Centre, PO Box 300, A-1400 Wien, Austria
Price: $4.00
Series: UNIDO guides to information sources
Geographical area: World.

Information sources on the meat processing industry* 105

Publisher: United Nations Industrial Development Organization
Address: Vienna International Centre, PO Box 300, A-1400 Wien, Austria
Price: $4.00
Series: UNIDO guides to information sources
Geographical area: World.
Content: List of professional, trade and research organizations.

Information sources on the non- 106 alcoholic beverage industry*

Publisher: United Nations Industrial Development Organization
Address: Vienna International Centre, PO Box 300, A-1400 Wien, Austria
Price: $4.00
Series: UNIDO guides to information sources
Geographical area: World.
Content: Lists professional, trade, and research organizations.

Information sources of the vegetable 107 oil processing industry*

Publisher: United Nations Industrial Development Organization
Address: Vienna International Centre, PO Box 300, A-1400 Wien, Austria
Price: $4.00
Series: UNIDO guides to information sources
Geographical area: World.
Content: List of professional, trade and research organizations.

International directory of agricultural 108 engineering institutions/repertoire international d'institutions de génie rural*

Editor: H. von Hulst
Publisher: Food and Agriculture Organization of the United Nations
Address: Via delle Terme di Caracalla, 1-00100 Roma, Italy
Year: 1983
Pages: 500
Price: $18.00
Content: The directory covers central government services as well as international and national institutes dealing with land and water development, farm power and machinery, rural electrification, farm buildings, and farm work organization. Entries include: institution name, scientific staff, training and research, recent publications, language of correspondence; entries in institution's chosen FAO correspondence language. Arrangement: geographical by English name of the country.

International directory of animal 109 health and disease data banks*

Editor: Jesse Ostroff
Publisher: National Agricultural Library
Address: Agriculture Department, 10301 Baltimore Boulevard, Beltsville, MD 20705, USA
Year: 1982
Pages: 100
Price: Free
Content: The directory covers about 115 data banks containing bibliographic, epidemiologic, laboratory and clinical, and research-in-progress data on animal health and disease. Entries include: data bank name, organization name, address, name and telephone of contact, objectives and types of information available, description of data elements, file size, thesaurus used, processing method, products and services, publications, data bank vendors, search services, future plans. Arrangement: by type of data, then geographical. Indexes: subject, organization name/title.

International directory of fish 110 technology institutes*

Publisher: Food and Agriculture Organization of the United Nations
Address: Via delle Terme de Caracalla 1-00100 Roma, Italy
Year: 1984
Pages: 125
Price: $8.50
Content: Entries for over 80 institutes in some 50 countries. Entries include: institute name, address, name of director, responsible authority, source of financing, date founded, major fields of interest, description of facilities, size of professional staff, description of programme publications. Arrangement: geographical.

International directory of forestry and 111 forest products libraries*

Editor: Helvi M. Bessenyei, Peter A. Evans
Publisher: Pacific Southwest Forest and Range Experiment Station, US Forest Service
Address: Agriculture Department, Box 245, Berkeley, CA 94701, USA
Year: 1983
Pages: 95
Price: $11.50 (paper); $4.50 (microfiche)
Geographical area: World.
Content: Entries for about 270 libraries whose collec-

tions and services are concerned with forestry and forest products. Entries include: location, name, mailing address, telephone, name and title of librarian, size of collection, number of serials received, document services used, major subjects of collection, types of users, services, supporting institution name. Arrangement: geographical. Indexes: country, personal name.

International directory of marine scientists* 112

Publisher: Fishery Data Centre
Address: Fisheries Department, Food and Agriculture Organization of the United Nations, 1-00100 Roma, Italy
Year: 1984
Pages: 500
Content: The directory covers 19 000 scientists involved in marine science research and development. Entries include: name, affiliation, address, and scientific field(s). Arrangement: geographical, then by institution. Indexes: personal name, scientific field.

List of research workers in the agricultural sciences in the Commonwealth 113

Editor: G. Philip Rimington
Publisher: Commonwealth Agricultural Bureaux
Address: Farnham House, Farnham Royal, Slough SL2 3BN, UK
Year: 1981
Pages: x + 658
Price: £24.20, $48.50
ISBN: 0-85198-485-1
Content: The list contains the names and postal addresses and, where supplied, the telephone numbers and abbreviated telegraphic addresses of research workers in the agricultural sciences, at government and state-aided institutions in countries contributing to the Commonwealth Agricultural Bureaux; and indicates generally the lines of study in which each worker is specially interested.
The volume is divided into four parts. Part A gives the names of the executive council, CAB liaison officers and staff at the Commonwealth Institutes and Bureaux; Part B gives the names and addresses of research workers by countries; Part C is an index of the institutions mentioned in the list; and Part D an alphabetical index of the names of research workers.
Next edition: July 1987.

Middle East and Africa food directory 1986-87* 114

Editor: Saadeddine Chehab
Publisher: Middle East Food
Address: PO Box 135121, Beirut, Lebanon
Year: 1985
Price: $110.00
Content: The directory covers public and private sector firms serving the food, beverage, and catering market. Volume one covers the Middle East and comprises over 4000 listings, volume two details over 3000 firms in Africa.

Poultry world international directory 1985 115

Publisher: Farmers Publishing Group
Address: Survey House, Sutton, Surrey SM1 4QQ, UK
Pages: x + 82
Price: £2.00
Geographical area: World.
Content: Names and addresses; buyers' guide; brand names; stock suppliers; services; UK trade associations; UK egg packers; UK egg processors; UK poultry processors; UK poultry wholesalers.
Next edition: August 1985; £2.00.

Sourcebook on food and nutrition, third edition* 116

Publisher: Marquis Who's Who
Address: c/o Henry Thompson, London Road, Sunningdale, Berkshire, UK
Year: 1982
Price: £54.45
ISBN: 0-8379-4503-8
Content: Includes a directory section giving information on grant programmes, libraries, associations, colleges and universities, and career opportunities in the field of nutrition.

Who's who in world agriculture, second edition 117

Publisher: Longman Group Limited
Address: Sixth Floor, Westgate House, The High, Harlow, Essex CM20 1NE, UK
Year: 1984
Pages: 1300 (in 2 vols)
Price: £195.00
ISBN: 0-582-90111-1
Series: Reference on research
Content: This directory provides professional biographical profiles of over 12 000 senior research and advisory scientists in agriculture and veterinary sciences from more than 120 countries. Biographical details include full address, higher education, present post, previous experience, and main professional and research interests. The subjects covered are agricultural engineering, animal production, botany, fisheries and aquaculture, food sciences, forestry and forest products, horticulture, plant production, soil science, veterinary science and zoology.

INTERNATIONAL DIRECTORIES

World directory of forest pathologists and entomologists* 118

Publisher: International Union of Forestry Research Organizations
Address: Schonbrunn, Tirolergarten, A-1131 Wien, Austria
Year: 1985
Content: Entries include: name, title, affiliation, address, research interest, specialty. Arrangement: geographical. Indexes: subject, personal name, tree host, pathogen, insect.

World food crisis: an international directory of organizations and information resources 119

Editor: Thaddeus C. Trzyna
Publisher: California Institute of Public Affairs
Address: PO Box 10, Claremont, CA 91711, USA
Year: 1977
Pages: xxvi + 140
Price: $20.00
ISBN: 0-912102-21-7
Series: Who's doing what series
Geographical area: World.
Content: Listing of over 800 organizations and 275 major programmes in nearly every country of the world.

World list of forestry schools* 120

Publisher: Food and Agriculture Organization of the United Nations
Address: Via delle Terme de Caracalla, 1-00100 Roma, Italy
Year: 1985
Content: The list covers university-level and non-university-level schools of forestry, worldwide. Entries include: school name, address. Arrangement: geographical within above two categories.

CHEMICAL AND MATERIALS SCIENCE

Achema-jahrbuch 1983-85 121

Editor: Dicter Behrens
Publisher: DECHEMA
Address: PO Box 97 01 46, Theodor-Heuss-Allee 25, 6000 Frankfurt am Main 97, German FR
Year: 1985
Price: DM80.00
ISSN: Volume 1 - 0340-3726; volume 2 - 0340-8310; volume 3 - 0340-8337
Geographical area: Mainly Europe and USA.
Content: Volume one comprises 415 institute reports from 23 countries. The reports have been arranged in alphabetical order by countries, by place names within each country, and finally by the institute directors' or authors' names.
Volume two covers industrial and laboratory equipment, machinery, plant, and processes, giving: alphabetical list of firms; alphabetical list of advertisers; company reports, alphabetically arranged by firm's names; advertisement section alphabetically arranged by firm's names; alphabetical subject index for the section on company reports.
Volume three covers chemical engineering from A to Z, including: index of exhibitors classified by halls; alphabetical list of societies, research institutes (as far as they are exhibitors) and firms participating in the Achema yearbook with their full addresses and particulars of their principal fields of activity; alphabetical list of trade names and picture marks with descriptive matter and names of trade name owners; alphabetical list of apparatus, machinery, plant, constructional materials, and general supplies for chemical science and technology.
Next edition: 1987-88; DM100.00.

Chemaddressbook 122
[Chemicals address book]

Editor: Friedrich Derz
Publisher: Walter de Gruyter
Address: Genthiner Strasse 13, 1 Berlin 30, German FR
Year: 1974
Pages: xxxii + 819
Availability: OP
ISBN: 311-00-4274-6
Series: Volume 3 of Chem Buydirect
Geographical area: World.
Content: The index of identification codes, listed according to countries, gives the full name and address of manufacturers and distributors. In a separate index the firm names are listed alphabetically with corresponding identification codes. It also contains an international postcode index, in which chemical firms are listed by national postcode (Zip-code) and city name.

Chemfacts: PVC 123

Publisher: Chemical Data Services
Address: Quadrant House, The Quadrant, Sutton, Surrey SM2 5AS, UK
Year: 1984
Price: £80.00
Geographical area: World.
Content: The main section contains entries on almost 200 countries, giving the PVC producers and their plant locations, with details of capacity, process and licensor, as well as available production, import and export figures and analyses of trade. A detailed introduction

describes the properties, production and uses of PVC and lists the licensors of PVC technology, and a directory section with the producer companies, head office addresses, telephone and telex numbers, completes the book.

Chemical company profiles: the Americas, second edition 124

Publisher: Chemical Data Services
Address: Quadrant House, The Quadrant, Sutton, Surrey SM2 5AS, UK
Year: 1981
Pages: 244
Price: £55.00
Content: Information on some 1700 firms in 34 countries, including more than 600 in the USA, 160 in Canada, nearly 350 in Brazil, and approximately 100 each in Argentina, Colombia, and Mexico. The description of each company includes its postal address, and where available telephone and telex numbers, cable address, history and current structure, manufacturing and trading activities, directorate, number of employees, turnover and other financial data, and domestic and foreign subsidiaries.

Chemical industry directory and who's who 125

Editor: Kim Robinson
Publisher: Benn Business Information Services Limited
Address: PO Box 20, Sovereign House, Sovereign Way, Tonbridge, Kent TN9 1RQ, UK
Year: 1985
Pages: vi + 1465
Price: £42.00
ISSN: 0-86382-018-2
Geographical area: World.
Content: ABC guide to chemical manufacturers and traders; buyers' guide to chemicals; ABC guide to plant and equipment; buyers' guide to laboratory apparatus and scientific instruments; trade names; who's who in the chemical industry; UK section; European chemical and oil storage depots; professional and trade organizations; independent consultants; administrative groups; overseas associations.
Next edition: March 1986; £55.00

Chemical industry yearbook, second edition 126

Publisher: Chemical Data Services
Address: Quadrant House, The Quadrant, Sutton, Surrey SM2 5AS, UK
Year: 1984
Price: £80.00
Geographical area: World.
Content: The second edition follows two years after the first and reviews the production of and trade in the same 50 chemicals over a wider range of countries. Latin America, North Africa and the Middle East are more fully represented, but some of the countries chosen for the first edition have been replaced by others. The chemicals comprise 15 inorganics, 25 organics, five fertilizers and five polymers, including SBR.
The Year Book is divided into five sections - the first gives, country by country, production figures where available for the latest years published, and brief notes outlining some of the factors affecting the chemical industry such as fuel and feedstocks, mineral resources and power. Section 2 tabulates by country import and export (volume and values) figures for the two latest available years. Section 3 lists by product, for the years 1978 to 1982, production figures for each country, and Section 4 lists by product, for the years 1980 to 1982, the trade statistics for each country which reports them. The fifth section lists existing and planned plants for 47 of these chemicals in the countries covered.
Used in conjunction, the two editions of the Year Book span up to seven years in production figures, five in trade.

Chemical plant contractor profiles, fourth edition 127

Publisher: Chemical Data Services
Address: Quadrant House, The Quadrant, Sutton, Surrey SM2 5AS, UK
Year: 1983
Price: £80.00
Geographical area: World.
Content: The fourth edition includes 234 profiles, more than twice as many as the third edition of 1981. The format of the book has been revised to give, wherever possible, a standard structure for each company profile. Full address details head a text describing the company's history, present structure and number of employees. Following this are subsections on the directorate, financial data (capital, turnover, r&d expenditure), subsidiaries/associates/offices and processes and services.
An up-to-date selection of the company's chemical and related projects, recently completed, under way or planned, is presented in a table at the end of the profile. In this the entries, arranged by project/product, include details of the client, plant location, plant capacity, licensor and date of completion. Aspects of the company's involvement - feasibility study, management, construction, engineering, design and procurement - are indicated in the majority of these entries.

Chemical research faculties: an international directory 128

Editor: Gisella Linder Pollock
Publisher: American Chemical Society
Address: 1155 Sixteenth Street, NW, Washington, DC

INTERNATIONAL DIRECTORIES

20036, USA
Year: 1984
Pages: xxxv + 523
Price: $129.95 (US/Canada); $155.95 (export)
ISBN: 0-8412-0817-4
Geographical area: World.
Content: International directory listing over 735 institutions and universities granting advanced degrees in chemistry, chemical engineering, biochemistry, and pharmaceutical/medicinal chemistry. It contains three indexes: to faculty; to research subjects; and to institutes.

Chemproductindex 129
[Chemical production index]
Editor: Friedrich Derz
Publisher: Walter de Gruyter
Address: Genthiner Strasse 13, 1 Berlin 30, German FR
Year: 1976
Pages: xxxvi + 1979 (in 2 parts)
Availability: OP
ISBN: 311-00-2141-2
Series: Volume 1 of ChemBuydirect
Geographical area: World.
Content: Each of the different names (synonyms) of a chemical refers to the same registry number, which is assigned to each compound. The names are listed alphabetically.

Chemsuppliers directory 130
[Chemical Suppliers Directory]
Editor: Friedrich Derz
Publisher: Walter de Gruyter
Address: Genthiner Strasse 13, 1 Berlin 30, German FR
Year: 1976
Pages: xxx + 980
Availability: OP
ISBN: 311-00-4273-8
Series: Volume 2 of ChemBuydirect
Geographical area: World.
Content: Identification codes for manufacturers and distributors are listed according to the registry number of the chemicals they supply.

Directory of metallurgical consultants and translators 131
Editor: William A. Weida, William G. Jackson, Gayle J. Anton
Publisher: Metals Information
Address: American Society for Metals, Metals Park, OH 44073, USA
Year: 1984
Pages: v + 78
Price: $45.00; £30.00
Geographical area: World.
Content: The directory contains entries for over 450 consultants, and 200 translators, listed in alphabetical and geographical order. Each listing includes: each individual or organization's name, address, telephone number and telex where applicable, as well as specific metallurgical disciplines. Titles of indexes: consultants-keyword index; consultants-corporate/individual name index; translators-language index; translators-corporate/individual name index.
Not available on-line.

Footwear, raw hides and skins and leather industry in OECD countries 1981-82* 132
Publisher: Organization for Economic Cooperation and Development
Address: Publications Office, 2 rue André-Pascal, 75775 Paris Cedex 16, France
Year: 1983
Price: £4.50
ISBN: 92-64-02542-1

Handbook of textile fibres, fifth edition 133
Editor: J. Gordon Cook
Publisher: Merrow Publishing Company Limited
Address: ISA Building, Dale Road Industrial Estate, Shildon, County Durham DL4 2QZ, UK
Year: 1984
Pages: Volume 1 - xxviii + 208; volume 2 - xxx + 723
Price: £26.00 (volume 1 - £7.50; volume 2 - £18.50)
ISBN: Volume 1 - 0-904095-39-8; volume 2 - 0-904095-40-1
Geographical area: World.
Content: Volume one covers natural fibres, and volume two man-made fibres. Each gives comprehensive coverage of fibres produced throughout the world with information on source, production and processing, structure and properties, and use on each fibre.

Index of reviews in polymer science 134
Editor: M.S.M. Alger
Publisher: Rapra Technology Limited
Address: Shawbury, Shrewsbury, Shropshire SY4 4NR, UK
Year: 1977
Pages: v + 81
Price: £20.00
Geographical area: World.
Content: Around 2000 entries comprising title, author and reference, listed under type of polymerization, molecular structure, properties, analytical methods, and cross referenced under polymer.

Information sources on building boards from woods and other fibrous materials* 135

Publisher: United Nations Industrial Development Organization
Address: Vienna International Centre, PO Box 300, A-1400 Wien, Austria
Price: $4.00
Series: UNIDO guides to information sources
Geographical area: World.

Information sources on leather and leather products* 136

Publisher: United Nations Industrial Development Organization
Address: Vienna International Centre, PO Box 300, A-1400 Wien, Austria
Price: $4.00
Series: UNIDO guides to information sources
Geographical area: World.

Information sources on natural and synthetic rubber* 137

Publisher: United Nations Industrial Development Organization
Address: Vienna International Centre, PO Box 300, A-1400 Wien, Austria
Price: $4.00
Series: UNIDO guides to information sources
Geographical area: World.

Information sources on utilization of agricultural residues for the production of panels, pulp and paper* 138

Publisher: United Nations Industrial Development Organization
Address: Vienna International Centre, PO Box 300, A-1400 Wien, Austria
Price: $4.00
Series: UNIDO guides to information sources
Geographical area: World.

Information sources on the ceramics industry* 139

Publisher: United Nations Industrial Development Organization
Address: Vienna International Centre, PO Box 300, A-1400 Wien, Austria
Price: $4.00
Series: UNIDO guides to information sources
Geographical area: World.
Content: List of professional, trade, and research organizations.

Information sources on the furniture and joinery industry* 140

Publisher: United Nations Industrial Development Organization
Address: Vienna International Centre, PO Box 300, A-1400 Wien, Austria
Price: $4.00
Series: UNIDO guides to information sources
Geographical area: World.
Content: List of professional, trade, and research organizations.

Information sources on the glass industry* 141

Publisher: United Nations Industrial Development Organization
Address: Vienna International Centre, PO Box 300, A-1400 Wien, Austria
Price: $4.00
Series: UNIDO guides to information sources
Geographical area: World.
Content: List of professional, trade, and research organizations.

Information sources on the paint and varnish industry* 142

Publisher: United Nations Industrial Development Organization
Address: Vienna International Centre, PO Box 300, A-1400 Wien, Austria
Price: $4.00
Series: UNIDO guides to information sources
Geographical area: World.
Content: List of professional, trade, and research organizations.

Information sources on the pesticides industry 143

Publisher: United Nations Industrial Development Organization
Address: Vienna International Centre, PO Box 300, A-1400 Wien, Austria
Year: 1982
Pages: xii + 77
Price: $4.00
Series: UNIDO guides to information sources
Geographical area: World.
Content: There are 658 entries arranged in the following order: professional, trade, and research organizations, learned societies and special information services; directories; sources of statistical and marketing data; basic books, handbooks, and monographs; proceedings, papers, and reports; periodicals; abstracting, indexing, and bibliographical material; dictionaries and encyclopaedias; other potential sources of information.

Information sources on the petrochemical industry*   144

Publisher: United Nations Industrial Development Organization
Address: Vienna International Centre, PO Box 300, A-1400 Wien, Austria
Price: $4.00
Series: UNIDO guides to information sources
Geographical area: World.

Information sources on the printing and graphics industry*   145

Publisher: United Nations Industrial Development Organization
Address: Vienna International Center, PO Box 300, A-1400 Wien, Austria
Price: $4.00
Series: UNIDO guides to information sources
Geographical area: World.
Content: List of professional, trade, and research organizations.

Information sources on the pulp and paper industry*   146

Publisher: United Nations Industrial Development Organization
Address: Vienna International Centre, PO Box 300, A-1400 Wien, Austria
Price: $4.00
Series: UNIDO guides to information sources
Geographical area: World.
Content: List of professional, trade, and research organizations.

Information sources on the soap and detergent industry*   147

Publisher: United Nations Industrial Development Organization
Address: Vienna International Centre, PO Box 300, A-1400 Wien, Austria
Price: $4.00
Series: UNIDO guides to information sources
Geographical area: World.
Content: List of professional, trade, and research organizations.

International ceramic directory*   148

Editor: B.G.R. Logan
Publisher: London and Sheffield Publishing Company Limited
Address: 5 Pond Street, Hampstead, London NW3 2PN, UK
Year: 1984
Price: £40.00
ISBN: 0-900091-03-7
Geographical area: World.
Content: Manufacturers of all types of ceramic products, excluding refractory products, are listed by countries.
Next edition: 1988.

International directory of acid deposition researchers, North American and European edition*   149

Editor: Steven F. Vozzo
Publisher: Environmental Protection Agency
Address: Corvallis, OR 97330, USA
Year: 1983
Pages: 160
Price: $16.00 (paper); $4.50 (microfiche)
Content: Entries for more than 1300 scientists involved in some aspect of acid precipitation research in North America and Europe. Entries include: scientist name, title, organization, address, telephone, areas of expertise related to acid precipitation. Arrangement: geographical. Indexes: subject, alphabetical.

International pulp and paper directory-IPPD   150

Publisher: Miller Freeman Publications Incorporated
Address: 500 Howard Street, San Francisco, CA 94105, USA
Year: 1982
Pages: 500
Price: $77.00
Geographical area: North and South America, Europe, Asia.
Content: Entries for over 2500 mills listed alphabetically by country, giving: mailing address, telephone, and telex; company ownership; officers and key management, production, technical, engineering, and marketing personnel; pulp grades produced and capacity; paper and paperboard grades produced and capacity; production facilities, equipment, paper machines, sizes and speeds; coating facilities and equipment; converting operations; power generation facilities. Worldwide index of pulp and paper mills and headquarters; alphabetical index of 8000 management, production, technical, engineering and marketing personnel at headquarters offices and mill sites; directory by grades; directory of producers by country; associations; research organizations.
Next edition: October 1985; $117.00.

Leather guide, sixteenth edition 1986*   151

Editor: J. Hedges, P. Bryant
Publisher: Benn Business Information Services
Address: Sovereign Way, Tonbridge, Kent TN9 1RW, UK

Year: 1985
Price: £39.00
ISBN: 0-86382-026-3
Geographical area: World.
Content: Listing, alphabetically by company name within countries, of tanners and merchants, hide and skin suppliers; chemical suppliers, machinery manufacturers, trade organizations, etc; buyers' guide.

Materials research centres: a world directory of organizations and programmes in materials science 152

Editor: E. Mitchell, E. Lines
Publisher: Longman Group Limited
Address: Sixth Floor, Westgate House, The High, Harlow, Essex CM20 1NE, UK
Year: 1983
Pages: 576
Price: £110.00
ISBN: 0-582-90013-1
Series: Reference on research
Content: Details of over 4000 industrial organizations, academic establishments, research associations, government laboratories and international advisory bodies which conduct, finance or promote research into materials science. Arranged alphabetically by country, the book is indexed by establishment and by subject.

Middle East and world water directory, second edition 1983-84* 153

Editor: Rene P. Kareh
Publisher: Arab Water World
Address: PO Box 135121, Beirut, Lebanon
Year: 1983
Price: $110.00
Content: Volume one contains entries for over 7000 firms serving the water and sewerage industry of the Middle East and Africa. Volume two covers the Americas, Europe, the far East, and Australasia, and lists over 3000 leading manufacturers of water and sewage equipment.
Next edition: July 1985; $110.00.

New trade names in the rubber and plastics industry 154

Editor: Richard Juniper
Publisher: Pergamon Press Limited
Address: Headington Hill Hall, Headington, Oxford OX3 0BW, UK
Year: 1983
Pages: vi + 220
Price: £30.00
Geographical area: World.
Content: Around 2250 entries are listed by trade name, company, and classification. Directory also gives company names and addresses.
Available on-line.
Next edition: May 1985; £35.00.

Refining, construction, petrochemical, and natural gas processing plants of the world 155

Publisher: Midwest Oil Register Incorporated
Address: PO Box 700597, Tulsa, OK 74170 USA
Year: 1985
Pages: 518
Price: $30.00
Content: Entries give: company name, address, telephone number, capacities, products, personnel showing titles. Also contains a section for companies who are engaged in engineering, construction and service of refineries, natural gas processing, and petrochemical plants.
Not available on-line.
Next edition: December 1985; $30.00.

Source journals in metallurgy 156

Publisher: Metals Information
Address: Metals Park, OH 44073, USA
Price: £23.50; $40.00
Content: Listing of the 1300 scientific, engineering and trade journals processed for abstracting and indexing purposes by Metals Information. This reference is divided into three sections: an alphabetical listing of all journals, with the standard ISO abbreviation, publisher, frequency of publication and ISSN number for each periodical; an alphabetical listing of all the publishing companies, including their complete address, and a list of their metallurgical journals; and the third section groups the journals by the major emphasis of their subject content.

World calendar of forthcoming meetings, metallurgical and related fields 157

Editor: Lynne Biggs
Publisher: Metals Information, Institute of Metals
Address: 1 Carlton House Terrace, London SW1Y 5DB, UK
Year: 1985
Pages: ii + 60
Price: £50.00; $80.00
Geographical area: World.
Content: The publication is issued quarterly and contains around 700 entries. Conferences are listed in chronological order, and entries include: location, subject matter, name of contact, sponsors where relevant, lan-

guage, and registration. Indexes are as follows: conference listing; subject index; sponsor index; location index.
Not available on-line.
Next edition: September 1985; $80.00/£50.00.

World directory of chemical producers, 1985-86 — 158

Publisher: Chemical Information Services Limited
Address: PO Box 61, Oceanside, NY 11572, USA
Year: 1985
Pages: 724
Price: $295.00, £268.00
ISSN: 0196-0555
Content: Nearly 50 000 alphabetically-listed product titles (including cross-references) manufactured by over 5000 chemical producers worldwide. There are two sections: a product section listing chemicals in alphabetical order with reference to all producers of each compound; and an address section listing manufacturers first by country and then alphabetically. Name, address, telephone, telex, and cable addresses are given.
Virtually all synonyms by which a chemical may be known in commerce have been inserted as cross-references in the alphabetical sequence of the product section. Substances in this section include basic organic and inorganic chemicals, including bulk pharmaceuticals, basic insecticides, herbicides, fungicides and other agricultural chemicals, research chemicals, and laboratory reagents. It does not include chemical and pharmaceutical specialities and formulated mixtures. Catalysts, dyes, and resins are included and listed by general categories.

World directory of crystallographers, sixth edition — 159

Editor: A.L. Bednowitz
Publisher: International Union of Crystallography
Address: PO Box 322, 3300 AH Dordrecht, The Netherlands Netherlands
Year: 1981
Pages: 228
Price: $10.00
ISBN: 90-27701310-3
Content: A who's who supported by the International Union of Crystallographers.
Next edition: 1985.

World directory of fertilizer products, fifth edition — 160

Publisher: British Sulphur Corporation
Address: Parnell House, 25 Wilton Road, London SW1V 1NH, UK
Year: 1981
Price: $150.00; £55.00
ISBN: 0-902777-52-1

Content: Entries for more than 300 international fertilizer producers, listing the products offered, their formulations and physical form. A synopsis of each trading company's activities is given, together with addresses, telephone and cable information. Likewise the major producers/suppliers of raw materials are listed with a short description of their activities.
A product guide to companies and organizations of fertilizers and fertilizer raw materials is included, which will prove invaluable for all those engaged in global trading. This information is presented in a tabulated form which will enable users to see at a glance which products each concern deals with, and for each product all the companies/organizations involved in its international trade. It will be noted from the contents listing that the types of fertilizer raw materials, fertilizer intermediates, and fertilizer products covered by each company/organization are given in detail.
The listing of the world's purchasing organizations by country together with their addresses, etc which appeared in the previous edition, has been substantially expanded.
Next edition: September 1986.

Worldwide chemical directory, fourth edition — 161

Publisher: Chemical Data Services
Address: Quadrant House, The Quadrant, Sutton, Surrey SM2 5AS, UK
Year: 1982
Price: £65.00
Content: The directory lists, country by country, some 10 500 companies which make and/or distribute all kinds of chemical products, including inorganic, organic and speciality chemicals, pharmaceuticals, plastics and fertilizers; it has a full alphabetical company index.
Company details given include postal and street addresses, cable addresses, telex and telephone numbers, and a short description of the company's activities. A separate section provides addresses and other details of the leading trade and professional associations, national and international, of the chemical and pharmaceutical industries.

EARTH AND SPACE SCIENCES

Arid lands research institution: a world directory 1977 — 162

Editor: Patricia Paylore
Publisher: University of Arizona Press
Address: 1615 East Speedway, Tucson, AZ 85719, USA

Pages: xv + 317
Price: $12.50
ISBN: 0-8165-0631-0
Content: Approximately 200 institutions arranged by continent, then country with name index of 1000 arid lands research scientists, and a subject index. Entries include information on the nature of the institution, geographica, location, field studies, climate and vegetation types; scope of interest; research programme; finances; staff and organization; facilities; publications in series; and history.

COSPAR directory of organization and members 163

Publisher: Committee on Space Research (COSPAR)
Address: 51 boulevard de Montmorency, 75016 Paris, France
Year: 1985
Pages: vi + 128
Price: F45.00
Geographical area: World.
Content: Members of COSPAR (national and union); COSPAR commissions and panels and their membership lists; some other ICSU organizations' officers; national committees for COSPAR; addresses of all persons appearing in the directory (alphabetical index).
Next edition: Late 1986; F50.00.

Directory of geophysical research 164

Editor: O' Reilly
Publisher: Elsevier Science Publishers bv
Address: PO Box 211, 1000 AE Amsterdam, Netherlands
Year: 1985
Pages: 200
Geographical area: World.
Content: Individuals are listed under such classifications as: exploration; resources; exploration methods; techniques and instrumentation; computational techniques; the oceans; etc. There is a geographical index.

Directory of geoscience departments in universities in developing countries, third edition 165

Editor: B.K. Tan, S. Chandra Kumar
Publisher: Association of Geoscientists for International Development
Address: c/o AIT, PO Box 2754, Bangkok 10501, Thailand
Year: 1983
Pages: iv + 108
Price: $8.00
ISBN: 974-8202-09-7
Geographical area: Africa, Asia, Latin America, and the Caribbean.
Content: Details of geoscience departments listed country by country giving university, faculty address, departments, heads and staff, regular courses and degree, average annual number of graduates, scholarships etc, job opportunities available during vacations, publications, language of instruction. University index; faculty index.
Next edition: 1987.

Directory of neotropical protected areas 166

Publisher: International Union for the Conservation of Nature and Natural Resources
Address: 1196 Gland, Switzerland
Pages: xxv + 465
Price: $35.00
ISBN: 0-907567-63-0
Geographical area: North America, South America, Caribbean.
Content: Approximately 352 entries, arranged by country subdivided by protected area. Entries are indexed, and cover: management category, biogeography, legal protection, date established, geographical location, altitude, area, land tenure, physical features, vegetation, noteworthy fauna, zoning, tourism, disturbances or deficiencies, scientific research, special scientific facilities, principal reference material, staff, local park or reserve administration, name of CNPPA coordinator, information dated.
Not available on-line.

Directory of palaeontologists of the world, fourth edition 167

Editor: Vivianne Berg-Madsen
Publisher: International Palaeontological Association
Address: E-206 Natural History Building, Smithsonian Institution, Washington, DC 20560, USA
Year: 1984
Pages: 193
Availability: OP
Content: Approximately 4400 palaeontologists are listed alphabetically along with a taxonomical index (which arranges workers by speciality: Mollusca, Mammalia), and a geographical index in which all addresses are listed alphabetically by country.
Not available on-line.
Next edition: 1989; $10.00.

Directory of sea-level research 168

Editor: I. Shennan, P. Pirazzoli
Publisher: Nils-Axel Mörner
Address: Geological Institute, Stockholm University, S-10691 Stockholm, Sweden
Year: 1984
Pages: 69

INTERNATIONAL DIRECTORIES

Price: Free
Series: IGCP-200 project
Geographical area: World.
Content: There are 465 entries covering 60 countries, arranged as follows: main directory with name, address, and research information; index of geographical areas of investigation; index of disciplines of study and methods/techniques used; index of home countries of participants.

Directory of Western palearctic wetlands   169

Editor: E. Carp
Publisher: International Union for the Conservation of Nature and Natural Resources
Address: 1196 Gland, Switzerland
Year: 1980
Pages: 550
Price: $27.50
ISBN: 2-88032-300-2
Series: Wetlands series
Geographical area: Northern hemisphere.
Content: There are entries for 44 countries, arranged by country subdivided by area. Entries cover: criteria for inclusion, geographical location, area, altitude, water depth, wetland type, ecology, legal status, tenure, management practices, threats, scientific threats, scientific research, principal reference material. There are no indexes.
Not available on-line.

Directory of world seismography stations*   170

Editor: Barbara B. Poppe
Publisher: National Oceanic and Atmospheric Administration
Address: Commerce Department, 325 Broadway, Boulder, CO 80303, USA
Year: Volume 1 1980; volume 2 1984
Pages: 465
Price: Volume 1 $13.00; volume 2 $15.00
Content: The directory is planned as a six-volume set covering all active and defunct stations. Volume 1, Part 1, covers more than 400 stations in the United States, Canada, and Bermuda, is the only section published to date. Volume 2 covers Japan, China, and Korea. Entries include: station name, address, telephone; name of operating organization; source name and address for station's records; site data (longitude, latitude elevation, date opened and closed, foundation matter, geological age); instrumentation description; timing system; history; system response curves. Arrangement: geographical. Indexes: station code (with state, name of operator). Volume 1, Part 2 will cover South America but publication plans are indefinite.

Doescher's directory of brachiopodologists*   171

Editor: Rex A. Doescher
Publisher: Doescher, Rex A.
Address: Natural History Building E-207, Smithsonian Institution, Washington, DC 20560, USA
Year: 1985
Pages: 56
Price: Free
Geographical area: World.
Content: There are 700 brachiopod workers listed from 39 countries. Individual data profiles of each worker was generated from coded responses of a questionnaire which classifies the brachiopodologist by superfamily, geological period, field of interest (taxonomy, anatomy, etc), and geographical areas (where he did research). A listout of workers by geological period and a telephone directory is included. The directory is updated monthly. This directory is currently stored on the Smithsonian's Honeywell (series 60) Model 66/05/ computer; however requests for the directory should be made to the editor.

Earth and astronomical sciences research centres: a world directory of organizations and programmes   172

Editor: Jennifer M. Fitch
Publisher: Longman Group Limited
Address: Sixth Floor, Westgate House, The High, Harlow, Essex CM20 1NE, UK
Year: 1984
Pages: 742
Price: £110.00
ISBN: 0-582-90020-4
Series: Reference on research
Content: Details on approximately 4800 organizations, laboratories and observatories throughout the world which conduct or finance research into any aspect of geology, cartography, surveying, ocean studies, meteorology and climatology, planetary and galactic observations, geochemistry, mineralogy and petrology, mining studies and earthquake control.
Arranged alphabetically by country, each entry contains, where appropriate, the address, type of organization and affiliation, senior staff, scope of interest and publications.
There is a titles of establishments index and a subject index.

Information sources in the earth sciences, second edition   173

Editor: J. Hardy, D.N. Wood, A. Harvey
Publisher: Butterworth and Company Limited
Address: Borough Green, Sevenoaks, Kent TN15 8PH, UK
Year: 1985

Pages: 576
Price: £45.00
ISBN: 0-408-01406-7
Series: Butterworths guide to information sources
Geographical area: World.

International directory of current research in the history of cartography and in carto-bibliography 174

Editor: Eila M.J. Campbell, P.K. Clark, A. Elizabeth Clutton
Publisher: Geo Books Norwich
Address: Regency House, 34 Duke Street, Norwich NR3 3AP, UK
Year: 1983
Pages: iv + 194
Price: £3.00
ISBN: 0-86094-144-2
Series: History of cartography
Geographical area: World.
Content: About 380 entries giving: name and address of research worker; area/topic of investigation; listings of recent publications (past 2 years). Index of places and subjects; index of personal names occuring in the listed research topics; index of contributors by country.
Next edition: June 1985; £3.50.

International directory of marine scientists, third edition 175

Publisher: Unesco; Food and Agriculture Organization of the United Nations (Joint publication)
Address: Unesco, 7 place de Fontenoy, 75700 Paris, France; FAO, Via delle Terme di Caracalla, 00100 Roma, Italy
Year: 1983
Pages: iv + 748
Price: Free
ISBN: 92-3-102243-1
Series: Aquatic sciences and fisheries information system
Content: Entries for 18 000 marine scientists, each giving name and subjects of specialization, arranged under 2500 institutions, each with postal address, some with telephone and telex numbers. Institution entries are arranged under approximately 100 countries and under a separate category for international organizations.
Alphabetical index of names giving country and page reference; index of names under approximately 100 subject areas, giving country and page reference to part 1.

International directory of mining 176

Editor: Jean Leroy
Publisher: McGraw Hill Publications
Address: 1221 Avenue of the Americas, Newport, NY 10020, USA
Year: 1985
Pages: 650
Price: $80.00
ISBN: 0-07-606924-9
Geographical area: World.
Content: Information on the industry and its support services, including listings of: mining company data; company data/headquarters listings; United States mine/plant units; Canadian mine/plant units; international mine/plant units; mineral sales organizations; ore shipping facilities; consultants and services directory; financial services; mine bureaux and geological surveys; industry surveys, tabulations; executive and operating; personnel index.

International hydrographic organization yearbook 177

Publisher: International Hydrographic Bureau
Address: BP 345, Monte Carlo, MC 98000, Monaco
Year: 1985
Pages: xiv + 214
Price: F55.00
Series: Periodical publications of IHO - PP-05
Geographical area: World.
Content: All information in bilingual form (English/French): lists names, addresses and general information on national hydrographic services throughout the world, (first the fifty-two member states of the IHO then about 43 other states which have supplied information). Biographical notes concerning heads of hydrographic offices of member governments and IHB directors; general information on IHO and IHB. Appendices: list of governments which have participated in work of IHB since its creation; table of reported tonnages for IHO member states; table of shares, contributions and votes; table of supplementary shares; former directing committees of the IHB; presidents and vice presidents of IH conferences; membership of finance committee; regional hydrographic commissions within the IHO.
Next edition: January 1986; F55.00

Meteorological services of the world 178

Publisher: World Meteorological Organization
Address: 41 Avenue Giuseppe Motta, Case postale No 5, CH-1211 Genève 20, Switzerland
Year: 1982
Pages: iv + 170
Price: SFr24.00
Content: The publication gives, for each country entered: state meteorological service (address, telex, telephone number, divisions); state meteorological organizations dependent on the above-mentioned service; state meteorological organizations independent of the above-mentioned service; meteorological institutes forming part of the service; meteorological institutes not forming

INTERNATIONAL DIRECTORIES

part of the service; publications; responsible authority for aeronautical meteorological services.
Next edition: September 1985.

Pollution research index: a guide to world research in environmental pollution, second edition   179

Editor: Dr Andrew I. Sors, David Coleman
Publisher: Longman Group Limited
Address: Sixth Floor, Westgate House, The High, Harlow, Essex CM20 1NE, UK
Year: 1979
Pages: x + 555
Availability: OP
ISBN: 0-582-90006-9
Content: A worldwide reference guide to over 2000 organizations, including government departments, universities, research institutes and manufacturing industry with separate sections on over 100 countries. It includes details of physical, chemical, and biological aspects of pollutants, their movement and transformations in air, freshwater, marine and terrestrial environments, their effect on man and nature, and techniques for their measurement and control.
Indexed by organization and by subject.
Next edition: Not planned.

Space activities and resources: review of United Nations international and national programmes   180

Publisher: United Nations Publications
Address: Room DCZ-853, New York, NY 10017, USA
Year: 1977
Price: $14.50
ISBN: 92-1-100158-7
Geographical area: World.

World directory of mineralogists   181

Editor: Dr Fabien Cesbron
Publisher: Bureau de Recherches Géologiques et Minières; International Mineralogical Association
Address: Editions du BRGM, PO Box 6009, 45060 Orléans Cedex, France
Year: 1985
Pages: 361
Price: $12.00
ISBN: 2-7159-0102-X
Content: The directory lists 7050 names of scientists from 34 countries working in the field of mineralogy, petrology, geochemistry, crystallography etc. It includes: list of countries (alphabetical order) with names, university degrees, present positions, professional addresses, telephone numbers, domains of major scientific interest; a general index of names (with corresponding country); addresses of national mineralogical associations with the names of their national representatives; names and addresses of chairmen, vice-chairmen and secretaries of the various commissions and working groups of the International Mineralogical Association. The complete list of addresses and names can be supplied on magnetic tape.

World palaeontological collections*   182

Editor: R.J. Cleevely
Publisher: Mansell Publishing Limited
Address: 950 University Avenue, Bronx, NY 10452, USA
Year: 1983
Pages: 450
Price: $75.00
Content: Covers collections of shells, fossils, insects and minerals made by British and European naturalists during the past two hundred years. Entries include: collection name, present location of collection, name of curator; name and title of collector, brief biography (with birth and death dates), memberships in major British natural history societies, biographical sources. Indexes: collector name, subject.

Worldwide directory of national earth-science agencies*   183

Publisher: US Geological Survey
Address: 12201 Sunrise Valley Drive, Reston, VA 22092, USA
Year: 1981
Pages: 80
Price: Free
Content: Entries cover governmental agencies in about 160 countries concerned with earth sciences; about 90 major international organizations involved in earth science research, study, and promotion are also listed. Entries include: for government agencies - name of agency, address, field of interest; for international organizations - name, address, field of interest. Agencies are arranged by country; organizations are by field of interest.

ELECTRONICS

Aerospace research index: a guide to world research in aeronautics, meteorology, astronomy, and space science   184

Editor: A.P. Willmore, S.R. Willmore
Publisher: Longman Group Limited

Address: Sixth Floor, Westgate House, The High, Harlow, Essex CM20 1NE, UK
Year: 1981
Pages: x + 597
Availability: OP
ISBN: 0-582-90009-3
Content: The directory aims to provide a comprehensive source of reference on government establishments, industrial laboratories, colleges and universities which are engaged on research in the related fields of aeronautics, meteorology, astronomy, and space science.
There is a title and a subject index.
Next edition: Not planned.

Asian and Pacific electrical and electronics directory* 185

Editor: Joachim Lee
Publisher: General Trade Directories Private Limited
Address: 2 Soon Wing Road, 03-09 Soon Wing Industrial Building, Singapore 1334, Singapore
Year: 1985
Pages: 1200
Price: $80.00
ISSN: 0129-7449
Content: The directory covers over 10 000 firms engaged in manufacturing, distributing, importing, exporting, and servicing electrical and electronic products. Entries include: company name, address, telephone, name of principal executive, products and services, trade and brand names. Arrangement: classified by type of product or service, then geographical. Indexes: product/service.
Next edition: Autumn 1986.

Computer books and serials in print 1985 186

Publisher: Bowker Publishing Company
Address: 58/62 High Street, Epping, Essex CM16 4BU, UK
Year: 1985
Pages: 1500
Price: £70.00
ISBN: 0-8352-2044-3
Geographical area: World.
Content: Listing of around 18 000 books and 1700 serials dealing with the specialized and growing field of computers, indexed within a new, special subject arrangement that stresses the applied uses of computers. All aspects of the technology are covered, including automation, robotics, personal computers and more, with full finding and ordering information for each entry.
Next edition: July 1986; £78.00.

Computer publishers and publications 1985-86: an international directory and yearbook, second edition 187

Editor: Efrem Sigel, Frederica Evan
Publisher: Communications Trends Incorporated
Address: 2 East Avenue, Larchmont, NY 10538, USA
Year: 1985
Pages: 450
Price: $95.00
ISBN: 0-8103-2137-8
Geographical area: World.
Content: The main section contains entries on publishers of popular magazines, trade and technical periodicals, scholarly journals, newsletters, looseleaf services, and general and technical books. Entries are arranged by publisher and include full name, address, and phone number; contact person and key personnel; year founded; brief description; details concerning periodicals and books published; and more. Indexes.
The Computer Publishers and Publications: Supplement provides updated information in this rapidly changing field on new periodicals, new book publishers, and significant changes at existing publishers. Index.
Next edition: July 1986.

Computers and information processing world index 188

Editor: Suzan Deighton, John Gurnsey, Janet Tomlinson
Publisher: Gower Publishing Company Limited
Address: Gower House, Croft Road, Aldershot, Hampshire GU11 3HR, UK
Year: 1984
Pages: x + 616
Price: £65.00
ISBN: 0-566-03410-7
Content: Organizations including research centres, libraries, user groups, trade associations, manufacturers, learned societies, professional associations, (subdivided by country); reference works - bibliographies, abstracts, data base guides, encyclopaedias, dictionaries, glossaries, and lexicons, directories, reports, handbooks, history, biography; computing applications - an indicative bibliography of key texts and major journals; communications; office systems; information storage and retrieval; artificial intelligence; management data processing; social and behavioural sciences; medicine and health care; natural sciences; engineering; humanities; simulation; graphics; social implications of computers; journals (subdivided by country); standardizing bodies; publishers' addresses; index.
Not available on-line.

INTERNATIONAL DIRECTORIES

Data world international 189

Publisher: Auerbach Publishers Incorporated
Address: 6560 North Park Drive, Pennsauken, NJ 08109, USA
Price: $845.00
Content: This four-volume continuous publication gives comprehensive EDP information including, in each volume: a directory of companies, product reports with user reactions, price data, and specification tables that describe and compare product features. Volume one covers general-purpose computers and peripherals; volume two, mini- and microcomputers; volume three, applications and systems software; volume four, data communications facilities, equipment, and services.

Directory of international broadcasting, fifth edition* 190

Editor: Christopher Surgenor
Publisher: BSO Publications Limited
Address: 3-5 St John Street, London EC1M 4AE, UK
Year: 1984
Price: £25.00
Geographical area: World.
Content: Listing of television and radio stations throughout the world, manufacturers of equipment, etc.
Next edition: June 1985.

Directory of United Nations information systems volume 1 - information systems and data bases* 191

Publisher: United Nations Organization
Address: Palais des Nations, CH-1211, Genève 10, Switzerland
Year: 1980
Price: $22.00
Geographical area: World.

Directory of United Nations information systems volume 2 - information sources in countries* 192

Publisher: United Nations Organization
Address: Palais des Nations, CH-1211, Genève 10, Switzerland
Year: 1980
Price: $13.00
Geographical area: World.

Directory of word processing systems 193

Editor: Denis J. Grimes, Brian W. Grimes
Publisher: Wiley, John, and Sons Limited
Address: Baffins Lane, Chichester, Sussex PO19 1UD, UK
Year: 1985
Pages: 634
Price: £41.75
ISBN: 0471-87820-0
Series: Kelly-Grimes computer buyers guide series
Geographical area: Europe and North America.
Content: This is one of a series of comprehensive guides for selecting the appropriate hardware, software, peripherals, and supplies for computing needs. Designed as a first-source computer product reference for new and experienced computer users, the book includes product specifications, ordering information, and manufacturers' recommendations. The Word Processing Systems Guide is a one-step source of information for buying word processing equipment.
Not available on-line.

IBM (R) PC compatible computer directory: hardware, software and peripherals 194

Editor: Brian W. Kelly, Dennis J. Grimes
Publisher: Wiley, John, and Sons Limited
Address: Baffins Lane, Chichester, Sussex PO19 1UD, UK
Year: 1984
Pages: 604
Price: £26.95
ISBN: 0471-87819-7
Series: Kelly-Grimes directories
Geographical area: Europe and North America.
Content: This book has been designed to be a first source computer product and service reference for both the neophyte and the experienced computer user. Brief articles at the beginning of this volume help readers understand the rapidly expanding world of personal computers, software, and their related products.
Not available on-line.

Information sources on the electronics industry* 195

Publisher: United Nations Industrial Development Organization
Address: Vienna International Centre, PO Box 300, A-1400 Wien, Austria
Price: $4.00
Series: UNIDO guides to information sources
Geographical area: World.

International defense directory 196

Editor: C.E. Howard
Publisher: Interavia SA
Address: PO Box 162, 86 Avenue Louis-Casai, 1216 Cointrin-Genève, Switzerland
Year: 1985
Pages: xciv + 600

Price: SFr500.00
Geographical area: World.
Content: There are 30 000 entries, country by country, split in fields of activity; alphabetical index of entries; alphabetical index of products and services; alphabetical index of product names.
Next edition: 1986

International directory of telecommunications: market trends, companies, statistics, products, and personnel   197

Editor: Steven Roberts
Publisher: Longman Group Limited
Address: Sixth Floor, Westgate House, The High, Harlow, Essex CM20 1NE, UK
Year: 1984
Pages: xv + 282
Price: £49.00
ISBN: 0-582-90021-2
Series: Companion for industry
Geographical area: World.
Content: An international guide outlining market developments in the telecommunications industry. It provides a description of market trends, statistics, and a summary of current and proposed national regulations, plus company profiles in the areas of telecommunications services, equipment manufacturers and broadcasting companies.
The directory is fully indexed by organization, product, and names of key personnel.
Next edition: 1986.

Marquis international who's who directory of optical science and engineering   198

Publisher: Marquis Who's Who Incorporated
Address: 200 East Ohio Street, Chicago, IL 60611, USA
Year: 1985
Pages: 750
Price: $125.00
ISBN: 0-8379-7001-6
Geographical area: World.
Content: Approximately 6500 professional profiles arranged alphabetically. Profiles include product and subject expertise, optical application areas, consulting experience, career history, professional memberships, education background, awards and grants, and home and/or office address.
Titles of indexes: area of expertise, application area, product expertise, and geographical.

Marquis who's who directory of computer graphics   199

Publisher: Marquis Who's Who Incorporated
Address: 200 East Ohio Street, Chicago, IL 60611, USA
Year: 1984
Pages: xvi + 549
Price: $125.00
ISBN: 0-8379-5901-2
Geographical area: World.
Content: The directory contains over 4500 entries, which are arranged alphabetically. Content of entries: name, current title and company name, job function, computer graphics function, computer graphics area(s) of interest and/or expertise, computer graphics industry application area(s), computer graphics product expertise, birth information, education, professional certification, career history, career-related activities, creative works, professional memberships, address(es), telephone number.
Titles of indexes: computer graphics area(s) of interest or expertise; computer graphics application area(s); computer graphics product expertise; and geographical.

Microwaves and rf product data directory*   200

Editor: Stacy V. Bearse
Publisher: Hayden Publishing Company Incorporated
Address: 50 Essex Street, Rochelle Park, NJ 07662, USA
Year: 1984
Pages: 930
Price: $25.00
Geographical area: World.
Content: The directory covers about 4500 manufacturers of high-frequency equipment components, including antennas and accessories, materials and accessories, passive components, semiconductors, solid-state components, systems and subsystems, test instruments and equipment, and tubes and amplifiers. Entries are arranged alphabetically, and include company name, address, phone, names of principal executives, and a list of sales offices is given in advertisers' listings.
Next edition: October 1985.

Optical industry and systems purchasing directory*   201

Publisher: Optical Publishing Company Incorporated
Address: Box 1146, Berkshire Common, Pittsfield, MA 01202, USA
Year: 1985
Pages: 1000 (in 2 vols)
Price: $64.00
ISSN: 0191-0647
Content: Over 1600 United States and over 250 foreign manufacturers and suppliers are listed in Volume 1, the

INTERNATIONAL DIRECTORIES

buyers' guide. (Volume 2 is an optical industry encyclopedia and dictionary). Entries include: company, name, address, telephone, names of executives and technical personnel, description of products and services. Arrangement: alphabetical. Indexes: geographical, product.
Next edition: March 1986.

Satellite directory (telecommunications)* 202

Publisher: Phillips Publishng Incorporated
Address: 7315 Wisconsin Avenue, Suite 1200N, Washington, DC 20014, USA
Year: 1985
Pages: 1000
Price: $197.00
Geographical area: World.
Content: The directory covers: 4000 hardware and technical service companies, programming services companies, consultants, communications attorneys, telecommunications services, publishers; 8000 earth stations; related trade associations, government agencies, and satellite telecommunications carriers. Entries include: for earth stations - city or site name, parent company name, call sign, operating frequency, data licensed, stations site coordinates, antenna size; for companies, consultants, and associations - name, address, telephone, key personnel, activities, products, or services; for government agencies and congressional committees - name, areas of responsibility, address, telephone, committee members, or contacts for specific areas of concern. Entries are arranged by activity; earth stations give same data by company name and geographically. The index is by company/organization name.
Next edition: January 1986.

ENERGY SCIENCES

Arab oil and gas directory, tenth edition* 203

Publisher: Arab Petroleum Research Centre
Address: 107 Park Street, London W1Y 3FB, UK
Year: 1985
Price: £180.00
Content: Information on all aspects of oil and gas production, exploration, and development in the Arab world.

Directory of energy information centres in the world 204

Editor: World Energy Conference
Publisher: Institut Français de l'Énergie
Address: 3 rue Henri Heine, 75016 Paris, France
Year: 1983
Pages: x + 278
Price: £29.00; F350.00
ISBN: 2-85933-009-7
Content: There are 400 entries from 69 countries. The directory is published in English and in French, and identifies by country the national and international organizations able to supply energy information (fields covered, nature of documents available, services offered, means of which they dispose, access to information networks).
Not available on-line.

Directory of gas utility companies 205

Publisher: Midwest Oil Register Incorporated
Address: PO Box 700597, Tulsa, OK 74170, USA
Year: 1985
Pages: 794
Price: $25.00
Geographical area: World.
Content: Entries give: company name, address, telephone number, compressor stations, if pipe line company or electric company also, personnel with titles.
Not available on-line.
Next edition: December 1985; $25.00.

Directory of industrial and technological research institutes: industrial conversion of biomass* 206

Publisher: United Nations Industrial Development Organization
Address: Vienna International Centre, PO Box 300, A-1400 Wien, Austria
Year: 1983
Price: £1.52
Geographical area: World.

Energy balances of developing countries, 1971-82 207

Publisher: International Energy Agency
Address: OECD Publications and Information Centre, 1750 Pennsylvania Avenue NW, Washington, DC 20006, USA
Year: 1984
Pages: 350
Price: $30.00
ISBN: 92-64-02543-X
Content: Shows production, trade, and consumption for each major type of energy for over 50 countries.

Energy policies and programmes of IEA countries 1983 review* — 208

Publisher: International Energy Agency
Address: OECD Publications and Information Centre, 1750 Pennsylvania Avenue NW, Washington, DC 2006, USA
Year: 1984
Pages: 400
Price: $36.00
ISBN: 92-64-12591-4
Content: Reports on and assessment of each member country's energy programmes and policies, including forecasts and statistics.

European petroleum directory/Africa - Middle East petroleum directory — 209

Publisher: PennWell Publishing Company
Address: PO Box 1260, Tulsa, OK 74101, USA
Year: 1984
Pages: 734 (in 2 vols)
Price: $120.00 (North America); $150.00 (elsewhere) (for the set)
Content: Gives addresses, telephone and telex numbers for companies in drilling, refining, exploration, pipelines, engineering, etc, totalling over 10 000 companies and 40 000 individuals. Worldwide production surveys. Available only as two volume set.
Next edition: September 1985; $120.00/$150.00.

Geothermal world directory, twelfth edition — 210

Editor: K. Darwin, A. Tratner
Publisher: Geothermal World
Address: 5762 Firebird Court, Camarillo, CA 93010, USA
Year: 1985
Pages: 500
Price: $55.00
ISSN: 0094-9779
Content: Information covering the spectrum of geothermal energy activities: research, technologies, industry, governmental programmes, educational curriculum and contacts, and information resources. Listings on international individuals and institutions including international geothermal developments' past, present and future in the areas of direct utilization, electrical energy production, and novel multi-purpose uses of this immense terrestrial heat source.
Next edition: June 1986; $55.00.

Information sources on non-conventional sources of energy* — 211

Publisher: United Nations Industrial Development Organization
Address: Vienna International Centre, PO Box 300, A-1400 Wien, Austria
Price: $4.00
Series: UNIDO guides to information sources
Geographical area: World.

International directory of new and renewable energy information resources and research centres* — 212

Publisher: Solar Energy Research Institute
Address: Energy Department, 1617 Cole Boulevard, Golden, CO 80401, USA
Year: 1982
Pages: 500
Price: $35.00
Content: Entries for about 3000 international, federal, regional, state, and local government agencies, universities, private organizations, associations, publishers of periodicals, directories, and data bases; and other groups or services which can supply research and information on solar and renewable energy. Entries include: for government, research, and information organizations - name, address, type of organization; name, title, and telephone of contact; description of work and areas of interests, expertise, services; for associations - name and address; for publishers - title of publication, year published, name and address of publisher. Arrangement: geographical, then by type of work performed. Indexes: organization name, solar technology, publication title.

Land drilling and oil well servicing contractors directory, eleventh edition — 213

Editor: William R. Leek
Publisher: PennWell Publishing Company
Address: PO Box 1260, Tulsa, OK 74101, USA
Year: 1984
Pages: 300
Price: $75.00 (USA and Canada); $93.50 (elsewhere)
Geographical area: World.
Content: Over 900 drilling contractors and 1350 oilwell servicing firms are listed. Included are companies involved in: contract drilling, workover and completion, stimulation, acidizing, fracing, logging, swabbing, perforating, bailing, directional drilling, rods and tubing, casing crews, plugging and wireline services. Over 12 000 key people at 3500 locations are listed, from officers and directors to key operating personnel. The total number of rigs with depths and areas of operation are listed for the drilling contractors in this directory.

Offshore contractors and equipment directory, 17th edition — 214

Editor: William R. Leek
Publisher: PennWell Publishing Company
Address: PO Box 1260, Tulsa, OK 74101, USA

Year: 1985
Pages: 500
Price: $85.00 (USA and Canada); $106.50 (elsewhere)
Geographical area: World.
Content: Names and descriptions of over 2500 firms in the drilling, offshore construction, geophysical, diving and transportation industry, along with their affiliates, subsidiaries, divisions, addresses, telephone, cable and telex numbers. There is a company index included for quick reference. The equipment index consists of mobile drilling rig specifications, workover and miscellaneous rig specifications, tender and fixed platform rig specifications, construction and pipelay barge specifications, and Offshore Magazine's 1985 marine transportation survey.

New features in the directory include the division of the construction section into companies that actually perform engineering and construction activities and a service/supply/manufacturers section naming supply firms that provide miscellaneous services or manufacture components used in the offshore industry. The drilling personnel section has been divided into two sections - one for drilling contractors, and workover and miscellaneous drilling contractors. The transport section has been conveniently subdivided into 'air' and 'marine' sections for the first time, and the drilling rig survey not only organizes rigs alphabetically by rig name, but also divides rigs by rig type. For the first time, PennWell's Electronic Rig Stats (ERS) on-line computer database service has prepared the mobile drilling rig specifications survey for this directory.
Next edition: April 1986; $85.00 and $106.50.

Oil directories of companies outside the USA and Canada 215

Publisher: Midwest Oil Register Incorporated
Address: PO Box 700597, Tulsa, OK 74170, USA
Year: 1985
Pages: 502
Price: $20.00
Geographical area: World, except USA and Canada.
Content: Entries give: company name, address, telephone number, kind of company, personnel with titles. Not available on-line.
Next edition: November 1985; $20.00.

Synthetic fuels and alternate energy worldwide directory* 216

Publisher: PennWell Publishing Company
Address: Box 21278, Tulsa, OK 74101, USA
Year: 1985
Price: $60.00
Content: The directory covers: over 2000 companies engaged in coal liquification and gasification, solar, geothermal, nuclear, oil shale, tar sands, heavy oil, and biomass; research organizations, engineering firms, government agencies, and other related groups are also listed. Entries include: for firms - company name, address, telephone, telex, cable address, names and titles of key personnel, projects in the field; for others - name, address, telephone, telex, cable address, name and title of contact, area of involvement or interest. Entries are classified by type of organization. Indexes: company name, subject. Former title: Worldwide Synthetic Fuels and Alternate Energy Directory.
Next edition: August 1986.

Who's who in world power generation markets, second edition* 217

Editor: C. Firth
Publisher: Martin, John, Publishers
Address: 24 Old Bond Street, London W1, UK
Year: 1985-86
Price: £50.00

World directory of energy information 218

Editor: Cambridge Information and Research Services Limited
Publisher: Gower Publishing Company Limited
Address: Gower House, Croft Road, Aldershot, Hampshire GU11 3HR, UK
Year: 1984
Pages: 328
Price: £35.00
ISBN: 0-566-02387-3
Content: Volume one covers 400 organizations in Western Europe; volume two 489 organizations of the Middle East, Africa, and Asia/Pacific; volume three 473 organizations in the Americas.
Each volume in the series is divided into four parts:
Part one provides an overview of the main features of energy in the region, with appropriate maps and tables.
Part two looks in detail at individual countries in a standard sequence. Key energy indicators are followed by sections dealing with energy market trends, the supply industries, energy trade, and energy policies.
Part three contains profiles of hundreds of organizations involved in the exploration, production, transmission and distribution of energy, as well as relevant official bodies.
Part four consists of a bibliography of publications on energy matters at either the national or transnational level.
Indexes are included which cross-reference publications by subject matter and country.

World energy directory: a guide to organizations and research activities in non-atomic energy 219

Editor: Wendy M. Smith
Publisher: Longman Group Limited

Address: Sixth Floor, Westgate House, The High, Harlow, Essex CM20 1NE, UK
Year: 1985
Pages: 582
Price: £120.00
ISBN: 0-582-90026-3
Series: Reference on research
Content: An international guide to 3000 industrial, official, academic and independent organizations and laboratories which conduct, or promote, research and development work in non-nuclear energy. The subjects covered include studies of new energy sources, fossil fuels, energy conservation, energy storage and strategic studies. It is fully indexed by title of establishment and by subject.

World pipelines and international directory 220

Editor: J.N.H. Tiratsoo
Publisher: Scientific Surveys Limited
Address: PO Box 21, Beaconsfield, Buckinghamshire HP9 1NS, UK
Year: 1983
Pages: viii + 120
Price: £25.00
ISBN: 0-87201-925-X
Content: Entries are arranged in chapters according to geographical area, and supported by 20 maps.
Next edition: March 1986; £30.00.

Worldwide petrochemical directory 221

Publisher: PennWell Publishing Company
Address: Box 21278, Tulsa, OK 74101, USA
Year: 1985
Pages: xxviii + 252
Price: $85.00 (USA and Canada); $106.50 (elsewhere)
ISSN: 0084-2583
Content: The purpose of this directory is to provide a complete listing of the companies active in the manufacture of petrochemicals throughout the world. Petrochemicals are defined as basic chemicals or intermediates made from crude oil, refinery products, natural gas or natural gas liquids. Facilities for manufacture may range from an installation devoted entirely to making petrochemicals out of these base stocks to perhaps only one unit within the framework of a crude oil refinery, gas processing or chemical plant.
Companies are listed alphabetically country-by-country within the major geographical regions of the world. A survey of all plants on stream, newly completed or under construction is included. Complete address information and lists of specialized personnel on company staffs are featured. An alphabetical company index is included.
Next edition: July 1986.

Worldwide refining and gas processing directory, 43rd edition 222

Publisher: PennWell Publishing Company
Address: Box 21278, Tulsa, OK 74101, USA
Year: 1985
Pages: 454
Price: $85.00
ISSN: 0277-0962
Content: The purpose of this directory is to provide a complete listing of the companies active in crude oil and gas processing throughout the world, as well as the firms with engineering/construction services available for such operations.
Statistical surveys show plant capacities, processes, and principal products. Construction reports tell of work planned, underway, and newly completed. Companies are listed alphabetically country-by-country within the large regions of the world. Complete address information and lists of specialized personnel on company staffs are featured. An alphabetical company index is included.
Next edition: July 1986.

ENGINEERING AND TRANSPORTATION

Bulk carrier register, 17th edition 223

Publisher: Clarkson, H., and Company Limited
Address: Publications Department, 12 Camomile Street, London EC3A 7BP, UK
Year: 1985
Price: £90.00 (surface mail); £112.00 (air mail)
Series: Clarkson's registers
Geographical area: World.
Content: Over 5000 bulk, ore, and combined carriers having a deadweight of 10 000 tons or more are listed alphabetically with extensive details. The statistical section contains analyses of size and age, dimensions, draught, and speed. Vessels are also listed under their respective owners.

Directory of hydraulic research institutes and laboratories* 224

Publisher: International Association for Hydraulic Research
Address: Rotterdamseweg 185, Box 177, Delft, Netherlands
Year: 1980
Pages: 600
Price: Dfl60.00
Geographical area: World.
Content: The directory covers about 300 laboratories,

INTERNATIONAL DIRECTORIES

worldwide. Entries include: laboratory name, address, telephone, director's name, number of staff, financial data, and type of services. Arrangement: geographical.
Next edition: 1986.

Directory of shipowners, shipbuilders and marine engineers 225

Editor: Keith Wilson
Publisher: Business Press International
Address: Quadrant House, The Quadrant, Sutton, Surrey SM1 1QH, UK
Year: 1985
Pages: 1600
Price: £40.00 (United Kingdom); £45.00 (elsewhere)
ISBN: 0-617-00384-X
Geographical area: World.
Content: The book is divided into sections as follows: shipowners and managers - about 1300 companies (all ships and senior personnel); ferry operators - about 286 companies (all ships and senior personnel); tugs and service craft - about 364 companies (all ships and senior personnel); shipbuilders and ship repairers - about 1200 companies (senior personnel and major facilities); marine engine builders - about 220 companies (senior personnel and product information); naval architects and consultants - about 900 companies (complete addresses); associations (international, British, and others) - about 400 entries (complete addresses and contact person); classification societies - 20-plus listed many with full personnel and outport details; government and official bodies - about 70 countries with marine body contacts listed. Indexes are as follows: general index; ship index; personnel index.
Not available on-line.
Next edition: June 1986; £42.00 (UK).

Engineering and industrial directory* 226

Editor: Anthony Walker
Publisher: Savory Milln and Company
Address: 3 London Wall Buildings, London EC2M 5PU, UK
Year: 1985
Price: £150.00 per volume
Content: Volume one covers mechanical engineering, contracting, aerospace, and defence; volume two covers motors, metals, general industrial topics, and industrial consumables. Entries give details of company activities and operating statements.

Engineering research centres: a world directory of organizations and programmes 227

Editor: T. Archbold, J.C. Laidlow, J. MacKechnie
Publisher: Longman Group Limited
Address: Sixth Floor, Westgate House, The High, Harlow, Essex CM20 1NE, UK
Pages: 1031
Price: £135.00
ISBN: 0-582-90018-2
Geographical area: Reference on research
Content: Details of over 7300 engineering research and development establishments throughout the world. Each entry includes, where available, the address, type of organization and affiliation, senior personnel, scope of interest and publications. Arranged alphabetically by country the directory is indexed by titles of establishments and by subject area.

Information sources in architecture 228

Editor: Valerie J. Bradfield
Publisher: Butterworth and Company Limited
Address: Borough Green, Sevenoaks, Kent TN15 8PH, UK
Year: 1983
Pages: 440
Price: £30.00
ISBN: 0-408-10763-4
Series: Butterworth guides to information sources
Geographical area: World.

Information sources in engineering, second edition 229

Editor: L.J. Anthony
Publisher: Butterworth and Company Limited
Address: Borough Green, Sevenoaks, Kent TN15 8PH, UK
Year: 1985
Pages: 704
Price: £38.00
ISBN: 0-408-11475-4
Series: Butterworth guides to information sources
Content: A guide to information sources in all major branches of engineering: mechanical, electrical, civil, and chemical.

Information sources on the cement and concrete industry* 230

Publisher: United Nations Industrial Development Organization
Address: Vienna International Centre, PO Box 300, A-1400 Wien, Austria
Price: $4.00
Series: UNIDO guides to information sources
Geographical area: World.
Content: List of professional, trade, and ressearch organizations.

Information sources on the machine tool industry* 231

Publisher: United Nations Industrial Development Organization
Address: Vienna International Centre, PO Box 300, A-1400 Wien, Austria
Price: $4.00
Series: UNIDO guides to information sources
Geographical area: World.
Content: List of professional, trade, and research organizations.

Information sources on woodworking industry machinery* 232

Publisher: United Nations Industrial Development Organization
Address: Vienna International Centre, PO Box 300, A-1400 Wien, Austria
Price: $4.00
Series: UNIDO guides to information sources
Geographical area: World.

Interavia aerospace directory 233

Editor: J. Didelot
Publisher: Interavia SA
Address: PO Box 162, 86 Avenue Louis-Casai, 1216 Cointrin-Genève, Switzerland
Year: 1985
Pages: cc + 1222
Price: SFr235.00
Geographical area: World.
Content: Over 50 000 entries, listed country by country, and under product sections; alphabetical index of all entries.
Next edition: April 1986.

International directory of building research, information and development organisations, fifth edition* 234

Editor: ICBRIDO
Publisher: Spon, E. & F.N.
Address: 11 New Fetter Lane, London EC4P 4EE, UK
Year: 1985
Price: £25.00
ISBN: 0-419-12990-1
Geographical area: World.
Content: Information on research institutes, government organizations, universities, and private companies, giving details of: address and telephone, history, finance, key personnel, number of staff, structure and organization, research, publications. Subject and personnel indexes.

International directory of consultants and technical sources, third edition 1985* 235

Editor: Bryan A. Denyer
Publisher: Institute of Marine Engineers
Address: 76 Mark Lane, London EC3R 7JN, UK
Year: 1985
Price: £17.50
Geographical area: World.
Content: Guide to consultings, testing services, and research organizations in the marine and offshore industries.
Next edition: 1986.

International directory of consulting engineers 236

Editor: Roderick Rhys Jones
Publisher: Rhys Jones Marketing
Address: The Lodge, Diamond Terrace, Greenwich, London SE10 8QN, UK
Year: 1985
Pages: xxvii + 232
Price: $35.00; £24.00
ISBN: 0-9508456-1-2
Geographical area: World.
Content: Entries for 199 subscribers (all members of FIDIC: International Federation of Consulting Engineers) each with full page giving: general description; names of key personnel; international experience; personnel; services; fields of specialization; typical projects. Index of international offices: about 1500 line entries. Index of fields of specialization: grid index covering over 80 fields for easy reference, prefaced by cross reference of headings in English, French, German, Spanish, and Arabic. Index of offices. List of other FIDIC publications.
Introduction, notes for guidance, introduction to FIDIC, in English, French, German, Spanish, and Arabic.
Next edition: January 1987; £25.00.

International directory of consulting environmental and civil engineers* 237

Editor: A. Shaw
Publisher: International Research Services Incorporated
Address: Box 225, Blue Bell, PA 19422, USA
Year: 1984
Pages: 150
Price: $25.00
Content: The directory covers 850 engineering firms and manufacturers active in environmental and civil engineering; worldwide coverage. Entries include: firm name, address, telephone, services or products. Arrangement: geograhical. Indexes: firm name, product/service.
Next edition: January 1986.

INTERNATIONAL DIRECTORIES

International robotics yearbook 238

Editor: Profesor I. Aleksander
Publisher: Kogan Page
Address: 120 Pentonville Road, London N1 9JN, UK
Year: 1983
Pages: 336
Price: £50.00
ISBN: 0-85038-691-8
Geographical area: World.
Content: Directory of manufacturers, suppliers, and services; directory of world research and development activities - details of about 200 academic and industrial groups, and details of grants to research establishments and fundings available; information on robot associations.

International who's who in engineering 239

Editor: Dr Ernest Kay
Publisher: Melrose Press Limited
Address: 3 Regal Lane, Soham, Ely, Cambridgeshire CB7 5BA, UK
Year: 1984
Pages: x + 589
Price: £52.50
ISBN: 0-900332-71-9
Geographical area: World.
Content: There are 3500 entries containing, personal, family, professional details, education, memberships, hobbies, publications, honours and address; list of engineering institutions.
Not available on-line.

Inventory of world commercial air carriers, second edition 240

Publisher: International Civil Aviation Organization
Address: 1000 Sherbrooke Street West, Suite 400, Montreal, Quebec H3A 2R2, Canada
Year: 1979
Pages: 61
Price: $2.75
Content: The inventory is quadrilingual in English, French, Russian, and Spanish.

Liquid gas carrier register, 20th edition 241

Publisher: Clarkson, H., and Company Limited
Address: Publications Department, 12 Camomile Street, London EC3A 7BP, UK
Year: 1985
Price: £55.00 (surface mail); £61.00 (air mail)
Series: Clarkson's registers
Geographical area: World.
Content: Over 700 operating vessels (having at least five per cent of cargo capacity available for carriage of gas in liquid form) are listed showing all major characteristics in alphabetical sequence. Listings of methane, ethylene, and other liquid gas carriers in ascending capacity order. Statistical analyses by type, capacity, flag, country, year of build, and future delivery dates.

Loadstar bulk handling directory, 1985 242

Editor: R.J. Miller
Publisher: Loadstar Publications
Address: 131 Salusbury Road, London NW6 6RG, UK
Pages: 288
Price: £24.00 (Europe); $45.00 (elsewhere)
ISBN: 0-9510103-0-1
Geographical area: World.
Content: The directory contains some 1300 entries subdivided under 64 category headings of bulk materials handling and ancillary equipment. Each entry provides brief technical details. There is an index of manufacturers' address as well as a general index. There is also tabular information divided on a country-by-country basis.
Not available on-line.
Next edition: January 1986; £30.00 and $55.00.

Middle East and Africa construction directory 1986-87* 243

Editor: Riyadh A. Chehab
Publisher: Arab Construction World
Address: PO Box 135121, Beirut, Lebanon
Year: 1985
Price: $110.00
Content: Volume one comprises listings of 8000 companies, importers, wholesalers, distributors, contractors, consulting engineers, etc. in the Middle East. Volume two covers Africa and contains 5000 entries.

Offshore drilling register 1983 244

Publisher: Clarkson, H., and Company Limited
Address: Publications Department, 12 Camomile Street, London EC3A 7BP, UK
Price: £55.00 (surface mail); £61.00 (air mail)
Series: Clarkson's registers
Geographical area: World.

Offshore service vessel register, eighth edition 245

Publisher: Clarkson, H., and Company Limited
Address: Publications Department, 12 Camomile Street, London EC3A 7BP, UK
Year: 1985
Price: £90 (surface mail); £105.00 (air mail)
Series: Clarkson's registers
Geographical area: World.

Content: Over 3000 vessels are listed alphabetically with dimensions, machinery details, carry capacities, specialization, etc. Lists of special types, composition of fleets, and statistical analyses of vessel types, age groups, national flags, loaded draughts, horsepower, and deck lengths.

Railway directory and year book   246

Editor: C.M. Bushell
Publisher: Business Press International Limited
Address: Quadrant House, Sutton, Surrey SM2 5AS, UK
Year: 1985
Pages: 695
Price: £26.00
ISBN: 0-617-00389-0
Geographical area: World.
Content: About 2500 entries giving details of organizations in the railway industry including personnel; indexes - general (organizations by name), and personnel (people by name).
Not available on-line.
Next edition: November 1985; £28.00.

Robotics: a worldwide guide to information sources   247

Publisher: Bowker Publishing Company
Address: 58/62 Epping High Street, Epping, Essex CM16 4BU, UK
Year: 1985
Pages: 239
Price: £59.00
ISBN: 0-8352-1820-1
Content: More than 4000 listings of reference materials, organizations, industrial research laboratories, manufacturers and scientific and technical specialists and will enable anyone, laypersons and professionals, to develop answers to questions concerning the mechanization of work. The volume includes handbooks, monographs, periodicals, journals, abstracts, conference and symposia proceedings worldwide, US and other government agency reports, dissertations, AV information, biographical information on specialists in related fields, product descriptions, and much more. Each section features a type of information or source and is preceded by an introduction. An index lists information included by category, author, and title.

Robotics, CAD/CAM market place 1985   248

Publisher: Bowker Publishing Company
Address: 58/62 High Street, Epping, Essex CM16 4BU, UK
Pages: 239
Price: £59.00

ISBN: 0-8352-1820-1
Geographical area: World.
Content: Listing of 2100 publications, 180 specialist associations, over 400 educational institutes with specialist courses, 700 research institutes, over 370 manufacturers of robot products, and manufacturers of allied products.

Sea technology buyers guide/directory   249

Editor: David M. Graham
Publisher: Compass Publications Inc
Address: 1117 N 19th Street, Suite 1000, Arlington, VA 22209, USA
Year: 1984
Pages: 184
Price: $20.50
Geographical area: World.
Content: Index; market overview; industrial firms; product buyers' guide; service buyers' guide; federal government; educational institutions; research vessels; geophysical survey vessels.
Next edition: September 1985; $20.50.

Ship and boat international guide to the small ship and workboat market   250

Editor: R.G. White
Publisher: Metal Bulletin Journals Limited
Address: Park House, Park Terrace, Worcester Park, Surrey KT4 7HY, UK
Year: 1985
Pages: 100
Price: £6.00
Geographical area: World.
Content: Section one lists ship and boat builders and repairers worldwide, with details of facilities; section two gives ships on order worldwide of less than 100 m length or 3000 grt; section three lists marine diesel engines for propulsion or auxiliary drive, by manufacturers and model.

Tanker register 1985, 25th edition   251

Publisher: Clarkson, H., and Company Limited
Address: Publications Department, 12 Camomile Street, London EC3A 7BP, UK
Price: £90.00 (surface mail); £107.00 (air mail)
Series: Clarkson's registers
Geographical area: World.
Content: Details of 3285 tankers and combined carriers having a deadweight of 10 000 tons and above. Analyses of size, age, dimensions; national fleet tables; listings of tonnage delivered, name changes, and removals; listings of special types. All vessels are listed alphabetically and all owners are tabulated with the names and sizes of ships.

INTERNATIONAL DIRECTORIES

Transport engineers handbook, second edition* 252

Editor: G. Montgomerie
Publisher: Kogan Page
Address: 12 Pentonville Road, London N1 9JN, UK
Year: 1985
Price: £12.00
ISBN: 0-85038-990-9
Content: Part two consists of a directory of manufacturers, distributors, and trade associations.

Tunnelling Directory 1985 253

Editor: Anna Way
Publisher: Morgan-Grampian plc
Address: 30 Calderwood Street, Woolwich, London SE18 6QH, UK
Pages: iv + 312
Price: £20.00
ISBN: 0-86213-068-9
Geographical area: World.
Content: There are 1600 entries in three main sections: alphabetical list of companies; company details by country; companies according to activity. Two sections of additional information cover: international tunnelling association; research and trade organizations. All information is cross-referenced and given in English, French, German, Spanish, and Italian.
Not available on-line.
Next edition: May 1986; £20.25.

World engine digest* 254

Editor: C. Firth
Publisher: Martin, John, Publishers
Address: 24 Old Bond Street, London W1, UK
Year: 1985
Price: £135.00
ISBN: 0-906237-35-1
Content: Includes: production/market data for diesel and petrol engines; profiles of world's leading manufacturers; company activities and develoments.
Next edition: 1986.

World yearbook of robotics research and development 255

Editor: Professor I. Aleksander
Publisher: Kogan Page
Address: 120 Pentonville Road, London N1 9JN, UK
Year: 1985
Pages: 456
Price: £38.00
ISBN: 0-85038-933-X
Content: Entries for 350 public and private establishments involved in robotics research giving areas of interest, current and future research, grants and sponsorship, publications. Indexes of research centres, key personnel, and research activities by subject.

INDUSTRY AND MANUFACTURING

Brown's directory of North America and international gas companies* 256

Editor: Dean Hale
Publisher: Energy Publications Incorporated
Address: Box 1589, Dallas, TX 75221, USA
Year: 1984
Pages: 300
Price: $145.00
Content: Entries cover operating gas companies (utilities, including municipal systems, and transmission companies), holding companies, and state public service commissions; includes major gas companies outside the United States. Entries include: for utilities and gas pipelines - company name, address, telephone, officers and other key personnel, source of gas supplies, communities served, technical and operating data; for holding companies - company name, address, names of officers and directors, subsidiaries, sales, employees, etc; for commissions - body name, address, names of staff members; some listings include commissioners' names. Arrangement: by activity, then alphabetical. Indexes: gas companies, communities with companies serving them.
Next edition: October 1985.

Directory of industrial information services and systems in developing countries* 257

Publisher: United Nations Industrial Development Organization
Address: Vienna International Centre, PO Box 300, A-1400 Wien, Austria
Year: 1982
Price: £1.03

Directory of United Nations databases and information systems* 258

Editor: Advisory Committee for the Coordination of Information Systems
Publisher: United Nations Publications
Address: Palais des Nations, CH-1211, Genève 10, Switzerland
Year: 1984
Pages: v + 323
ISBN: 92-9048-295-8

Geographical area: World.
Content: There are 615 systems, services and data bases listed for 38 organizations in the UN systems. Indexes, in English, French and Spanish, are name/acronym, subject, and geographical (English only). May become available as an on-line data base in the future.

East-West business directory* 259

Editor: Professor Carl H. McMillan
Publisher: Duncan Publications
Address: 3 Colin Gardens, Hendon, London NW9 6EL, UK
Year: 1985
Price: £49.50
ISBN: 0-946093-04-0
Geographical area: Europe and North America.
Content: Listing of companies in OECD countries with Soviet or East European investment; also of Eastern commercial activity in OECD countries. An updating service is available for £25.00 per annum.

European and North American scrap directory 260

Editor: J. Bailey, B. Robinson, Milton Nurse, Paula Read
Publisher: Metal Bulletin Books Limited
Address: Park House, Park Terrace, Worcester Park, Surrey KT4 7HY, UK
Year: 1981
Pages: 597
Price: £35.00
ISBN: 0-900542-55-1
Geographical area: Europe and North America.
Content: Detailed information on approximately 1500 companies engaged in trading and physical processing of iron and steel and non-ferrous scrap.

Fairplay world shipping year book 261

Editor: W. Peach
Publisher: Fairplay Publications Limited
Address: 52-54 Southwark Street, London SE1 1UJ, UK
Year: 1985
Pages: xx + 940
Price: £30.00
ISBN: 0-905045-73-4
Geographical area: World.
Content: There are 1700 individual company entries. Each entry lists: company name, address, telephone, telex, telegram, directors, senior management personnel, type of company, UK representatives, financial accounts, balance sheet, financial results. Within each of the main chapters covering each sector of the shipping industry, and under the particular company entry, pertinent extras are included, eg shipowner entries also include comprehensive fleet lists plus routing; towage and salvage operators also list tyes of service offered, number of tugs, other equipment; marine equipment suppliers also list the products they specialize in.
Editorial index (based on company name entry); personnel index (based on directors/staff entries); ship index (based on each ship entry in fleet lists).
Statistics: tables covering merchant fleets, shipbuilding, shipbuilding prices, freight rate comparisons.
Not available on-line.
Next edition: June 1986; £34.00.

Ferro-alloy directory 262

Editor: Danielle Donougher
Publisher: Metal Bulletin Books Limited
Address: Park House, Park Terrace, Worcester Park, Surrey KT4 7HY, UK
Year: 1984
Pages: 367
Price: £27.50
ISBN: 0-900542-89-6
Geographical area: World.
Content: Listing of ferro-alloy producers and traders with product guides, totalling over 500 entries.

Financial Times industrial companies year book 1985 263

Publisher: Longman Group Limited
Address: 6th Floor, Westgate House, The High, Harlow, Essex CM20 1NE, UK
Year: 1985
Pages: xx + 660
Price: £46.00
ISBN: 0-582-90331-9
Series: Financial Times international year books
Geographical area: World.
Content: There are 759 entries on the world's principal industrial companies. Includes addresses; directors/officers/senior management; business; recent acquisitions; subsidiary/associate companies; details of operations; breakdown of three years' turnover by business segment/geographical area; share capital details; shareholder information; financial figures for last three years. Section arranged alphabetically by name.
Company index (all entries and secondary references within text); geographical index (where companies based); sector index.
Next edition: March 1986; £50.00.

Financial Times mining international year book 1986 264

Publisher: Longman Group Limited
Address: 6th Floor, Westgate House, The High, Harlow, Essex CM20 1NE, UK

Pages: xxxii + 536
Price: £46.00
ISBN: 0-582-90339-4
Series: Financial Times international year books
Geographical area: World.
Content: There are 720 entries on mining companies (principally extracting metals and minerals; also refiners/smelters and marketers). Entries give: addresses; directors/officers/senior management; subsidiary/associate companies; recent acquisitions; description of operations (where company operates, mines owned, joint ventures, exploration programmes etc); production and reserve figures for last three years (where available); shareholder information; financial figures for last three ears (where available). Section arranged alphabetically by name.
109 entries on mining/metals associations/institutes. Entries give details of addresses; officers; objectives; eligibility for membership; publications. Section arranged alphabetically by name.
Company index (all entries and secondary references within text). Product index (selected metals/minerals produced - cross reference to main company section); geographical index (areas where companies are exploring/mining) with maps shaded to show areas of mining activity mentioned in text of index; tables giving countries' output of selected metals for last three years; international professional services (business card ads for consultants etc); suppliers' directory and buyers' guide.
Next edition: October 1985; £50.00.

Financial Times oil and gas international year book 1985   265

Publisher: Longman Group Limited
Address: 6th Floor, Westgate House, The High, Harlow, Essex CM20 1NE, UK
Year: 1984
Pages: xxxii + 612
Price: £49.00
ISBN: 0-582-90330-0
Series: Financial Times international year books
Geographical area: World.
Content: There are 493 entries on oil and gas companies whose interests include upstream activities (exploration and production); 299 entries on those involved in downstream activities only (refining, marketing, pipelines); 107 oil brokers and traders; 69 associations.
Upstream and downstream entries give addresses; details of directors/officers; subsidiary/associate companies; description of operations: where company operates, landholdings, wells drilled, refinery throughput etc; recent acquisitions; production and reserves figures for last two years (where available); share capital details; shareholder information; financial figures for last three years (where available).
Oil brokers/traders entries give addresses; officers; brief description of business; subsidiary/associate companies; ownership of company.
Associations give addresses; officers; objectives; eligibility for membership; publications.
All these sections are arranged alphabetically by name. Company index (all entries and secondary references within text); geographical index (areas of oil/gas exploration/production by company); tables on oil production/refining/tanker tonnage/oil consumption; country by country; international professional services (business cards for consultants etc); suppliers' directory and buyers' guide.
Next edition: December 1985; £52.00.

Financial Times who's who in world oil and gas 1982-83   266

Publisher: Longman Group Limited
Address: 6th Floor, Westgate House, The High, Harlow, Essex CM20 1NE, UK
Year: 1982
Pages: xxiv + 636
Price: £32.00
ISBN: 0-582-90313-0
Series: Financial Times international year books
Geographical area: World.
Content: Biographical details of approximately 3500 top people involved in the oil and gas industry. Includes key personnel from commercial, academic, and government fields. Entries give personal details (nationality, date of birth, married/single, children), clubs, leisure interests etc); academic qualifications; current and past appointments; publications; languages; contact address.
Index of organizations: names of personnel with entries in book are listed by company/government dept etc.

Industrial minerals directory   267

Editor: Brian Coope
Publisher: Metal Bulletin Books Limited
Address: Park House, Park Terrace, Worcester Park, Surrey KT4 7HY, UK
Year: 1982
Pages: 640
Price: £35.00
ISBN: 0-900542-66-7
Geographical area: World.
Content: Detailed coverage on a country basis, of non-metallic mineral producers plus a buyers' guide listing, under mineral headings, suppliers of a given mineral with country of origin. There are over 1800 entries.
Next edition: Autumn 1986.

Information business 1984: a guide to companies and individuals: the annual directory of the Information Industry Association 268

Editor: Faye Henderson, Fred Rosenau
Publisher: Elsevier Scientific Publishers
Address: PO Box 211, 1000 AE Amsterdam, Netherlands
Year: 1984
Pages: xxii + 528
Price: Dfl165.00
ISBN: 0-444-87514-X
Geographical area: USA, Europe, Japan.
Content: The main section of this guide contains 355 pages of descriptive information on 214 leading information-sector firms. Most of these are US-based, but there are numerous entries also for major Japanese, British, and European companies. The Professional Members section provides descriptive and contact information on more than sixty important consultants, industry analysts, information scientists, marketing specialists, writers and editors, investors, and attorneys specializing in the information sector. The international section lists several hundred non-domestic subsidiaries, trading partners, agents and representatives, with addresses, telephone, telex and cable details. The subject and services index contains well over 200 terms pointing to the main descriptive listing. Also useful is the alphabetical list of more than 150 trade names and their owners. The volume closes with an index of close to 2000 personal names, with complete addresses and telephone numbers.

Information sources on industrial maintenance and repair* 269

Publisher: United Nations Industrial Development Organization
Address: Vienna International Centre, PO Box 300, A-1400 Wien, Austria
Price: $4.00
Series: UNIDO guides to information sources
Geographical area: World.

Information sources on industrial quality control* 270

Publisher: United Nations Industrial Development Organization
Address: Vienna International Centre, PO Box 300, A-1400 Wien, Austria
Price: $4.00
Series: UNIDO guides to information sources
Geographical area: World.

Information sources on industrial training* 271

Publisher: United Nations Industrial Development Organization
Address: Vienna International Centre, PO Box 300, A-1400 Wien, Austria
Price: $4.00
Series: UNIDO guides to information sources
Geographical area: World.

Information sources on the canning industry* 272

Publisher: United Nations Industrial Development Organization
Address: Vienna International Centre, PO Box 300, A-1400 Wien, Austria
Price: $4.00
Series: UNIDO guides to information sources
Geographical area: World.
Content: List of professional, trade, and research organizations.

Information sources on the clothing industry* 273

Publisher: United Nations Industrial Development Organization
Address: Vienna International Centre, PO Box 300, A-1400 Wien, Austria
Price: $4.00
Series: UNIDO guides to information sources
Geographical area: World.

Information sources on the foundry industry* 274

Publisher: United Nations Industrial Development Organization
Address: Vienna International Centre, PO Box 300, A-1400 Wien, Austria
Price: $4.00
Series: UNIDO guides to information sources
Geographical area: World.

Information sources on the iron and steel industry* 275

Publisher: United Nations Industrial Development Organization
Address: Vienna International Centre, PO Box 300, A-1400 Wien, Austria
Price: $4.00
Series: UNIDO guides to information sources
Geographical area: World.

Information sources on the packaging industry * 276

Publisher: United Nations Industrial Development Organization
Address: Vienna International Centre, PO Box 300, A-1400 Wien, Austria
Price: $4.00
Series: UNIDO guides to information sources
Geographical area: World.

Information trade directory 1983 277

Publisher: Learned Information Limited
Address: Besselsleigh Road, Abingdon, Oxford OX13 6EF, UK
Year: 1982
Pages: x + 282
Price: £25.00
ISBN: 0-904933-36-9
Geographical area: World.
Content: There are sections on: information production; information distribution; information retailing; support services and suppliers; associations and government agencies; conferences and courses; sources of information.

International directory of market research organizations 1985 278

Publisher: Market Research Society; British Overseas Trade Board
Address: Market Research Society - 15 Belgrave Square, London SW1X 8PF, UK
Pages: x + 441
Price: £35.00; $90.00
ISBN: 0-906117-03-8
Geographical area: World.
Content: There are 1300 entries giving details of research agencies in sixty countries, arranged alphabetically.
Not available on-line.
Next edition: 1987.

International federation of cotton and textile industries directory, twelfth edition, 1985-86 279

Publisher: International Textile Manufacturers Federation
Address: Am Schanzeagraben 29, Postfach, CH-8039 Zürich, Switzerland
Pages: 30
Geographical area: World.
Content: Details of ITMF organization and personnel; member associations (name, address, telephone, telex, key personnel); associate members; publications.
A biennial publication.

International zinc and galvanizing directory, third edition 280

Editor: Norman Connell
Publisher: Metal Bulletin Books Limited
Address: Park House, Park Terrace, Worcester Park, Surrey KT4 7HY, UK
Year: 1983
Pages: 98
Price: £20.00
ISBN: 0-900542-78-0
Geographical area: World.
Content: A detailed directory covering zinc producers, traders, and galvanizers. Also a buyers' guide (divided into producing and trading companies). There are approximately 500 entries.

Iron and steel works of the world 281

Editor: Richard Serjeantson, Raymond Cordero, Henry Cooke
Publisher: Metal Bulletin Books Limited
Address: Park House, Park Terrace, Worcester Park, Surrey KT4 7HY, UK
Year: 1983
Pages: 840
Price: £60.00
ISBN: 0-900542-82-9
Geographical area: World.
Content: Information on major iron and steel producers and re-rollers, tube makers, iron powder producers, strip coaters, cold rolled section makers; over 1900 entries arranged alphabetically by country plus buyers' guide covering a broad range of product headings.
Next edition: 1987.

Licht's, F.O., internationales zuckerwirtschaftliches jahr- und adressbuch 282

[Licht's, F.O., international sugar economic yearbook and directory]
Editor: Dr Helmut Ahlfeld
Publisher: Licht, F.O., GmbH
Address: POB 1220, Am Mühlengraben 22, 2418 Ratzeburg, German FR
Year: 1984
Pages: 608
Price: DM132.00
Geographical area: World.
Content: Names and addresses of companies, organizations, and institutions concerned with sugar in all parts of the world; locations and capacities of sugar factories in over one hundred countries; product report and buyers' guide; world sugar statistics 70-page removable supplement.
Next edition: September 1985; DM135.00.

List of company directories and summary of their contents, second edition*   283

Publisher: United Nations Industrial Development Organization
Address: Vienna International Centre, PO Box 300, A-1400 Wien, Austria
Year: 1983
Price: $16.00
Geographical area: World.

Major companies of the Arab world, ninth edition*   284

Editor: G.F. Bricault
Publisher: Graham and Trotman
Address: 66 Wilton Road, London SW1V 1DE, UK
Year: 1985
Price: £110.00
ISBN: 0-86010-600-4
Content: Entries for over 6000 Arab companies, giving: address, telephone, and telex; names of key personnel; financial information and bankers; activities; subsidiaries.
Next edition: March 1986; £120.00

Marquis who's who directory of online professionals   285

Publisher: Marquis Who's Who Incorporated
Address: 200 East Ohio Street, Chicago, IL 60611, USA
Year: 1984
Pages: 852
Price: $85.00
ISBN: 0-8379-6001-0
Geographical area: World.
Content: Approximately 6000 entries; professionals listed in alphabetical format; sketch or profile content includes career, education, professional memberships, address, telephone number, on-line function, on-line experience, on-line systems used, equipment used, data bases used, subject expertise, consulting expertise, electronic mail systems used, etc.
Indexes: by on-line function; by data base subject expertise; by geographical location.
Available on-line as File 235 on DIALOG.

Metal traders of the world   286

Editor: David Gilbertson
Publisher: Metal Bulletin Books Limited
Address: Park House, Park Terrace, Worcester Park, Surrey KT4 7HY, UK
Year: 1983
Pages: 736
Price: £40.00
ISBN: 0-900542-80-2
Geographical area: World.
Content: Detailed information including head office address, trading personnel, ownership, trading specialization, products handled plus a classified guide of ores, metals, and semis showing companies active in their trading. There are over 1800 entries.
Next edition: 1987.

Metallurgical plant makers of the world   287

Editor: Richard Serjeantson
Publisher: Metal Bulletin Books Limited
Address: Park House, Park Terrace, Worcester Park, Surrey KT4 7HY, UK
Year: 1980
Pages: 751
Price: £33.00
ISBN: 0-900542-47-0
Geographical area: World.
Content: International guide to ferrous and non-ferrous plant and equipment designers, manufacturers, and overall engineering and contracting companies. A worldwide directory giving detailed information on approximately 1100 of these companies, with a buyers' guide listing the plant they produce/design.

Mines and mining equipment and service companies worldwide, third edition*   288

Editor: Don Nelson
Publisher: Spon, E. & F.N.
Address: 11 New Fetter Lane, London EC4P 4EE, UK
Year: 1985
Price: £35.00
ISBN: 0-419-13260-0
Content: Includes: principal mining companies; mining equipment, services, and consulting companies; equipment and services directory; geographical and personal indexes.

Minor metals survey   289

Editor: Norman Connell
Publisher: Metal Bulletin Books Limited
Address: Park House, Park Terrace, Worcester Park, Surrey KT4 7HY, UK
Year: 1981
Pages: 118
Price: £15.00
ISBN: 0-900542-63-2
Geographical area: World.
Content: Feature articles spanning the industry; world directory of producers and traders; specialist product buyers' guide.

INTERNATIONAL DIRECTORIES

Non-ferrous metal works of the world 290

Editor: Richard Serjeantson
Publisher: Metal Bulletin Books Limited
Address: Park House, Park Terrace, Worcester Park, Surrey KT4 7HY, UK
Year: 1982
Pages: 643
Price: £42.00
ISBN: 0-900542-71-3
Geographical area: World.
Content: Guide to the world's non-ferrous metal smelters, refiners, semi-fabricators and secondary ingot makers. There are over 2000 entries.
Next edition: November 1985.

Phillips paper trade directory 1985: mills of the world 291

Editor: George Hutton
Publisher: Benn Business Information Services Limited
Address: PO Box 20, Sovereign Way, Tonbridge, Kent TN9 1RQ, UK
Pages: ix + 520
Price: £50.00 (United Kingdom); £54.00 (elsewhere)
ISBN: 0-86382-022-0
Geographical area: World.
Content: Details of 3600 mills in 82 countries; 9500 company entries in products section; alphabetical list of 2000 merchants of paper board and paper products; 1300 manufacturers in A-Z buyers' guide of UK companies; alphabetical list of European suppliers of machinery, equipment, and materials; product guide; A-Z of UK waste paper merchants and processors; country-by-country guide to world exporters and importers; A-Z of brand names, watermarks and their owners; alphabetical who owns whom of European groups and consortia; sales agents and organizations listed by country; trade service specialists; UK paper and allied trade associations of the world.
Next edition: December 1985; £55.00 (United Kingdom); £60.00 (elsewhere).

Reference book for world traders* 292

Publisher: Croner Publications Incorporated
Address: 211-03 Jamaica Avenue, Queens Village, NY 11428, USA
Price: $85.00 per annum
Content: Basic data and details on world trade and research. Volume one covers information sources in the USA, basic rules of exporting, etc. Volumes two and three cover general trade information worldwide. Updated regularly; available on subscription.

Securitech, the annual international guide to security equipment and services 293

Editor: R. Morris
Publisher: UNISAF Publications Limited
Address: Queensway House, Redhill, Surrey, UK
Year: 1985
Pages: 300
Price: £17.50
ISBN: 0-86108-184-6
Geographical area: World.
Content: There are 290 entries, as well as an index of products.
Next edition: April 1986; £20.00.

Stainless steel international survey and directory 294

Editor: Milton Nurse, Claire Miller
Publisher: Metal Bulletin Books Limited
Address: Park House, Park Terrace, Worcester Park, Surrey KT4 7HY, UK
Year: 1982
Pages: 119
Price: £20.00
ISBN: 0-900542-74-8
Geographical area: World.
Content: Feature articles on the industry; world list of stainless steel refining plant; international directory of stainless steel producers; product buyers' guide. There are approximately 220 stainless steel entries.
Next edition: November 1985.

Standard and Poor's register of corporations, directors and executives* 295

Publisher: Standard and Poor's Corporation
Address: 45 boulevard Bischoffstein, Box 2, 1000 Bruxelles, Belgium
Year: 1985
Price: £330.00 (3 volume set)
Geographical area: World.
Content: Facts and figures on more than 45 000 public and private corporations including lists of 40 000 senior personnel; comprehensive indexes including index of executives and classified industrial index.

Steel traders of the world 296

Editor: Raymond Cordero
Publisher: Metal Bulletin Books Limited
Address: Park House, Park Terrace, Worcester Park, Surrey KT4 7HY, UK
Year: 1984
Pages: 876
Price: £40.00

ISBN: 0-900542-79-9
Geographical area: World.
Content: World listing of approximately 2000 steel traders and the products handled.
Next edition: 1988.

Sugar industry buyers' guide*   297

Editor: D. Leighton
Publisher: International Sugar Journal Limited
Address: 23A Easton Street, High Wycombe, Buckinghamshire HP11 1NX, UK
Year: 1985
Pages: 30
Price: $2.00
Geographical area: World.
Content: Display advertisements, announcements of new products and classified entries comprising lists of manufacturers and suppliers under product headings, followed by an address section listing these firms alphabetically with postal and cable addresses, telephone, and telex numbers.

Sugar year book   298

Publisher: International Sugar Organization
Address: Haymarket House, 28 Haymarket, London SW1Y 4SP, UK
Year: 1983
Pages: viii + 338
Price: £10.00
Geographical area: World.
Content: Lists of exporting and importing members of the International Sugar Organization. General tables on sugar statistics including world production, exports and imports.
Not available on-line.
Next edition: October 1985; £10.00.

Trade directories of the world, 31st edition*   299

Publisher: Croner Publications Incorporated
Address: 211-03 Jamaica Avenue, Queens Village, NY 11428, USA
Year: 1985
Price: £65.00
Content: Entries for approximately 3300 trade, industrial, and professional directories with listings in English, French, Spanish, and German. Indexes to: trades and professions; countries; general import/export directories.

Trade directory information in journals, fifth edition*   300

Publisher: Science Reference Library
Address: 25 Southampton Buildings, London WC2A 1AW, UK
Year: 1985
Pages: £2.50
Content: A list of trade directories appearing in journals.

Trade names dictionary company index*   301

Editor: Donna Wood
Publisher: Gale Research Company
Address: Book Tower, Detroit, MI 48226, USA
Price: $265.00
ISBN: 0-8103-0699-9
Content: Alphabetical name and address list of the 36 400 companies mentioned in the Trade Names Dictionary, with trade names of products they manufacture.

Trade shows and professional exhibits directory   302

Editor: Robert J. Elster
Publisher: Gale Research Company
Address: Book Tower, Detroit, MI 48226, USA
Year: 1985
Pages: 549
Price: $85.00
ISBN: 0-8103-1109-7
Geographical area: World.
Content: Focusing on those shows with the largest attendance and greatest number of exhibitors, this directory provides detailed entries on a wide range of conferences, conventions, trade and industrial shows, expositions, and similar events that utilize exhibits. Altogether, more than 2100 such events are listed, with entries furnishing details on attendance, cost and number of exhibits, audience, dates and locations for five years, and much more. There are five indexes: Geographical; chronological; subject; organizations; and show/exhibition name and keyword indexes. A softbound supplement is published between editions.

Wire industry yearbook 1985   303

Editor: Brenda Mitchell
Publisher: Magnum Publications Limited
Address: 110/112 Station Road East, Oxted, Surrey RH8 0QA, UK
Pages: 448
Price: £16.25
ISSN: 0084-0424
Series: Wire industry (international monthly publication)
Geographical area: World.
Content: Entries cover: 900 product groups, each one listed in five languages; approximately 3000 manufacturers/suppliers worldwide are listed under appropriate

products groups; alphabetical address section; trade names section; metric/imperial conversion tables.
Next edition: January 1986; £18.00.

World aluminium survey 304

Editor: Norman Connell
Publisher: Metal Bulletin Books Limited
Address: Park House, Park Terrace, Worcester Park, Surrey KT4 7HY, UK
Year: 1981
Pages: 275
Price: £20.00
ISBN: 0-900542-62-4
Geographical area: World.
Content: This survey is divided into a producers directory and products directory, listed alphabetically by country and product. There are approximately 1000 entries.

World directory of fertilizer manufacturers, fifth edition 305

Publisher: British Sulphur Corporation
Address: Parnell House, 25 Wilton Road, London SW1V 1NH, UK
Year: 1981
Pages: 126
Price: $200.00; £75.00
ISBN: 0-902777-76-7
Content: The directory presents a factual description of each manufacturing company's history, structure, affiliations, product range, plant locations, manufacturing capability, and financial data, and provides a balanced view of the part played by fertilizers in the overall operations of the company concerned. Countries have been listed in alphabetical order, which gives users an immediate picture of the scale of the industry in each country, and the alphabetical index of companies makes it simple to locate individual entries.
Next edition: September 1986.

World directory of institutions offering courses in industrial design 306

Editor: International Council of Societies of Industrial Design and the Secretariat of the United Nations Industrial Development Organization
Publisher: United Nations Industrial Development Organization
Address: Vienna International Centre, PO Box 300, A-1400 Wien, Austria
Year: 1976
Pages: 76
Price: Free
Geographical area: World.
Content: The directory lists 146 institutions, arranged in alphabetical country order, in tabular form giving for one or several institutions the following data: name, general policy, departments related to design, main source of finance, size of staff and student body, admission requirements, length and description of courses, approximate fees and scholarship possibilities.

World directory of wood-based panel producers 307

Publisher: Miller Freeman Publishing Incorporated
Address: 500 Howard Street, San Francisco, CA 94105, USA
Year: 1977
Availability: OP

World guide to fertilizer plant equipment* 308

Publisher: British Sulphur Corporation
Address: Parnell House, 25 Wilton Road, London SW1V 1NH, UK
Year: 1977
Pages: 198
Price: $40.00, £20.00

World precious metals survey and directory 309

Editor: Norman Connell
Publisher: Metal Bulletin Books Limited
Address: Park House, Park Terrace, Worcester Park, Surrey KT4 7HY, UK
Year: 1982
Pages: 118
Price: £15.00
ISBN: 0-900542-73-X
Geographical area: World.
Content: In-depth articles covering most aspects of precious metals production/trading plus two directories detailing producers with their respective operations and products and trading companies with their products handled. There are approximately 500 entries.

MEDICAL AND BIOLOGICAL SCIENCES

Animal health international directory* 310

Editor: Susan Thompson
Publisher: IMS World Publications
Address: 37 Queen Square, London WC1N 3BL, UK
Year: 1984
Price: $500.00
Geographical area: World.

Content: Entries for over 26 000 veterinary trade names; names and addresses of animal healthcare companies.

Biotechnology International* 311

Editor: Carmel Barnard
Publisher: IMS World Publications
Address: 37 Queen Square, London WC1N 3BL, UK
Year: 1985
Pages: $500.00 (set)
Geographical area: World.
Content: Three-part survey of biotechnology markets of the world with information on leading companies, university research, key organizations; research index.

Compendium of bryology: a world 312
listing of herbaria, collectors,
bryologists, current research

Publisher: Cramer, J.
Address: In den Springäckern 2, D-3300 Braunschweig, German FR
Year: 1985
Pages: 355
Price: DM60.00
ISBN: 3-7682-1434-6
Geographical area: World.

Current awareness in biological 313
sciences (CABS)

Editor: Professor Harry Smith
Publisher: Pergamon Press Limited
Address: Headington Hill Hall, Oxford OX3 0BW, UK
Year: 1985
Pages: 11 000
Price: $1865.00 per annum
ISSN: 0733-4443
Geographical area: World.
Content: Worldwide index of current research in the biological sciences. Approximately 150 000 references per annum listed of authors active in the fields of biochemistry, cell and developmental biology, ecology, endocrinology, genetics and molecular biology, immunology, microbiology, neuroscience, pharmacology and toxicology, physiology, and plant science. Entries give authors' names, addresses, title of paper published, source. The directory is fully classified, updated monthly and is available on-line. It has been published monthly since January 1983.

Dental schools of the world* 314

Publisher: Council of International Relations
Address: American Dental Association, 211 E. Chicago Avenue, Chicago, IL 60611, USA
Year: 1984

Pages: 100
Price: Free
Content: Entries include: institution name, address, name of director.

Directory of biomedical engineers* 315

Editor: Patricia I. Homer
Publisher: Alliance for Engineering in Medicine and Biology
Address: 4405 East-West Highway, Suite 42, Bethesda, MD 20814, USA
Year: 1983
Pages: 55
Price: $25.00
ISSN: 0740-6843
Geographical area: World.
Content: The directory covers over 1300 biomedical engineers and others concerned with the involvement of engineering in the physical, biological, and medical sciences. Entries include name, highest degree earned and/or certification, title and department, organizational affiliation, address, telephone, principal occupation, memberships in professional associations, principal areas of interest, and are arranged geographically.
Next edition: 1985.

Directory of cancer research 316
information resources*

Editor: International Cancer Research Data Bank
Publisher: United States Department of Health, Education and Welfare
Address: Blair Building, 8300 Colesville Road, Silver Spring, MD 20910, USA
Year: 1979
Pages: 250

Directory of international and national 317
medical related societies*

Publisher: Pergamon Press
Address: Headington Hill Hall, Oxford OX3 0BW, UK
Year: 1982
Price: £41.00
ISBN: 0-08-027991-0
Geographical area: World.
Content: Entries for 4000 societies giving name, address, membership details, conferences, etc.

Directory of major medical libraries 318
worldwide*

Publisher: US Directory Service
Address: PO Box 011565, Miami, FL 33101, USA
Year: 1980
Price: $86.00
ISBN: 0-916524-07-8

Content: This directory is a reference source for the names and addresses of 2794 important major medical libraries in over 1500 urban centres in 108 countries throughout the world. All the libraries listed are associated with one or more of the following: universities, hospitals, medical centres, medical research laboratories, medical associations, experimental institutions, scientific institutions, health research institutes, educational societies, and medical societies.

Directory of medical schools worldwide, third edition 1983-84   319

Publisher: US Directory Service
Address: PO Box 001565, Miami, FL 33101, USA
Pages: 191
Price: $24.95
ISBN: 0-916524-17-5
ISSN: 0160-6468
Content: Listing of medical schools in over 100 countries. A detailed introduction provides vital facts and information on admissions, statistics, language and curricula. Entries are arranged alphabetically by nation for quick reference.

Directory of on-going research in cancer epidemiology*   320

Editor: C.S. Muir, G. Wagner
Publisher: Oxford University Press
Address: Walton Street, Oxford OX2 6DP, UK
Year: 1984
Pages: xx + 728
Price: £18.00
ISBN: 0-19-723062-8
Series: IARC scientific publications
Geographical area: World.
Content: Entries are arranged alphabetically by country.
Next edition: November 1985; £20.00.

Environmental education - list of institutions in the region of Asia and the Pacific   321

Publisher: Unesco Regional Office for Education in Asia and the Pacific
Address: PO Box 1425 GPO, Bangkok 10500, Thailand
Year: 1983
Pages: v + 38
Geographical area: Asia-Pacific.
Content: There are 157 entries in which institutions are categorized as national or regional, formal or non-formal, within the alphabetical listing of countries. No index, not available on-line.

Foreign medical school catalogue 1977*   322

Editor: C.R. Modica
Publisher: Foreign Medical School Information
Address: Bay Shore, NY, USA
Year: 1976
Pages: 165
Price: $9.95
Content: Medical schools in 65 countries are listed with figures on admissions.

Genetic engineering and biotechnology firms worldwide directory*   324

Editor: Marshall Sittig, Robert Noyes
Publisher: Sittig and Noyes
Address: Box 592, Kingston, NJ 08528, USA
Year: 1985
Pages: 400
Price: $177.00
Content: The directory covers: about 1200 firms, including major firms with biotechnology divisions as well as small independent firms; worldwide coverage. Entries include: company name, address, division name (where applicable), research laboratory locations, number of employees engaged in biotechnology and genetic engineering, equity interests held by others, areas of research activity, currently available products.
Next edition: Autumn 1986.

Guide to parasite collections of the world*   324

Editor: Dr J.R. Lichtenfels, M.H. Pritchard
Publisher: American Society of Parasitologists
Address: 1041 New Hampshire Street, Lawrence, KS 66044, USA
Year: 1982
Pages: 80
Price: $5.00
Content: The guide covers nearly 70 collections. Entries include: collection name, address, names of curators, kinds of specimens, associated libraries, loan and accession procedure, services, history, published lists of specimens. Arrangement: geographical.

Information sources in biotechnology   325

Editor: Anita Crafts-Lighty
Publisher: Macmillan Press
Address: 4 Little Essex Street, London WC2R 3LF, UK
Year: 1985
Pages: 306
Price: £45.00
ISBN: 0-333-39290-6

Geographical area: World.
Content: What is biotechnology? the science and the business; information sources in biotechnology - monographs, book series, and textbooks; conferences and their proceedings; trade periodicals and newsletters; research and review periodicals; abstracting and secondary sources; computer data bases; patents and patenting; market surveys.

Information sources in the biological sciences, third edition 326

Editor: H.V. Wyatt
Publisher: Butterworth and Company Limited
Address: Borough Green, Sevenoaks, Kent TN15 8PH, UK
Year: 1985
Pages: 250
Price: £18.00
ISBN: 0-408-11472-X
Series: Butterworths guides to information sources.
Geographical area: World.

Information sources in the medical sciences, third edition 327

Editor: L.T. Morton, S. Godbolt
Publisher: Butterworth and Company Limited
Address: Borough Green, Sevenoaks, Kent TN15 8PH, UK
Year: 1983
Pages: 552
Price: £38.00
ISBN: 0-408-11473-8
Series: Butterworths guides to information sources

Information sources on the pharmaceutical industry* 328

Publisher: United Nations Industrial Development Organization
Address: Vienna International Centre, PO Box 300, A-1400 Wien, Austria
Price: $4.00
Series: UNIDO guides to information sources
Geographical area: World.
Content: List of professional, trade, and research organizations.

International biotechnology directory 1985 329

Editor: J. Coombs
Publisher: Macmillan Press
Address: 4 Little Essex Street, London WC2R 3LF, UK
Year: 1984
Pages: 464

Price: £65.00
ISBN: 0-333-36682-4
Geographical area: World.
Content: The directory covers not only companies, research organizations, societies and associations, journals, newsletters and data bases but also includes an invaluable buyers' guide of products and services. Part one is an introduction to biotechnology, including overview, international organizations, information services; data bases; journals; newsletters. Part two contains national profiles - country by country guide to government departments; national profiles; societies. Part three covers products and areas of research - a country by country guide to: non-commercial organizations; companies - products guide. Classification index; product-buyers' guide; buyers' guide alphabetic index.

International cytogenetics laboratory directory* 330

Editor: Barbara Kaplan
Publisher: Association of Cytogenetic Technologists
Address: 616 S Orchard Drive, Burbank, CA 91506, USA
Year: 1984
Pages: 40
Price: $10.00
Content: Entries for about 450 laboratories (with at least one association member on the staff) which study heredity using genetic and cellular biology techniques. Entries include: laboratory name, address, telephone, areas of specialization, techniques, numbers and types of laboratory tests performed, and names of director and cytogenetic technologists. Arrangements: geographical. Indexes: alphabetical. Former title: Association of Cytogenetic Technologists - Laboratory Directory. Also includes list of members, laboratory directors, and certified cytogenetic technologists.
Next edition: December 1985.

International directory of genetic services 331

Editor: Dr Henry T. Lynch
Publisher: March of Dimes Birth Defects Foundation
Address: 1275 Mamaroneck Avenue, White Plains, NY 10605, USA
Year: 1983
Pages: 57
Price: $2.00
Geographical area: World.
Content: Identification of genetic units and their services worldwide. The data has been cross-referenced as follows: directory of genetic units; directors with unit number, country, and genetic services rendered; genetic services rendered by country and unit numbers; and availability of genetic services.

International directory of organizations concerned with aging*   332

Publisher: United Nations Industrial Development Organization
Address: Vienna International Centre, PO Box 300, A-1400 Wien, Austria
Year: 1978
Price: $10.00
Geographical area: World.

International directory of psychologists: exclusive of the USA, fourth edition*   333

Editor: Kurt Pawlik
Publisher: Elsevier Science Publishers
Address: PO Box 211, 1000 AE Amsterdam, Netherlands
Year: 1985
Pages: viii + 1182
Price: $64.75; Dfl175.00
ISBN: 0-444-87774-6
Content: The directory provides - in a standardized form - professional, biographical, and address information on over 32 000 psychologists from 43 countries outside the USA. The following information is included in the entries: name and title, sex, date and place of birth, mailing address, highest academic degree, present position, institution of employment, membership of psychological associations and primary fields of interest or research in psychology.

International directory of specialized cancer research and treatment establishments*   334

Publisher: International Union against Cancer
Address: 1 rue du Conseil-General, CH-1205 Genève, Switzerland
Year: 1982
Pages: 700
Price: SFr170.00
Content: Entries for about 700 major institutions with specialized competence in the field of cancer research and/or treatment in 50 countries. Entries include: institution name, address, telephone, telex; affiliations; names of directors and department heads; number of personnel; amount of annual budget; annual cancer-patient statistics; description of activities; availability of postgraduate training posts; and data on cancer library facilities, cancer registry, information services. Arrangement: geographical. Indexes: institution name, director name, department head name.
Next edition: August 1986; SFr220.00.

International index of laboratory animals   335

Editor: Dr M.P.W. Festing
Publisher: Medical Research Council
Address: Woodmansterne Road, Carshalton, Surrey SM5 4EF, UK
Year: 1980
Pages: 142
Price: £5.00
ISBN: 0-901053-04-X
Geographical area: World.
Content: About 4000 entries of names and locations of stocks of laboratory vertebrates, arranged alphabetically by strain name.
Next edition: December 1985; £10.00.

International mycological directory   336

Publisher: Commonwealth Mycological Institute
Address: Ferry Lane, Kew, Surrey TW9 3AF, UK
Year: 1971
Pages: v + 23
Price: £3.50
ISBN: 0-85198-239-5
Geographical area: World.
Content: International and national mycological societies and organizations, including specialist, amateur, medical and veterinary, phytopathological and microbiological groups; lists of mycological journals and newsletters; major herbaria and culture collections arranged on a national basis.
Next edition: 1987; International Mycological Association.

International medical who's who, second edition   337

Publisher: Longman Group Limited
Address: Sixth Floor, Westgate House, The High, Harlow, Essex CM20 1NE, UK
Year: 1985
Pages: 1372
Price: £195.00
ISBN: 0-582-90112-X
Series: Reference on research
Content: Provides, in two volumes, professional biographical profiles of over 9000 senior medical scientists and biochemists from about 100 countries. Details given include full address, present post, directorships and professional appointments, major publications, and main professional and research interests.
Subjects covered are anatomy and physiology, biochemistry, biophysics, dental sciences, immunology and transplantation, clinical medicine, microbiology, neoplasia, pharmacology and therapeutics, clinical psychology and surgery.
There is an index of names listed by country and subject.

International register of specialists and current research in plant systematics* 338

Editor: R.W. Kiger, T.D. Jacobsen, R.M. Lilly
Publisher: Hunt Institute for Botanical Documentation
Address: Carnegie-Mellon University, Pittsburgh, PA 15213, USA
Year: 1981
Pages: 350
Price: $5.00
Content: Entries for over 1500 botanists engaged in plant classification. Entries include: name, address, botanical specialties, current projects, alphabetically arranged. Indexes: taxonomic; geographical; methodology/general subject; geological age.

International zoo yearbook 339

Editor: P.J.S. Olney
Publisher: Zoological Society of London
Address: Regent's Park, London NW1 4RY, UK
Year: 1984
Pages: xii + 395
Price: £26.50 (hardback); £19.50 (softback)
ISSN: 0074-9664
Geographical area: World.
Content: Species of wild animals bred in captivity: listed under fishes (175 species); amphibians (40 species); reptiles (85 species); birds (950 species); mammals (750 species).
Census of rare animals in captivity: listed under amphibians (8 species); reptiles (54 species and subspecies), birds (89 species and subspecies), mammals (315 species and subspecies).
Studbooks and world registers for rare species of wild animals in captivity - 61 listings.
The above three sections are all available separately at £5.00 each.
Cumulative author index; cumulative subject index.
List of Zoos and Aquaria of the World (900 entries - 70 pages) is also published in alternative volumes (last published vol 22 1982 : next published Vol 24/25 1985).
Next edition: Autumn 1985; £30.00

Medical and healthcare books and serials in print 1985: an index to literature in the health sciences 340

Publisher: Bowker Publishing Company
Address: 58/62 Epping High Street, Epping, Essex CM16 4BU, UK
Year: 1985
Pages: 2365 (in 2 vols)
Price: £106.50
ISBN: 0-8352-2048-6
Geographical area: World.
Content: Listing of 63 000 in-print books and 11 000 serials in 5800 subject areas. Books are listed by subject, author, and title, and serials by subject and title. Full bibliographic and ordering information is provided for each title.
Separate indexes give contact information for publishers and distributors, abstracting and indexing services, and micropublishers.

Medical research centres: a world directory of organizations and programmes, sixth edition 341

Editor: Leslie T. Morton, Jean F. Hall
Publisher: Longman Group Limited
Address: Sixth Floor, Westgate House, The High, Harlow, Essex CM20 1NE, UK
Year: 1983
Pages: 1112
Price: £175.00
ISBN: 0-582-90017-4
Series: Reference on research
Content: Provides a comprehensive guide to over 3000 organizations controlling 10 000 departments and laboratories throughout the world which conduct or finance medical and biological research. The entries are arranged alphabetically by country in two volumes. The directory provides for each entry, where available: title; full postal address; telephone and telex numbers; parent or affiliation; product range for industrial companies; name of director; names of senior scientific staff; activities and major current research programmes.
The subject matter extends across the medical specialties but excludes veterinary medicine, botany and zoology.
There is a titles of establishments index and a subject index.

Naturalists' directory and almanac (international)* 342

Editor: Dr Ross H. Arnett
Publisher: Flora and Fauna Publications
Address: 2406 NW 47th Terrace, Gainesville, FL 32606, USA
Pages: 310 (in 3 parts)
Price: $12.95 (per part)
Content: The directory covers 5000 active amateur and professional naturalists, worldwide, who will correspond. Includes separate listings of natural history periodicals, societies, museums, and sources of natural history supplies. Published in three parts: Insect Collectors and Identifiers, Plant Collectors and Identifiers, and General Naturalists. Entries include: name, address, interests, services. Arrangement: alphabetical in each part. Indexes (all in part three): geographical, specialty, institution.

INTERNATIONAL DIRECTORIES

Nature's directory of biologicals* 343

Publisher: Macmillan Publishing Company
Address: Houndsmill, Basingstoke, Hampshire, UK
Year: 1983
Price: £22.50
ISBN: 0-333-34937-7
Geographical area: World.
Content: Listing of over 400 companies selling laboratory products and nearly 2000 products. Buyers' guide to enzymes, antibodies, cell lines, fine biochemicals, etc.

Pharmacology and pharmacologists: an international directory 344

Publisher: Oxford University Press
Address: Walton Street, Oxford OX2 6DP, UK
Year: 1981
Pages: xxii + 387
Price: £50.00
ISBN: 0-19-200101-9
Geographical area: World.
Content: There are 3200 entries arranged in alphabetical order (surname), giving addresses and telephone numbers, designations and fields of work, specific research activities, qualifications, past appointments, and publications; research activities index.
Not available on-line.

World directory of collections of cultures of microorganisms 345

Editor: V.F. McGowan, V.B.D. Skerman
Publisher: World Data Center, University of Queensland
Address: Department of Microbiology, University of Queensland, St Lucia 4067, Queensland, Australia
Year: 1982
Pages: xxxi + 641
Price: $30.00
Content: There are 358 collection entries from 53 countries with information provided in the following format: directory of institutions; lists of species of microorganisms; geographical index of collections; index of collections; index of main interests of the collections; list of personnel.
Next edition: 1986; $30.00.

World directory of medical schools, fifth edition 346

Publisher: World Health Organization
Address: Avenue Appia, 1211 - Genève 27, Switzerland
Year: 1979
Pages: 358
Availability: OP
ISBN: 92-4-150006-9
Content: Information on undergraduate medical education, and the institutions that provide it, in more than 100 countries. For each country there is a general statement on the pattern and administrative structure of undergraduate medical education. Each country statement is followed by a tabulation of the undergraduate medical schools in the country concerned. The names and addresses are, for the most part, given in the national language. Other columns in the table list the year instruction started, the number of teaching staff, total enrolment, the number of admissions, and the number of graduates. In the three last-mentioned columns, the student body has been subdivided into nationals and foreigners and males and females wherever the information has been made available.
A series of annexes presents the following information: data on registration and licensure for medical practice in a number of countries without medical schools; an example of a medical school's general objectives; definition of 'integration' in the medical curriculum; and a comprehensive list of the members of medical schools in 1955, 1960, 1970, and 1975, by country and area.
Next edition: 1986.

World directory of pharmaceutical manufacturers* 347

Editor: B. Bell
Publisher: IMS WORLD Publications Limited
Address: York House, 37 Queen Square, London WC1N 3BL, UK
Year: 1984
Pages: 300
Price: $200.00
Content: Entries for over 2000 leading drug companies in over 30 major markets. Entries include: company name, address, telephone, telex number, list of products or services, parent company and subsidiaries, company's leading products listed by trade name and therapeutic category (including approximate total number of products marketed). Arrangement: by country. Indexes: subsidiary name.
Next edition: 1986.

World directory of schools of public health and postgraduate training in public health 348

Editor: World Health Organization
Address: Avenue Appia, 1211 - Genève 27, Switzerland
Year: 1985
Pages: 189
Price: SFr20.00
ISBN: 92-4-150007-7
Next edition: Late 1985

World list of family planning addresses 349

Publisher: International Planned Parenthood Federation
Address: 18-20 Lower Regent Street, London SW1Y 4PW, UK
Year: 1985
Pages: ii + 22
Price: Free
ISSN: 0535-1774
Content: There are 130 entries, listed alphabetically by name of country.
Next edition: January 1986; free.

World meetings: medicine 350

Publisher: Macmillan Publishing Company
Address: 200D Brown Street, Riverside, NJ 08370, USA
Price: $140.00 (for four quarterly issues)
ISBN: 0-02-695300-5
Content: The publication covers the meetings and conferences to be held during the next two years in medicine, and the medical sciences. It presents complete details on the location, date, content, and goals of every upcoming meeting dealing with the scientific, behavioural, managerial, and applied aspects of medicine and allied health care.
Indexes: keyword index; location index; date index; deadline index; sponsor directory and index.

PHYSICS, MATHEMATICS AND NUCLEAR SCIENCES

Information sources in physics, second edition 351

Editor: D.F. Shaw
Publisher: Butterworth and Company Limited
Address: Borough Green, Sevenoaks, Kent TN15 8PH, UK
Year: 1984
Pages: 456
Price: £35.00
ISBN: 0-408014-74-1
Series: Butterworths guides to information sources
Geographical area: World.
Content: A review of the sources of information in all major branches of physics. Each chapter is written by a specialist in his chosen field. There are fourteen chapters treating special subject areas following a modified version of the ICSU and INSPEC classifications. The remaining six chapters include an introduction by John Ziman, a discussion of the scope of physics and the control of the literature and other sources of information, and a treatment of abstracting and indexing services, references material, general treatises, patent and 'grey' literature.

International nuclear energy guide, thirteenth edition 1985* 352

Publisher: Enercom
Address: 65-67 avenue des Champs Elysées, 75008 Paris, France
Year: 1984
Price: F490.00
Geographical area: World.
Content: Listing of administrative and professional organizations, research centres, protection and safety, nuclear fuel cycle, design and construction and allied nuclear firms, equipment and services; world guide; bibliography.

International who's who in energy and nuclear sciences 353

Publisher: Longman Group Limited
Address: Sixth Floor, Westgate House, The High, Harlow, Essex CM20 1NE, UK
Year: 1983
Pages: 532
Price: £105.00
ISBN: 0-582-90110-3
Series: Reference on research
Geographical area: World.
Content: Biographical details of over 3800 research chemists, research physicists and development engineers in over 70 countries involved in the generation, storage and efficient use of energy. The first part lists individuals alphabetically giving personal and professional information, publications and public appointments. The second part is a country and topic list of the same individuals.

Nuclear engineering international buyers guide 354

Editor: J. Varley
Publisher: Electrical - Electronic Press
Address: Quadrant House, The Quadrant, Sutton, Surrey SM2 5AS, UK
Year: 1985
Pages: 80
Price: £6.00
Geographical area: World.
Content: Not available on-line.
Next edition: April 1986; £6.50.

INTERNATIONAL DIRECTORIES

Nuclear industry almanac - volume 2 Asia/Pacific region*   355

Publisher: Nuclear Energy Intelligence
Address: Teal House, Moat Lane, Prestwood, Great Missenden, Buckinghamshire HP16 9DA, UK
Year: 1986

World directory of mathematics   356

Editor: Professor G.D. Mostow
Publisher: American Mathematical Society
Address: PO Box 6248, Providence, RI 02940, USA
Year: 1982
Pages: 725
Price: $23.00
Content: List of important mathematical organizations; alphabetical list of mathematics; geographical list of mathematics.
Next edition: 1986.

World nuclear directory: a guide to organizations and research activities in atomic energy, seventh edition   357

Editor: C.W.J. Wilson
Publisher: Longman Group Limited
Address: Sixth Floor, Westgate House, The High, Harlow, Essex CM20 1NE, UK
Year: 1985
Pages: 387
Price: £95.00
ISBN: 0-582-90025-5
Series: Reference on research
Content: An international guide to over 1500 organizations and laboratories which conduct or promote research, development or substantial manufacturing work in the atomic energy field. Subjects covered range from high energy nuclear physics, plasma physics and fusion technology, to radioactive waste management, economics and regulatory developments. A new feature for this edition is the inclusion of profiles of around 30 countries. Fully indexed by title and by subject.

2 AFRICA

GENERAL SCIENCES

African book world and press: a directory, third edition 1

Editor: Hans M. Zell, Carol Bundy
Publisher: Saur Verlag, K.G.
Address: PO Box 56, 14 St Giles, Oxford OX1 3EL, UK
Year: 1983
Pages: xx + 295
Price: £46.00; $78.00; DM188.00
ISBN: 0-905450-10-8 (UK); 3-598-10439-1 (Germany)
Geographical area: Africa.
Content: The directory provides information, in both English and French, on libraries, publishers and the retail book trade, research institutions with publishing programmes, magazines and periodicals, major newspapers, and printing industries throughout Africa. It contains over 4600 entries arranged in 51 sections country by country.

African international organizations directory 1984-85 2

Publisher: Saur Verlag, K.G.
Address: Postfach 711009, D-8000 München 71, German FR
Year: 1985
Pages: 604
Price: DM248.00
ISBN: 3-598-21650-5
Series: Guides to international organizations
Geographical area: Africa.
Content: Detailed descriptive entries for international organizations active within 55 African countries, grouped in sections according to organization type: intercontinental membership organizations; regionally defined membership organizations; organizations emanating from places, persons, and other bodies; organizations having special form; internationally-orientated organizations grounded in a particular country; recently reported or proposed organizations. There is a multi-access, multi-lingual index system.

Africa south of the Sahara 1984-85* 3

Publisher: Europa Publications
Address: 18 Bedford Square, London WC1B 3JN, UK
Year: 1984
Price: £48.00
Geographical area: Central and Southern Africa.
Content: Details of geography, history and economics, statistics, constitution, and judicial system as well as government, diplomative corps, political parties, communications, finance, and industry, arranged country by country.
Next edition: September 1985; £52.00.

Directory for scientific research organizations in South Africa* 4

Editor: Marian de Wind
Publisher: Council for Scientific and Industrial Research
Address: PO Box 395, Pretoria 0001, South Africa
Year: 1984
Content: Entries cover South African organizations conducting research in science and technology, giving address, head of research, subject of research, special facilities, publications, etc.

Directory of African experts* 5

Publisher: United Nations Industrial Development Organization
Address: PO Box 300, A-1400 Wien, Austria
Price: $31.00
Geographical area: Africa.

Content: A main index contains the full bibliographical data, qualifications, specializations, and experience of over 3600 experts. A subject index is arranged by field of specialization.
This is the first edition of the directory published by the Pan African Documentation and Information System (PADIS) as part of a computerized personnel common register file concerning African specialists.

Directory of African universities/ Repertoire des universités Africaines (third edition) — 6

Editor: Daniel M'Boungou-Mayengué
Publisher: Association of African Universities
Address: PO Box 5744, Accra, Ghana
Year: 1985
Price: $20.00
Content: 72 entries under 35 African countries. The entries include information about African universities, on postal addresses, principal officers, academic, technical and administrative staff, faculties, departments and institutes, degrees, diplomas and certificates, admission requirements, library services, etc. The publication is in two parts: one containing anglophone, the other francophone universities.

Directory of development research and training institutes in Africa* — 7

Publisher: Organization for Economic Cooperation and Development
Address: Publications Office, 2 rue André-Pascal, 75775 Paris Cedex 16, France
Year: 1982
Price: £5.00
ISBN: 92-64-02353-4

Directory of scientific and technical societies in South Africa* — 8

Editor: Marian de Wind
Publisher: Council for Scientific and Industrial Research
Address: PO Box 395, Pretoria 0001, South Africa
Year: 1984
Content: Details of around 500 scientific and technical societies in South Africa, arranged alphabetically with names and addresses of officials, membership conditions, fees, publications etc.

Directory of southern African libraries, 1983 — 9

Editor: J.A. Fourie
Publisher: Pretoria State Library
Address: PO Box 397, 0001 Pretoria, South Africa
Pages: 553
Price: R45.00
ISBN: 0-7989-1193-X
Series: Contributions to library science
Geographical area: Southern Africa.
Content: The directory includes libraries and information centres not only in the Republic of South Africa, but also those in the national states, South West Africa, as well as the independent states of Bophuthatswana, Botswana, Ciskei, Lesotho, Malawi, Swaziland, Transkei, Venda and Zambia.
Full directory information is given for each library, including street and postal address, telephone number, telex number, head of library, loan and reference facilities, hours, services, service aids (catalogues, classification systems), automated operations, specializations, salaries and acquisitions, statistics of staff and holdings, branch libraries, and history.
Within each territory, arrangement is according to category of library: national (sub-arranged by name); special, university and college libraries, arranged geographically by town; central government libraries arranged by catchword in an alphabetical sequence after the entry for the Library Services Branch of the Department of National Education.
These arrangements are complemented by geographical, subject specialization and bilingual name indexes.

Egyptian directory of scientific centres and organizations: a directory of organizations in science, technology, agriculture, medicine and social sciences, second edition — 10

Editor: National Information and Statistics on Science and Technology
Publisher: Academy of Scientific Research and Technology
Address: 101 Kasr El-Eini Street, Cairo Egypt
Year: 1979
Pages: 578
Availability: OP
Series: Directories of scientific centres and organizations
Content: The directory provides detailed information on about 265 university research institutes and faculties, ministries and affiliated institutes, organizations and laboratories conducting research. Each entry includes date of foundation, key personnel, divisions and number of qualified staff, and research and development activities.
Not available on-line.
Next edition: 1985.

Guide des services d'information en Tunisie, 1984-85   11

[Guide to Information Services in Tunisia, 1984-85]
Editor: Rabii Bannouri
Publisher: Centre de Documentation Nationale
Address: 77 rue Ibn Khaldoun, 1001 Tunis RP, Tunisia
Year: 1985
Pages: xii + 180
Content: The guide is published in two language versions: Arabic and French. It contains around 450 entries alphabetically arranged. The first edition was published in 1983, since when it has been completely revised. It is not available on-line.

Guide to the museums of Southern Africa   12

Editor: Hans Fransen
Publisher: Southern African Museums Association
Address: College of Careers, PO Box 2081, Cape Town, South Africa
Year: 1978
Pages: xii + 219
Price: R4.00
Content: Approximately 262 entries, The museums are grouped in chapters according to countries (and, for South Africa, provinces). In each chapter, they appear in alphabetical order of towns. Text follows a uniform pattern: history of institution and of building, scope of collections and other activities (research, publications, educational services, etc). Index and thematical index. Not available on-line. It is hoped to print an Afrikaans edition in the near future.

Inventaire - instituts de recherche et de formation en matière de développement - Afrique   13

[Directory -development research and training institutes - Africa]
Editor: OECD Development Centre
Publisher: Organization for Economic Cooperation and Development
Address: 2 rue André-Pascal, 75775 Paris Cedex 16, France
Year: 1982
Pages: xv + 156
Price: £10.00
ISBN: 92-64-02353-4
Series: Liaison bulletin between development research and training institutes
Content: The directory gives information on 290 development research and training institutes in 40 African countries. The database includes general information relating to the institutes (name, address, etc), a short description of their postgraduate training and research programmes, of other activities (publications, documentation, etc) and of the facilities available (library, computer, conference rooms).
Selective information searches can be made on specific areas of interest.
Next edition: 1986.

Libraries in West Africa: a bibliography   14

Editor: Helen Davies
Publisher: Saur Verlag, K.G.
Address: PO Box 56, 14 St Giles, Oxford OX1 3EL, UK
Year: 1982
Pages: 186
Price: £16.00; DM56.00
ISBN: 3-598-10440-5
Content: The publication lists almost 1400 items published between 1930 to the end of 1979, including books, journal articles, theses, reports, and conference papers. Arrangement is geographical and there is an author index as well as a select list of library journals published in West Africa included as an appendix.

Marine research centres: Africa   15

Publisher: Food and Agriculture Organization
Address: Via delle Terme di Caracalla, 00100 Roma, Italy
Year: 1982
Pages: xii + 254
Series: UNEP regional seas directories and bibliographies
Content: Compiled jointly by the United Nations Environmental Programme, the United Nations Economic Commission for Africa, and the United Nations Educational, Scientific and Cultural Organization, the directory lists 80 research centres in 24 countries, and is bilingual in French and English.

National register of research projects - part II: natural sciences: physical, engineering, and related sciences   16

Publisher: Science and Planning Directorate, Department of Constitutional Planning
Address: Private Bag X644, 0001 Pretoria, South Africa
Year: 1984
Pages: xxii + 282
Price: Free
ISBN: 0-7988-2972-9
Series: National register of research projects
Geographical area: South Africa.
Content: 5215 entries. The publication contains: a user guide; table of contents of research fields; individual project entries; keyword index; code list of research organizations; index of interdisciplinary research fields;

code list of fields of application; code list of magisterial districts and region.
Next edition: 1985; free.

SCOLMA directory of libraries and special collections on Africa, in the United Kingdom and Western Europe, fourth edition   17

Editor: Harry Hannam
Publisher: Saur Verlag, K.G.
Address: PO Box 56, 14 St Giles, Oxford OX1 3EL, UK
Year: 1983
Pages: 183
Price: £19.50; DM80.00
ISBN: 0-905450-11-6 (UK); 3-598-10502-9 (Germany)
Geographical area: Africa.
Content: Published on behalf of the Standing Conference on Library Materials on Africa (SCOLMA), this fourth edition of the SCOLMA directory has been extensively revised and updated. For the new edition the scope of the directory has been widened, and it now includes details not only of Africana collections in the United Kingdom and the Republic of Ireland, but also of significant holdings of African material in the libraries of Western Europe.
The new edition contains 275 entries, each entry providing full name and address, telephone, name of chief librarian and/or person in charge of the African collection, hours of opening, conditions of access, loan and reference facilities, size and description of the collection, and details of any relevant publications issued. A particular effort has also been made for this edition to identify holdings of audio-visual material relating to Africa. Indexes cover persons, places, and subjects.

Statistics Africa: sources for social, economic and market research   18

Editor: Joan M. Harvey
Publisher: CBD Research Limited
Address: 154 High Street, Beckenham, Kent BR3 1EA, UK
Year: 1978
Pages: xii + 374
Price: £25.00
ISBN: 0-900246-26-X
Content: Bibliography of sources of statistical information for each country, showing publisher, frequency, price, contents, etc. The publication also lists central statistical offices, and other organizations collecting statistics. It contains alphabetical indexes of organizations and of titles, as well as a subject index.
Not available on-line.

Zimbabwe research index   19

Publisher: Scientific Liaison Office
Address: PO Box 8510, Causeway, Harare, Zimbabwe
Year: 1980
Availability: OP
Content: This research index has been under review; publication expected to resume after November 1985.

AGRICULTURE AND FOOD SCIENCE

Directory of fertilizer facilities: Africa*   20

Publisher: United Nations Industrial Development Organization
Address: PO Box 300, A-1400 Wien, Austria
Price: $2.50
Geographical area: Africa.

Official list of professional and research workers, lecturing staff, extension and other workers in the agricultural field   21

Editor: N.S. Steenkamp
Publisher: Directorate of Agricultural Information, Department of Agriculture and Water Supply
Address: Private Bag X144, Pretoria 0001, South Africa
Year: 1985
Pages: 74
Price: Free
Geographical area: Southern Africa.
Content: List of institutes, university faculties, government departments, with address, telephone, telex, etc. Full list of the people who work there and their particular agricultural subjects. There is an alphabetical index of names, and a list of abbreviations.
Next edition: April 1986; Free.

INDUSTRY AND MANUFACTURING

Current African directories*   22

Publisher: CBD Research Publications
Address: 154 High Street, Beckenham, Kent BR3 1EA, UK
Year: 1972
Price: £27.00

ISBN: 0-900246-11-1
Content: Bibliography of trade directories published in or about Africa.

CZI register and buyer's guide 23

Publisher: Thompson Publications
Address: 1683, Harare, Zimbabwe
Year: 1984
Pages: iv + 200
Price: Z$10.00
Geographical area: Zimbabwe.
Content: Includes list of manufacturers arranged geographically; alphabetically arranged list of products; brand and trade names; chief executives.
Next edition: 1985; Z$12.00.

Directory of industrial and technological research institutes in Africa* 24

Publisher: United Nations Industrial Development Organization
Address: PO Box 300, A-1400 Wien, Austria
Year: 1982
Pages: 68

Directory of manufacturing establishments 25

Publisher: Central Statistics Office
Address: Tower Hill, Freetown, Western Area, Sierra Leone
Year: 1984
Pages: 25
Geographical area: Sierra Leone.
Content: The manufacturing directory consists of two main tables: number of establishments and persons engaged by provinces and by kind of activity; number of establishments and persons engaged by provinces and by size of employment. It contains the name of establishments and their addresses and the number of people employed.

Kompass register: Morocco* 26

Publisher: Kompass Maroc
Address: Boîte Postale 11-100, MA Casablanca, Morocco
Year: 1984
Price: £60.00
Content: Names and addresses of individual products arranged by Standard Industrial Classification.

Major companies of Nigeria, fourth edition, 1983* 27

Editor: M. Lawn, Jennifer Carr
Publisher: Graham and Trotman Limited
Address: Sterling House, 66 Wilton Road, London SW1V 1DE, UK
Year: 1983
Pages: 350
Price: £49.00, $78.00 (hard cover); £40.00, $64.00 (paperback)
ISBN: 0-86010-405-2 (hard cover); 0-86010-404-4 (paperback)
ISSN: 0144-2740
Series: Major companies
Content: This directory provides details of 2500 top companies in 74 business sectors, including company name, address, telephone, telex and cable numbers; names of directors and senior executives by job title; business activities; brand names; agencies; branches and subsidiaries; bankers; financial details; date of establishment.

Sociétés et fournisseurs d'Afrique noire 1984-85 28

[Companies and suppliers in Africa, 34th edition]
Publisher: EDIAFRIC-la documentation Africaine
Address: 10 rue Vineuse, 75116 Paris, France
Price: F390 (one volume); F700 (two volumes); F950 (three volumes)
Geographical area: French-speaking black Africa and North Africa.
Content: Published in three volumes:
L'annuaire des exportateurs (The exporter's directory): part 1 - index of companies, trade marks, who's who; part 2 - manufacturers and suppliers of French and foreign goods.
L'annuaire d'Afrique noire (African trade directory): companies, banks and public institutions of the twelve countries in French-speaking black Africa, listed by country; composition of the governments of the twelve states; index of companies; index of trade marks; who's who.
L'annuaire d'Afrique du Nord (North African trade directory): companies, banks, and public institutions of Algeria, Tunisia, and Morocco, listed by country; composition of the governments of the three countries; index of companies; index of trade marks; who's who.

Zambia directory 1985 29

Publisher: Directory Publishers of Zambia
Address: PO Box 30963, Lusaka, Zambia
Price: K30
Content: Numerical box numbers of towns in Zambia; telegraphic addresses; numerical phone numbers; towns of Zambia alphabetically listed; classified section.

Zambia industrial and commercial directory 1984-85 — 30

Publisher: Zambian Industrial and Commercial Association
Address: 30844 Lusaka, Zambia
Year: 1985
Pages: iv + 144
Price: $10.00
Content: The directory has an alphabetical list of companies; a classified section with index; general information on taxation, import and export procedures, investment incentives and diplomatic missions both in Zambia and abroad.
Next edition: 1987; $15.00

MEDICAL AND BIOLOGICAL SCIENCES

National register of research projects - part I: natural sciences: biological, medical and related sciences — 31

Publisher: Science Planning Directorate, Department of Constitutional Development and Planning
Address: Private Bag X644, 0001 Pretoria, South Africa
Year: 1984
Pages: xxi + 319
Price: Free
ISBN: 0-7988-2971-0
Series: National register of research projects
Geographical area: South Africa.
Content: 5616 entries. The publication contains: a user guide; table of contents of research fields; individual projects entries; keyword index; code list of research organizations; index of inter-disciplinary research fields; code list of fields of applications; code list of magisterial districts and region.
Next edition: 1985; Free.

3 ASIA

GENERAL SCIENCES

Arabian government and public services 1985　　1

Editor: Shahrukh Husain
Publisher: Beacon Publishing
Address: Jubilee House, Billing Brook Road, Weston Favell, Northamptonshire NN3 4NW, UK
Pages: xii + 308
ISBN: 0-906358-54-X
Series: Beacon business handbooks
Geographical area: Countries of the Gulf Cooperation Council.
Content: Approximately 800 entries arranged by country, giving up to 100 names per entry.
Next edition: December 1985; £47.00.

Association of development research and training institutes of Asia and the Pacific (ADIPA): a directory of members　　2

Editor: Elizabeth W. Ng
Publisher: University of Hong Kong Centre of Asian Studies
Address: Pokfulam Road, Hong Kong
Year: 1979
Price: HK$25.00
Geographical area: Asia.

Bangladesh trade directory*　　3

Publisher: Intertrade Publications Private Limited
Address: 55 Gariahat Road, Calcutta 700019, India
Year: 1984
Price: £65.00
Content: Details of 9000 distributors, manufacturers, banks, trade and professional associations, universities, and other organizations.

Brief directory of museums of India　　4

Editor: Usha Agrawal
Publisher: Museums Association of India
Address: National Museum, Janpath, New Delhi, India
Year: 1985
Pages: iv + 148
Price: Rs10.00
Geographical area: India.
Content: Lists information on 369 museums of India including: controlling authority; types of collections; publications; availability of guides series.

China phone book and address directory*　　5

Editor: Douglas J. Rasmussen
Publisher: China Phone Book Company Limited
Address: GPO Box 11581, Hong Kong
Year: 1985
Pages: 616
Price: $50.00
ISBN: 0250-4170
Content: Includes sections on: government and foreign affairs; research and education; media, communications, and culture; industry; affiliated professional groups; trade and finance; foreign government representatives and foreign businesses; hotels, restaurants, and travel facilities.

Directory of Asian museums　　6

Publisher: Unesco
Address: Place Fontenoy, 75007 Paris, France
Year: 1985
Pages: 295
Price: Free
Content: There are 700 entries, covering museums in 23 Asian countries. Each entry covers, as far as information has been provided, name, address, name of director or

curator, opening hours and admission charges, legal status, historical background, nature of collections, publications, services, and a bibliography of articles or books on the museum (compiled from the resources of the Unesco-ICOM Documentation Centre).
The countries are arranged by alphabetical order, and within each country, the cities in alphabetical order. There is an index of provinces by country.
Not available on-line.

Directory of current Hong Kong research on Asian topics   7

Editor: Elizabeth W. Ng
Publisher: University of Hong Kong Centre of Asian Studies
Address: Pokfulam Road, Hong Kong
Year: 1978
Pages: 197
Price: HK$40.00
ISBN: 0441-1900
Content: The directory is intended to be a comprehensive coverage of Asian research for all disciplines. Contains 1100 entries, index of researchers, subject index, glossary of Chinese names.

Directory of environmental organizations in India   8

Editor: S.K. Kesarwani, S.G. Bhat
Publisher: National Environmental Engineering Research Institute
Address: Nehru Marg, Nagpur 440 020, India
Year: 1976
Pages: 469
Content: This directory lists 208 Indian research and development agencies with activities in environmental science and technology. It is divided into four parts: research and teaching organizations; commissions, committees, directorates, bureaus and similar agencies, including standards organizations; water pollution prevention and control boards; and, societies and associations. Each entry includes the following information: name and address of organization; name of the person in charge, telephone number and telegraphic address; year of establishment; administrative structure; activities; special facilities; and publications. A subject index, a location index and a geographical index facilitate access to the directory.

Directory of international science organizations   9

[Kokusai gakujutsu dantai soran]
Publisher: Science Council of Japan
Address: 22-34 Roppongi 7 chome, Minato-ku, Tokyo 106, Japan
Year: 1979

Pages: 954
Price: Y9800
Geographical area: Japan.
Content: There are 926 non-governmental organizations and 69 inter-governmental organizations are listed. Data is obtained by questionnaires from the Science Council of Japan. Another edition is planned.

Directory of libraries in Singapore   10

Editor: Sng Yok Foong, Lau Siew Kheng, Khoo Guan Fong
Publisher: Library Association of Singapore
Address: c/o National Library, Stamford Road, Singapore 0617, Singapore
Year: 1983
Pages: vii + 194
Availability: OP
ISBN: 9971-83-620-3
Content: There are 137 entries, including list of members, type of libraries, and subject and library index.
Next edition: January 1986; S$30.00.

Directory of national systems of technicians education in south and central Asia*   11

Publisher: Unesco, Regional Office of Science and Technology for South and Central Asia
Address: 7 place de Fontenoy, 75700 Paris, France
Year: 1983
Pages: 100

Directory of non-governmental organizations in environment*   12

Publisher: Department of the Environment, Environmental Information System
Address: New Delhi, India
Year: 1984
Pages: xi + 213

Directory of officials and organizations in China 1968-83*   13

Editor: Malcolm Lamb
Publisher: Sharpe, M.E.
Address: 3 Henrietta Street, London WC2E 8LU, UK
Year: 1984
Price: £64.95
ISBN: 0-87332-277-0
Content: Details of Communist Party, ministries and commissions, armed forces, science academies and institutions, trade unions, regional government, etc, indexed by name and by characters.

Directory of research institutes and industrial laboratories in Israel 14

Editor: Naomi Shanin-Cohen
Publisher: National Centre of Scientific and Technological Information
Address: PO Box 20125, Tel-Aviv 61201, Israel
Year: 1980
Pages: 230
Price: $35.00
ISSN: 0334-3197
Content: There are 1152 entries, with alphabetical lists of, and subject index to research institutes and industrial laboratories.

Directory of research institutes in Israel 15

Editor: Geula Gilat, Dan Bry
Publisher: National Centre of Scientific and Technological Information
Address: PO Box 20125, Tel-Aviv 61201, Israel
Year: 1982
Pages: 138
Availability: OP
ISBN: 965-228-001-1
Content: There are 188 entries, with an alphabetical list of research institutes and a subject index.

Directory of scientific libraries in Thailand, third edition 16

Editor: Karnjana Chanyarak
Publisher: Thai National Documentation Centre
Address: Bang Khen, Bangkok, Thailand
Year: 1983
Pages: viii + 165
ISBN: Free
Content: The directory includes 79 libraries, which consist of 32 government libraries, 39 faculty libraries affiliated to academic institutions, 8 libraries belong to international associations and organizations. The directory is alphabetically arranged by name of libraries.
Next edition: 1986

Directory of scientific research institutes in the People's Republic of China 17

Editor: Susan Swannack-Nunn
Publisher: National Council for US-China Trade
Address: 1050 Seventeenth Street NW, Suite 350, Washington DC 20036, USA
Year: 1977
ISSN: OP
Content: Volume 1 covers agriculture, fisheries and forestry; volume 2, chemicals and construction; volume 3, deals with electrical and electronics, energy, light industry, machinery, including metals and mining, and transportation.

Directory of sci-tech r&d institutions in ROC 18

Publisher: National Science Council, Science and Technology Information Centre
Address: 128-1 Yen Chiu Yuan Road, Section 2, Nankang, Taipei, 115, Taiwan
Year: 1985
Pages: 380
Price: NT$250.00
Geographical area: Taiwan.
Content: Details of Academia Sinica, scientific and technical research institutes of the executive Yuan and provincial government; technical institutes of Taipei and Kaohsiung city government; universities and colleges; statutory bodies; government utilities and private enterprises. Indexed by organization.

Directory of special and research libraries in India 19

Publisher: Indian Association of Special Libraries and Information Centres
Address: P-291 CIT Scheme No 6M, Kankurgachi, Calcutta 700 054, India
Year: 1985
Pages: vi + 90
Price: £10.00
Content: There are 525 entries, arranged alphabetically under the name of the institutions. Entries include: address; phone; year of foundation; name of librarian; subjects; holdings; subscriptions; services; publications; staff. Indexes include: list of libraries by location; list of libraries by subject; name index.

Directory of special libraries and information sources in Indonesia 1981 20

[Direktori perpustakaan khusus dan sumber informasi di Indonesia 1981]
Editor: Hendrarta Kusbandarrumsamsi, Sudarisman Dwinarto
Publisher: Indonesian National Scientific Documentation Centre
Address: PO Box 3065/Jkt Jl Jendral Gatot, Subroto, Djakarta, Indonesia
Year: 1982
Pages: xxii + 308
Price: $20.00 (book); $9.00 (microfiche)
ISSN: 0216-2164

Content: 295 entries alphabetically under location. Details include: address; date founded; person in charge; staff; facilities; hours; users; resources; subjects; services provided; budget; publications; lending policy; subject index to libraries.
Next edition: September 1986; $25.00.

Directory of special libraries in Israel, fifth edition   21

Publisher: National Centre of Scientific and Technological Information
Address: PO Box 20125, Tel Aviv 61201, Israel
Year: 1980
Pages: 182
Price: $50.00
ISBN: 965-228-000-3
Content: Descriptions of 322 special libraries in institutes of higher education, technical schools, local councils, and government institutions. Each listing gives details on address, head librarian, hours of service, types of service, subject content, and collection information. Bilingual.

Directory of the cultural organization of the Republic of China*   22

[Chung hua min-kuo hsueh-shu chi-ko lu]
Publisher: National Central Library, Bureau of International Exchange of Publications
Address: 43 Nan-hai Road, Taipei 107, Taiwan
Year: 1985
Price: $15.00
Geographical area: Taiwan.
Content: Over 900 entries covering libraries, schools, learned societies, cultural centres, museums, and institutes of Chinese studies. There is a name index and subject index. This edition will be available on-line.

Directory of the learned societies in Japan   23

[Zenkoku gakkyokai soran]
Publisher: Science Council of Japan
Address: 22-34 Roppongi 7 chome, Minato-ku, Tokyo 106, Japan
Year: 1981
Pages: 683
Price: Y5200
Content: 1003 learned societies are listed. Data is obtained by questionnaires from the Science Council of Japan. A new edition is planned.

Directory of training research and information - producing centres in Iran*   24

Publisher: Iranian Documentation Centre
Address: PO Box 51-1387, 1188 Enqelab Avenue, Tehran, Iran
Year: 1983
Pages: 2 vols
Content: In Persian and English, the directory lists approximately 450 institutions, providing official name, parent organization, address and telephone number, subject areas and publications. Entries are classified by type of organization, eg universities and institutions of higher education, scientific associations, professional bodies, government. There are subject and alphabetical indexes.

Environmental education: list of institutions in the region of Asia and the Pacific*   25

Publisher: Unesco Regional Office for Education in Asia and the Pacific
Address: PO Box 1425, General Post Office, Bangkok 10500, Thailand
Year: 1983
Pages: 38

Government research institutes in Japan*   26

Publisher: Science and Technology Agency
Address: 2-2-1 Kasumigaseki, Chiyoda-ku, Tokyo 100, Japan
Year: 1981
Pages: 295

Handbook: southeast Asian institutions of higher learning   27

Editor: Dr Ninnat Olanvoravuth
Publisher: Association of Southeast Asian Institutions of Higher Learning
Address: Ratasastra Building, Chulalongkorn University, Henri Dunant Road, Bangkok 10500, Thailand
Year: 1982
Pages: xii + 643
Availability: OP
Next edition: 1986; $30.00.

Higher education in Malaysia - a bibliography   28

Editor: University of Malaysia Library
Publisher: Regional Institute of Higher Education and Development
Address: Room 803, 8th floor, RELC Building, 30

Orange Grove Road, Singapore 1025, Singapore
Year: 1984
Pages: xvii + 111
Price: S$32.00
ISBN: 9971-911-01-9
Series: Asean higher education bibliography series
Geographical area: East and West Malaysia.
Content: There are 1573 entries in two sections: A - the study (an overview of history and trends of higher education in Malaysia); B - the bibliography which lists universities and colleges, with details of their governance, finance, staff, methods, students, research, etc.

India who's who* 29

Publisher: INFA Publications
Address: Jeevah Deep Building, Parliament Street, New Delhi 110001, India
Year: 1984
Price: £34.75
Content: Entries for about 4000 professional persons in India, including list of Indian professional associations and institutions.

Inventaire descriptif des unités de recherche et de formation en sciences sociales - Asie 30
[Directory of social science research and training units - Asia]

Editor: OECD Development Centre
Publisher: Organization for Economic Cooperation and Development
Address: 2 rue André-Pascal, 75775 Paris Cedex 16, France
Year: 1975
Pages: xix + 218
Availability: OP
ISBN: 92-64-01426-8
Series: Liaison bulletin
Content: The directory contains information on 509 social science research and training units in 30 Asian countries. The data base includes general information relating to the units (name, address, etc), a short description of their postgraduate training and research programmes, and details of periodical titles and working links.
Selective information searches can be made on specific areas of interest.
Updating of the directory is to be carried out by the Association of Development Research and Training Institutes of Asia and the Pacific, Kuala Lumpur, in cooperation with the OECD Development Centre.

Japan directory of professional associations* 31

Publisher: Intercontinental Marketing Group
Address: PO Box 5056, Tokyo 100-31, Japan
Year: 1984
Price: $150.00
Content: The directory lists Japanese business, academic, trade, and professional organizations.

Japanese research institutes funded by ministries other than education 32

Editor: Seikoh Sakiyama
Publisher: Office of Naval Research
Address: American Embassy, APO San Francisco, CA 96503, USA
Year: 1981
Pages: 110
Price: Free
Content: The publication covers about 160 research centres, universities, and other similar organizations in Japan which are funded by government agencies other than the Ministry of Education and quasi-government agencies. Entries include institute name, address, telephone, divisions; number of staff and budget for three years, 1976-78. Entries are arranged by ministry (department or agency). There are indexes to subject of research, geographical location, and institute name.

Japanese research institutes funded by the ministry of education 33

Editor: Seikoh Sakiyama
Publisher: Office of Naval Research
Address: American Embassy, APO San Francisco, CA 96503, USA
Year: 1980
Pages: 135
Price: Free
Content: Lists nearly 320 institutes and laboratories with name, address, telephone, speciality, staff, and budget (for 1974-76). Entries are arranged by type of control. There are indexes to institute, location, and subject.

Korea directory* 34

Editor: S.T. Kim
Publisher: Korea Directory Company
Address: PO Box 3955, Seoul 100, Korea
Year: 1985
Price: $68.00
Content: The directory lists details of over 6000 companies in Korea, as well as government and industrial organizations.

Marine environmental centres: East Asian Seas   35

Publisher: Food and Agriculture Organization
Address: Via delle Terme di Caracalla, 00100 Roma, Italy
Year: 1984
Pages: vii + 138
Series: UNEP regional seas directories and bibliographies
Content: Compiled jointly by the United Nations Environmental Programme and the Food and Agriculture Organization, the directory lists 56 research centres in 5 countries.

Marine environmental centres: Indian Ocean and Antarctic   36

Publisher: Food and Agriculture Organization
Address: Via delle Terme di Caracalla, 00100 Roma, Italy
Year: 1985
Pages: ix + 226
Series: UNEP regional seas directories and bibliographies
Content: Compiled jointly by the United Nations Environmental Programme and the Food and Agriculture Organization, the directory lists 105 research centres in 30 countries.

New biotechnology marketplace: Japan*   37

Publisher: EIC Intelligence
Address: Public Affairs Department, 48 West 38th Street, New York, NY 10018, USA
Year: 1983
Price: $395.00
Content: Details of companies specializing in biotechnology.

Research institutes on Asian studies in Japan, 1981   38

Editor: Rokuro Kono
Publisher: Centre for East African Cultural Studies
Address: c/o The Tokyo Bunko, Honkomagome 2-chome, 28-21 Bunkyo-ku, Tokyo 113, Japan
Year: 1981
Pages: xix + 213
Price: $12.00
ISBN: 4-89656-208-9
Series: Directories, number 9
Content: There are 82 research institutes and research centres; 1184 university research centres; six libraries and archives; and 158 learned societies. For research institutes/centres and university research centres, each entry contains the following data: name of the institute (both in Japanese and English), address, telephone, cable address, name of its head, year of establishment, brief description, number of staff members, research divisions, research activities, research staff (including position and speciality), library holdings, access to the library, publications, and admission of foreign researchers. For learned societies, membership and qualifications are listed. There are indexes to: Japanese names of research institutes, research centres, university research centres, libraries, and archives; Japanese names of learned societies; keywords in English names of university research centres.

Research institutes on social sciences and humanities in the Republic of Korea, 1982-83   39

Editor: Masao Mori
Publisher: Centre for East Asian Cultural Studies
Address: c/o The Tokyo Bunko, Honkomagome 2-chome, 28-21 Bunkyo-ku, Tokyo 113, Japan
Year: 1983
Pages: viii + 240
Price: $12.00
ISBN: 4-89656-209-7
Series: Directories, number 10
Content: Two academies; 293 research institutes; eight museums and art galleries; nine foundations; and 96 learned societies. For research institutes, each entry contains the name of the institute (both in Korean and in English), address, telephone, the name of its head, the date of establishment, purpose or description, number of staff members, organization, activities, the names of research staff (including their posts and speciality), library holdings, access to the library, publications, and admission of foreign researchers.
For academies, museums and art galleries, and foundation, the names of research staff are not listed; for learned societies, membership and qualifications are listed. There are indexes to: Korean names of academies, research institutes, museums and art galleries, and foundations; and Korean names of learned societies.

Science and technology education in Malaysian universities*   40

Editor: A.S.H. Ong, L. Su Eng
Publisher: Regional Institute of Higher Education and Development
Address: 15 Grange Road, Singapore 0923, Singapore
Year: 1983
Pages: 54
Content: RIHED occasional paper 14 - not strictly a directory this, but a review of the human, financial and spatial resources available to Malaysian universities.

Science and technology in China   41
Editor: Tong B. Tang
Publisher: Longman Group Limited
Address: Sixth Floor, Westgate House, The High, Harlow, Essex CM20 1NE, UK
Year: 1984
Pages: 269
Price: £49.00
ISBN: 0-582-90056-5
Series: Longman guide to world science and technology
Content: Provides a comprehensive and up-to-date analysis of the current state of Chinese science policy, describing the structure of science organization and funding, and covers research and development in six major subject areas. There is a subject index, an establishment index and a directory of major institutions and research establishments.

Science and technology in Japan   42
Editor: Alun M. Anderson
Publisher: Longman Group Limited
Address: Sixth Floor, Westgate House, The High, Harlow, Essex CM20 1NE, UK
Year: 1984
Pages: 421
Price: £49.00
ISBN: 0-582-90015-8
Series: Longman guide to world science and technology
Content: Charts the major research programmes now underway in Japan and provides a detailed directory of the research institutes, universities and industrial companies where they are being carried out.

Science and technology in the Middle East   43
Editor: Ziauddin Sardar
Publisher: Longman Group Limited
Address: Sixth Floor, Westgate House, The High, Harlow, Essex CM20 1NE, UK
Year: 1982
Pages: 324
Price: £49.00
ISBN: 0-582-90052-2
Series: Longman guide to world science and technology
Content: This guide covers the Middle East Arab states, Turkey, Iran and Pakistan. It describes the state of science in the Middle East, including science policy and planning, manpower, science education, Arabization and Islamic science, and deals in detail with research and development activities in the fields of agriculture, medicine, nuclear science and industrial science and technology. There is a subject index, an establishment index and a directory of major institutions.

Science research institutes under the jurisdiction of the Ministry of Education, Science and Culture*   44
Publisher: Japan Society for the Promotion of Science
Address: 5-3-1, Kojimachi, Chiyoda-ku, Tokyo 102, Japan
Year: 1980
Pages: 318
Geographical area: Japan.

Scientific directory of Hong Kong, fourth edition, 1978   45
Publisher: Committee for Scientific Co-ordination Hong Kong, Hong Kong Government Printer
Address: Cornwall House, Hong Kong
Pages: iv + 144
Price: HK$35.00
Content: List of abbreviations for professional qualifications; list of abbreviations for universities; scientists; scientific facilities; index of scientists.

Standard trade index of Japan, 29th edition, 1985-86*   46
Publisher: Japan Chamber of Commerce and Industry
Address: c/o Alan Armstrong and Associates, 76 Park Road, London NW1 4SH, UK
Year: 1985
Price: £167.00
Content: Entries are categorized in three alphabetical sections - forms, commodities, and services. They cover manufacturers, organizations, products, government agencies, and trade and industrial organizations.

Taiwan buyers' guide*   47
Publisher: China Productivity Centre
Address: PO Box 769, 201-226 Tun Hua North Road, Taipei, Taiwan
Year: 1984
Price: £105.00
Content: Entries cover 15 000 manufacturers, exports, importers, services, representatives, government organizations, etc, in alphabetical and classified sections.

Universities handbook 1983-84   48
Publisher: Association of Indian Universities
Address: AIU House, 16 Kotla Marg, New Delhi 110 002, India
Year: 1984
Price: $90 (airmail); $80 (surface mail)
Geographical area: India.
Content: A comprehensive guide to all the 147 universities and university level institutions in India: 108 tradi-

tional universities; 23 agricultural universities; four medical institutions; and twelve technical institutions.
Next edition: 1986.

Universities of Pakistan year book 49
Publisher: University Grants Commission
Address: Sector H-9, Islamabad, Pakistan
Year: 1985
Content: List of universities and study centres in Pakistan.

Who's who in Indian engineering and industry 50
Editor: H. Kothari
Publisher: Kothari Publications
Address: 12 India Exchange Place, Calcutta 700001, India
Year: 1972
Pages: xxiv + 260
Availability: OP
Series: Who's who series in India
Content: Lists biographical details of leading engineers and industrialists of India. Also gives full information about organizations of interest to engineers.
Next edition: 1986; $50.00.

Who's who in Indian science 51
Editor: H. Kothari
Publisher: Kothari Publications
Address: 12 India Exchange Place, Calcutta 700001, India
Year: 1969
Pages: xii + 200
Availability: OP
Series: Who's who series in India
Content: List of biographical details of leading scientists of India and details of important scientific organizations.
Next edition: 1986; US$25.00.

AGRICULTURE AND FOOD SCIENCE

Asian agribusiness buyers' guide* 52
Publisher: International Trade Publications
Address: Queensway House, 2 Queensway, Redhill, Surrey RH1 1QS, UK
Year: 1984
Price: £45.50
Geographical area: Asia.
Content: Details of supply companies to Asia's agricultural, food processing, storage, and distribution industries.
Next edition: October 1985.

Assam directory and tea areas handbook 53
Editor: B. Mukherjee
Publisher: Assam Review Publishing Company
Address: 29 Waterloo Street, Calcutta - 700 069, India
Year: 1984
Pages: cc + 400
Price: Rs50.00
Geographical area: India, Bangladesh, Napal, Bhutan.
Content: Includes details of government agencies, societies and institutions as well as tea companies.
Next edition: June 1985; Rs50.00.

Botanical gardens of the People's Republic of China* 54
Editor: R.A. Howard
Publisher: Journal: Bulletin American Association of Botanical Gardens, Arbor 13, 33-44
Address: Box 206, Swarthmore, PA 19081, USA
Year: 1979

Environmental training programmes and policies in ASEAN: an overview 55
Editor: T. Chelliah
Publisher: Regional Institute of Higher Education and Development
Address: Room 803, 8th Floor, RELC Building, 30 Orange Grove Road, Singapore 1025, Singapore
Year: 1985
Price: S$21.00
ISBN: 9971-911-13-2
Series: RIHED-FNS studies in environmental education
Geographical area: South east Asia.
Content: Environmental concerns in ASEAN: an overview of regional environmental problems; development of environmental consciousness; volunteer citizen groups/non-governmental organizations; teaching and training programmes; research activities; relationship of teaching and research to national and regional problems.

Handbook of agricultural education 56
Publisher: Association of Indian Universities
Address: AIU House, 16 Kotla Marg, New Delhi 110 002, India
Year: 1983
Pages: 141
Price: $8.25
Next edition: April/May 1986.

Saudi Arabian agriculture guide, first edition, 1985 — 57

Editor: Gerald Pierce
Publisher: Beacon Publications
Address: Jubilee House, Billing Brook Road, Weston Favell, Northamptonshire NN3 4NW, UK
Year: 1985
Pages: xii + 272
Price: £27.00
ISBN: 0-906358-58-2
Series: Beacon business handbooks
Content: The reference section consists of a comprehensive alphabetical listing of companies involved in every aspect of agriculture in Saudi Arabia, a classified listing of products and services, a brand index and finally complete information on all government and public sector organizations concerned with the industry. The guide concludes with a comprehensive index.
The publication is in English and Arabic.
Next edition: April 1986; £27.00

CHEMICAL AND MATERIALS SCIENCE

Chemfacts: Japan, first edition — 58

Publisher: Chemical Data Services
Address: Quadrant House, The Quadrant, Sutton, Surrey SM2 5AS, UK
Year: 1981
Pages: 196
Price: £70.00
Content: Surveying the present position of the industry in Japan, this entirely new addition to the CDS Chemfacts series covers 103 major industrial chemicals and 235 chemical manufacturers.

Chemistry and chemical engineering in the People's Republic of China, a trip report on the US delegation in pure and applied chemistry — 59

Editor: John D. Baldeschwieler
Publisher: American Chemical Society
Address: 1155 Sixteenth Street NW, Washington, DC 20036, USA
Year: 1979
Pages: xix + 167
Price: $12.95 (USA and Canada); $15.95 (export) - paperback
ISBN: 0-8412-0502-7
Geographical area: People's Republic of China.
Content: The book contains information about the institutional structure of chemical research and development in China, with sections devoted to science policy, education, and technology. It contains an appendix of institutions that were visited and their programmes. Indexed.

China's chemicals and petrochemicals — 60

Editor: Douglas J. Rasmussen
Publisher: China Phone Book Company
Address: GPO Box 11581, Hong Kong
Year: 1985
Pages: 300
Price: $90.00
ISBN: 962-7081-07-8
Content: The publication lists: company names; addresses; telephone numbers, telex numbers, cable numbers of local firms; local factories; foreign firms.

Japan chemical directory — 61

Editor: Japan Chemical Week
Publisher: Chemical Daily Company Limited
Address: 19-16, Shibaura 3-chome, Minato-ku, Tokyo 108, Japan
Year: 1985
Pages: 800
Price: $110.00
ISSN: 0075-3203
Content: The main content covers Japan, listing manufacturers, equipment, traders, foreign capital enterprises; associations; and other organizations. The data covers company name, address, telephone number, cable address, telex and facsimile numbers, year of establishment, capital, number of employees, managing personnel, sales, business line, branch office, overseas offices, foreign business partners and foreign subsidiaries (only Janpan). The Asian chemical firms are listed by nation, accompanied by information such as company name, address, and main products.
An appendix lists Asian chemical firms in Australia, China, Hong Kong, Indonesia, Korea, Malaysia, New Zealand, Philippines, Singapore, Taiwan, and Thailand.
Next edition: May 1986.

ELECTRONICS

Asian computer yearbook — 62

Editor: Euan Barty
Publisher: Computer Publications Limited
Address: 22 Wyndham Street, 7th floor, Hong Kong
Year: 1983
Pages: 1000
Price: $72.00

Content: Entries cover: about 8000 computer installations; suppliers of computers and computer peripherals; software vendors; educational programs in data processing; consultants; data processing recruitment specialists; data preparation services; and other data processing services. Coverage includes Brunei, Bangladesh, Hong Kong, India, Indonesia, Korea, Macau, Malaysia, Pakistan, Philippines, Singapore, Sri Lanka, Taiwan, and Thailand. Generally, listings include company or organization name, address, phone, name of contact or principal executive, number of employees, trade names, products or services. Listings for computer installations have more detail. Entries arearranged geographically; regional computer societies and related organizations are listed separately.
Next edition: January 1986.

China's electronics and electrical products 63

Publisher: China Phone Book Company Limited
Address: GPO Box 11581, Hong Kong
Year: 1983
Pages: 480
Price: $95.00 (cloth back); $80.00 (paperback)
Content: Entries cover: about 4000 manufacturers of telecommunications equipment, household appliances, electro-optical apparatus, electronic components and accessories, electrical measuring and controlling devices, elevators, watches and clocks, and other electrical and electronic products; about 200 trading corporations, retail outlets, distributors; related companies and organizations. All material is entered in both English and Chinese in adjoining columns. Entries include company name, physical and mailing addresses, telephone,telex, cable address, products, brand names, trademarks, and logos, with alphabetical, brand name, or product indexes.
Next edition: 1986.

China's instruments and meters 64

Editor: Douglas J. Rasmussen
Publisher: China Phone Book Company
Address: GPO Box 11581, Hong Kong
Year: 1985
Pages: 400
Price: $90.00
ISBN: 962-7081-04-3
Content: The publication covers: about 2000 manufacturers in the People's Republic of China producing pressure and temperature control and measurement devices; cameras and other optical products and components; lasers; sound recorders and reproducers; navigation and surveying instruments; medical and dental equipment and supplies; about 300 trading corporations, distributors, and scientific and technical institutes; related companies and organizations. All material is in both English and Chinese in adjoining columns. Entries include company name, physical and mailing addresses, telephone, telex, cable address, products, brand names, trademarks, logos. Arrangement is by product category with alphabetical, brand name, or trademark indexes.

Japan aviation directory* 65

Editor: Hitoshi Ohashi
Publisher: Wing Aviation Press
Address: 2 Gomi Building 6th Floor, 1-14-5 Ginza 1-chome, Chuo-ku Tokyo 104, Japan
Year: 1984
Price: $45.00
ISBN: 0286-0635
Content: The directory gives information on Japan's aerospace industry, defence, space activities, general and government non-military aviation, and air transport and trading companies.

ENERGY SCIENCES

Arab oil and gas directory 1985 * 66

Editor: Arab Petroleum Research Center
Publisher: Graham and Trotman Limited
Address: Sterling House, 66 Wilton Road, London SW1V 1DE, UK
Year: 1985
Pages: 550
Price: £180.00; $190.00
Geographical area: Middle East, North Africa, and Iran.
Content: The tenth edition of this 'consultants' work book' provides detailed statistics and information on all aspects of oil and gas production, exploration and developments in the 24 Arab countries of the Middle East and North Africa and in Iran. Information includes details such as the texts of all relevant new laws and official documents, official surveys, current projects and developments, up-to-date statistics covering OPEC and OAPEC member countries, and has 26 maps in full colour. This book was prepared by a group of international oil experts and consultants who deal exclusively with the oil and gas industry in the Arab World.

ENGINEERING AND TRANSPORTATION

Arabian transport guide 1985   67

Editor: David Harrison, Garry Gimson
Publisher: Beacon Publishing
Address: Jubilee House, Billing Brook Road, Weston Favell, Northamptonshire NN3 4NW, UK
Year: 1984
Pages: xx + 400
Price: £27.00
ISBN: 0-906358-55-8
Series: Beacon business handbooks
Geographical area: Countries of the Gulf Cooperation Council and North Yemen.
Content: There are 3751 entries in country sections which are divided into four main fields: air transport; sea transport; land transport; freight forwarding and customs clearing. These are followed by an international section, arranged alphabetically by country.
Next edition: December 1985; £27.00.

Handbook of engineering education, 1985   68

Publisher: Association of Indian Universities
Address: AIU House, 16 Kotla Marg, New Delhi 110 002, India
Year: 1985
Geographical area: India.

INDUSTRY AND MANUFACTURING

Arabian computer guide, 1985   69

Editor: Robin Warner
Publisher: Beacon Publishing
Address: Jubilee House, Billing Brook Road, Weston Favell, Northamptonshire NN3 4NW, UK
Year: 1984
Pages: xii + 428
Price: £32.00
ISBN: 0-906358-52-3
Series: Beacon business handbooks
Geographical area: Arabian Peninsula and Jordan.
Content: Approximately 350 entries in Middle East, and 650 in the International section. Available on-line.
Next edition: October 1985; £32.00.

Arabian construction 1985   70

Editor: Peter Found
Publisher: Beacon Publishing
Address: Jubilee House, Billing Brook Road, Weston Favell, Northamptonshire NN3 4NW, UK
Year: 1984
Pages: iv + 696
Price: £37.00
ISBN: 0-906358-53-1
Series: Beacon business handbooks
Geographical area: Countries of the Gulf Cooperation Council.
Content: There are 6137 entries divided into the following sections: materials supply - manufacturers of every kind of building material product with specialization and agency details; plant and machinery - suppliers of all types of equipment; contractors - covering every type of specialization from air conditioning to well drilling; sub-contractors - details of carpentry, electrical, furnishing, glazing, painting and decorating, plastering, plumbing, and roofing specialists; industry services - professional and specialist organizations providing services to the construction industry, with an expanded surveying section.
Next edition: December 1985; £37.00.

Asia's 7500 largest companies   71

Publisher: ELC International
Address: 227-229 Chiswick High Road, London W4 2DW, UK
Year: 1985
Price: £85.00
Series: Largest companies
Geographical area: Asia and the Pacific.
Content: This directory of Asia-Pacific business gives financial analysis and rankings for over 2250 publicly quoted companies on the Stock Exchanges of nine countries, including Hong Kong, Indonesia, Japan, Korea, Singapore, Philippines, Thailand, Malaysia, and Taiwan. There is a listing of over 5000 privately owned companies by country and business activity, a trade index of companies analysed by ISIC, and an alphabetical index of companies with full postal addresses.

Chinese manufacturers association of Hong Kong directory of members 1985   72

Publisher: Chinese Manufacturers Association of Hong Kong
Address: CMA Building, 64-66 Connaught Road C, Hong Kong
Year: 1985
Pages: cl + 800
Price: HK$120.00
Content: 3000 entries giving the following information: name of company; address; telephone number; telex

number; products; trade marks. There are three indexes to: industry classification; Chinese characters; and brand names.
Next edition: 1987.

CMERI directory of indigenous engineering products 73

Publisher: Central Mechanical Engineering Research Institute
Address: Mahatma Gandhi Avenue, Durgapur 713209, West Bengal, India
Year: 1984
Pages: cxxii + 630
Price: Rs175.00
Geographical area: India.
Content: The directory covers products, manufactured in India only, and includes about 4000 products with about 3500 manufacturers with addresses, telephone numbers and telex number, alphabetically cross-indexed.

Diamond's Japan business directory* 74

Publisher: Diamond Lead Company
Address: 1-4-2 Kasumigaseki 1-chome, Chiyoda-ku, Tokyo 100, Japan
Year: 1984
Price: £210.25
Content: Details of 100 companies giving name, background, employees, officers, stock, financial data, etc.

Directory: affiliates and offices of Japanese firms in the ASEAN countries* 75

Publisher: JETRO
Address: c/o North Oxford Academic, 242 Banbury Road, Oxford OX2 7DR, UK
Year: 1982
Price: £34.00
Geographical area: Southeast Asia.
Content: Details of 2326 Japanese companies operating in Indonesia, Malaysia, the Philippines, Singapore, and Thailand.

Directory of Chinese foreign trade 1985-86 76

Publisher: Longman Professional/China Council for the Promotion of International Trade
Address: 21-27 Lambs Conduit Street, London WC1N 3NJ, UK
Year: 1985
Price: £65.00
ISBN: 085121-050-3
Content: This annual directory shows the names and addresses of all Chinese Foreign Trade Corporations with details of key personnel, trading activities, size of the organization, telex and telephone numbers, branch offices and subsidiaries, bank account numbers and overseas representatives.

Directory of Hong Kong industries 1985 77

Publisher: Hong Kong Productivity Centre
Address: TST PO Box 99027, Hong Kong
Year: 1985
Pages: 804
Price: $65.00 (air mail); $47.00 (surface mail)
Content: Provides a detailed reference to over 4000 major manufacturing companies in Hong Kong, including lists and indexes of: major trade and industrial associations, consulates and trade commissions; licensed banks and bank representatives; ISIC codes of products index; brand names used by companies; manufacturing companies in the directory; logos of manufacturing companies.
Next edition: February 1986; $70.00 (air mail); $50.00 (surface mail).

Directory of industrial laboratories in Israel 78

Editor: Geula Gilat
Publisher: National Centre of Scientific and Technological Information
Address: PO Box 20125, Tel-Aviv 61201, Israel
Year: 1984
Pages: 130
Availability: OP
ISBN: 965-228-002-1
Content: There are 300 entries, with alphabetical list of industrial laboratories and subject index. Not available on-line.

Directory of inter-Arab organizations* 79

Publisher: Committee for Middle East Trade
Address: c/o Coet, 33 Bury Street, St James's, London SW1Y 6AX, UK
Pages: 112
Price: £12.50
Content: The directory lists and describes 63 of the most important inter-Arab organizations in the Arab world. Inter-Arab agencies coordinate and encourage development of major industry operations; light industry development is typified by such organizations as the Arab Company for Drug Industries and Medical Appliances.

Handbook of indigenous manufacturers (engineering stores) 80
Publisher: Government of India Department of Publications, Delhi
Address: Civil Lines, Delhi 110054, India
Year: 1980
Pages: xiii + 226
Price: Rs25.00; £2.91; $9.00
Geographical area: India.

Japan company handbook 81
Editor: Hiroshi Takahashi
Publisher: Oriental Economist
Address: 1-4 Hongokucho, Nihonbashi, Chuo-ku, Tokyo 103, Japan
Year: 1985
Pages: 1192
Price: $30.00
ISSN: 0288-9307
Content: Entries are listed according to subject; the publication includes an index to the companies. Subjects covered include most areas from fishery and forestry to commerce, banking, and publishing.

Japan's iron and steel industry 82
Editor: S. Kawata
Publisher: Kawata Publicity Incorporated
Address: Hatori Building 5-6, Koishikawa 5-chome, Bunkyo-ku, Tokyo 112, Japan
Year: 1984
Pages: xlii + 180
Availability: OP

Kompass register: Indonesia* 83
Publisher: Kompass-Indonesia
Address: PO Box 615/DAK, Jelan Palmera Selatan 22, Djakarta, Indonesia
Year: 1984
Price: £60.00
Content: Names and addresses of individual products, arranged by Standard Industrial Classification.

Kompass register: Malaysia* 84
Publisher: Berita Kompass sdn Bhd
Address: 31 Jalan Riang, Kuala Lumpur 22-03, Malaysia
Year: 1985
Price: £60.00
Content: Names and addresses of individual products arranged by Standard Industrial Classification.

Kompass register: Singapore* 85
Publisher: Kompass Embassy Information Proprietary Limited
Address: 328C King George's Avenue, Singapore 0820, Singapore
Year: 1984
Price: £60.00
Content: Names and addresses of individual products arranged by Standard Industrial Classification.

Kompass trade information book 86
[Kompass buku merah]
Editor: Putri Zanina
Publisher: Nik Ibrahim Kamil
Address: Berita Kompass Sdn Bhd, 31 Jalan Riong, Kuala Lumpur 22-03, Malaysia
Year: 1984-85
Pages: xxiv + 1262
Price: M$120.00
Geographical area: Malaysia.
Content: List of 13 000 Malaysian companies listed by states, locality, and alphabetical order, and by products and services; company listings with corporate information; alphabetical index of products and services; alphabetical index of firms; alphabetical index of trade marks and foreign agencies. The publication is available on-line in Kuala Lumpur only.
Next edition: November 1985; M$120.00.

Major companies of Saudi Arabia 1985* 87
Editor: G.C. Bricault
Publisher: Graham and Trotman Limited
Address: Sterling House, 66 Wilton Road, London SW1V 1DE, UK
Pages: 176
Price: £45.00; $72.00
ISBN: 0-86010-710-8
Series: Major companies
Content: This title gives full details of over 1000 of the largest companies in Saudi Arabia, and information includes company name, address, telephone, telex; names of directors and senior management; business activities; agencies; branches; subsidiaries; bankers; financial details. Index included.

Major companies of the Arab world 1975-85* 88
Publisher: Graham and Trotman Limited
Address: Sterling House, 66 Wilton Road, London SW1V 1DE, UK
Pages: 6950; 9 vols on microfiche
Price: £445.00; $712.00
Series: Major companies

ASIA

Content: This ten-year record provided by the volumes of the directory Major Companies of the Arab World published between 1975 and 1985, gives a valuable insight into the development of the companies which implemented a large part of the decade's tremendous economic growth in the region. This record is now available on microfiche.

Major companies of the Arab world 1985, ninth edition*   89

Editor: G.C. Bricault
Publisher: Graham and Trotman Limited
Address: Sterling House, 66 Wilton Road, London SW1V 1DE, UK
Year: 1985
Pages: 1000
Price: £120.00, $192.00 (hard cover); £110.00, $176.00 (paperback)
ISBN: 0-86010-599-7 (hard cover); 0-86010-600-4 (paperback)
ISSN: 0144-1594
Series: Major companies
Geographical area: Middle East.
Content: The book provides details of over 6000 major Arab companies in 20 Arab countries. It lists over 35 000 directors and senior executives by job title in 73 business sectors. Information also includes: company name, address, telephone, telex and cable numbers; business activities; brand names; agencies; branches and subsidiaries; bankers; financial details; date of establishment. The directory is also available on microfiche.

Major companies of the Far East 1985, second edition*   90

Editor: J. Carr
Publisher: Graham and Trotman Limited
Address: Sterling House, 66 Wilton Road, London SW1V 1DE, UK
Year: 1984
Pages: 750 (2 vols)
Price: Vol 1: £55.00, $88.00 (hard cover); £45.00, $72.00 (paperback); vol 2: £55.00, $88.00 (hard cover); £45.00, $72.00 (paperback)
ISBN: Vol 1: 0-86010-605-5 (hard cover); 0-86010-606-3 (paperback); vol 2: 0-86010-607-1 (hard cover); 0-86010-608-X (paperback)
ISSN: Vol 1: 0267-2251; vol 2: 0267-226X
Series: Major companies
Content: This two-volume directory provides details of the major companies of the Far East, including company name, address, telephone, telex and cable numbers; names of directors and senior executives by job title; business activities; brand names; agencies; branches and subsidiaries; bankers; financial details; date of establishment.
Next edition: November 1985; £75.00

Pharmaceutical manufacturers of Japan, 1983-84*   91

Publisher: Yakugyo Jiko
Address: c/o Marnzen Company Limited, PO Box 5050, Tokyo 100-31, Japan
Year: 1984
Price: $88.00
Content: Details including address, telephone, facilities, results, sales, etc of Japans 330 leading drug manufacturers.

MEDICAL AND BIOLOGICAL SCIENCES

Directory of medical, public health and ayurvedic institutions in Himachal Pradesh   92

Publisher: Directory of Health Services
Address: Simla, Himachal Pradesh, India
Year: 1985
Pages: 94
Content: The publication lists: areas and populations; numbers of medical, public health, and ayurvedic institutions; numbers of beds available; names of medical, public health, and ayurvedic institutions in Himachal Pradesh.

Handbook of medical education, 1985   93

Publisher: Association of Indian Universities
Address: AIU House, 16 Kotla Marg, New Delhi 110 002, India

Indian pharmaceutical guide   94

Editor: Mohoan C. Bazaz
Publisher: Pamposh Publications
Address: 506 Ashok Bhavan, 93 Nehru Place, New Delhi 110 019, India
Year: 1985
Pages: 1568
Price: Rs200.00; $60.00; £40.00
Content: List of medical and pharmaceutical journals published in India; pharmaceutical directory; pharmaceutical distributor and chemists and druggists directory (25 000 entries); information on allied manufacturers and dealers, and analytical laboratories; list of products, packings, composition, and prices; list of pharmaceutical manufacturers and their products; list of cosmetic manufacturers and their products; drug and chemical index; manufacturers and advertisers index; general index to products.
Next edition: April 1986; $60.00; £40.00.

List of registered medical and surgical practitioners 95

Publisher: Medical and Health Department, Hong Kong
Address: Cornwall House, Hong Kong
Year: 1985
Price: HK$20.00; US$3.00
Geographical area: Hong Kong.
Content: The list includes all medical and surgical practitioners registered in Hong Kong on the first day of January 1985 (pursuant to Section 15 of the Medical Registration Ordinance) who resided in/outside Hong Kong. This list can be obtained from the Hong Kong Government Information Services Department by mail orders at the following address: Publications (Sales) Office, Information Services Department, Beaconsfield House, 4/F., Queen's Road Central, Hong Kong.

Modern pharmaceuticals of Japan VII 1985* 96

Publisher: Japan Pharmaceutical, Medical and Dental Supply Exporters Association
Address: 3-6 Nihonbashi-Honcho 4-chome, Chuo-ku, Tokyo, Japan
Pages: 124
Content: Contains specific trade names of pharmaceuticals used in Japan, several descriptions of original products, an index to pharmaceuticals, and a list of member firms with their addresses. The appendix presents general statistics of the pharmaceutical industry in Japan.

4 AUSTRALIA AND THE PACIFIC

GENERAL SCIENCES

Art galleries and museums of New Zealand 1

Editor: Keith W. Thomson
Publisher: Reed, A.H. and A.W., Limited
Address: 68-74 Kingsford-Smith Street, Wellington 3, New Zealand
Year: 1981
Pages: vii + 202
ISBN: 0-589-01364-5
Geographical area: New Zealand.
Content: Hundreds of entries of museums, mostly concerned with local history or art, with regional (north to south) index.

Australian Academy of Science yearbook 2

Publisher: Australian Academy of Science
Address: GPO Box 783, Canberra City, ACT 2601, Australia
Year: 1984-85
Pages: 240
Price: A$20.00
ISSN: 0067-1584
Content: The publication gives details of the Academy's council, fellowship, list of fellows, committees, science and industry forum, official representatives. Not available on-line.
Next edition: November 1985; A$22.00.

Australian libraries 3

Editor: Peter Biskup, Doreen Goodman
Publisher: Clive Bingley Limited
Address: 7 Ridgmount Street, London WC1E 7AE, UK
Year: 1982
Pages: 222
Price: £12.50
ISBN: 0-85157-326-6

Australian reference books* 4

Publisher: Dun and Bradstreet
Address: 27 Paul Street, London EC2, UK
Price: £925.00
Content: Details on over 130 000 companies giving name, address, year of formation, credit rating, etc. Published every three months.

Coral reef researchers: Pacific 5

Editor: Food and Agriculture Organization
Address: Via delle Terme di Caracalla, 00100 Roma, Italy
Year: 1984
Pages: v + 101
Series: UNEP regional seas directories and bibliographies
Content: Compiled jointly by the United Nations Environmental Programme, the Pacific Science Association, SPREP and the University of Guam, the directory lists names of 494 scientists from 27 countries. The 4 indexes cover special interest, taxonomy, geography, and residence.

CSIRO directory 1985 6

Publisher: Commonwealth-Scientific and Industrial Research Organization, Australia
Address: 314 Albert Street, East Melbourne, Victoria 2003, Australia
Year: 1985
Pages: vi + 100
Price: A$3.00
ISSN: 0157-7204

Geographical area: Australia.
Content: The directory lists CSIRO's institutes, divisions, and units, with sections on the Bureau of Scientific Services, planning and evaluation advisory unit, and state and territory advisory committees. Four indexes give locations by town, divisions/units, alphabetical listing of staff, and state locations.
Next edition: March 1986; A$4.00.

Database: Australian companies*   7

Publisher: Data Base Asia
Address: 5th floor, Arion Commercial Centre, 2 Queens Road West, Hong Kong
Year: 1983
Price: HK$500.00
ISBN: 962-09-0008-1
Geographical area: Hong Kong.
Content: Details on all major Australian companies operating in Hong Kong.

Directory of CSIRO research programs 1985   8

Editor: Elizabeth Odgers
Publisher: Commonwealth Scientific and Industrial Research Organization, Australia
Address: 314 Albert Street, East Melbourne, Victoria 3002, Australia
Year: 1985
Pages: xvi + 580
Price: A$30.00
ISSN: 0727-6753
Geographical area: Australia.
Content: The directory contains descriptions of all current research programmes, lists their associated projects, and provides particulars of locations, staff numbers, and expenditure. Names, addresses and telephone numbers of the people to contact for further information about any research topic are also included.
The problems being tackled and the possible implications of research findings are described, as well as the research itself. The introductory pages provide information on CSIRO's structure, the organization of its research, and the allocation of its resources. The three indexes comprise a personal names index, a subject index, and a combined divisional programme index that lists details of the divisions and units within CSIRO and identifies the programmes undertaken by each.
Next edition: May 1986; A$30.00.

DISLIC: directory of special libraries and information centres in New Zealand   9

Editor: Paul Szentirmay, Thiam Ch'ng Szentirmay
Publisher: New Zealand Library Association
Address: PO Box 12-212, Wellington, New Zealand
Year: 1984
Pages: 111
Price: NZ$19.90
Content: There are 385 entries, including indexes to the following: parent organizations; type of organization; subject; publications; named and special collections; names of personnel; localities; library symbols. Not available on-line.

Guide to DSIR   10

Editor: Q.W. Ruscoe
Publisher: Science Information Publishing Centre, Department of Scientific and Industrial Research.
Address: PO Box 9741, Wellington, New Zealand
Year: 1984
Pages: 40
Price: Free
ISBN: 0-477-06744-1
Series: DSIR information series
Geographical area: New Zealand.
Content: Activities and achievements of DSIR's biological industries group, resources group, and industrial group; directory of addresses and contacts - head office, DSIR research stations, and grant-aided research organizations; locations map.
Next edition: 1986; free.

Information consultants, freelancers and brokers directory 1985   11

Editor: Enid Hsieh, Margaret Wanklyn, Jennifer Goddard
Publisher: Information Management and Consulting Association
Address: GPO Box 2128T, Melbourne 3001, Victoria, Australia
Year: 1985
Pages: vii + 57
Price: A$20.00
ISBN: 0-9591-2001-7
Geographical area: Australasia.
Content: The directory contains 41 entries arranged alphabetically by business name. Each entry consists of the name of the business (this, in many cases is the name of an individual), address, brief description of services, experience, etc, in the words of the entrant, and the names and expertise of the key personnel. There are three indexes: geographical, personnel, and subject. It is published annually and updates during the year are incorporated in the 'IMCA Newsletter'. It is not currently available on-line.
Next edition: January 1986; A$25.00.

Marine environmental centres: South Pacific 12

Publisher: Food and Agriculture Organization
Address: Via delle Terme di Caracalla, 00100 Roma, Italy
Year: 1985
Pages: vii + 147
Series: UNEP regional seas directories and bibliographies
Content: Compiled jointly by the United Nations Environmental Programme and the Food and Agriculture Organization, the directory lists 68 research centres in 13 countries.

Pacific research centres: a directory of organizations in science, technology, agriculture and medicine 13

Publisher: Longman Group Limited
Address: Sixth Floor, Westgate House, The High, Harlow, Essex CM20 1NE, UK
Year: 1985
Pages: 350
Price: £110.00
ISBN: 0-582-90028-X
Content: Provides detailed information on about 1800 organizations which conduct or finance research and development in the Pacific region including Australia, China, Indonesia, Japan, Malaysia, New Zealand, and the Philippines. Organizations included are industrial and government laboratories, and academic institutions. Details given include address, size of organization, names of key personnel and an overview of main research activities.
There is a titles of establishments index and a subject index.

Scientific and technical research centres in Australia 14

Editor: A. Ermers
Publisher: Commonwealth Scientific and Industrial Research Organization (microfiche); Australian Scientific Industries Association (hard back)
Address: PO Box 89, East Melbourne, Victoria 3002, Australia
Year: 1984
Pages: vi + 427
Price: A$5.00 (microfiche); A$39.00 (hardback)
ISBN: 0643-03722-7 (microfiche); 0643-03721-5 (hardback)
Geographical area: Australia.
Content: The 1984 edition contains more than 1500 entries including some 500 from research centres not covered by the 1981 edition of this directory. It includes details of the research activities of commonwealth and state government departments and authorities; universities, institutes of technology and other advanced colleges; other organizations including those in private industry, non-profit-research institutions and hospitals. Full addresses are provided for each centre together with information on research directors, publications, numbers of staff and outlines of the scope of research with brief descriptions of current research projects and facilities. The subject index contains both broadly based and specific terms, and the research centres index gives information on all locations where research is carried out.
The information contained in the directory is also available for on-line interrogation via CSIRONET. Items can be retrieved by a computer search based on index and text words or by broad subject codes.
Next edition: 1986; $50.00.

South Pacific research register 15

Editor: Judith Titoko
Publisher: Pacific Information Centre, in association with University of the South Pacific Library
Address: PO Box 1168, Suva, Fiji
Year: 1984
Pages: iv + 78
Price: $13.00 (developed countries); $6.60 (developing countries)
Geographical area: South Pacific.
Content: Approximately 280 entries including: author list with detailed subject index. The following information, where available, is given for each researcher: name; private or permanent address; work address; official position; statement of research. Not available on-line.
Next edition: April 1986; same price as 1984.

AGRICULTURE AND FOOD SCIENCE

Australian fishing industry directory 1985* 16

Publisher: Department of Primary Industry
Address: Edmund Barton Building, Broughton Street, Barton, ACT 2600, Australia
Year: 1985
Pages: 200
Price: Free
Geographical area: Australia.
Content: 1000 entries.

Australian sugar year book 17

Publisher: Strand Publishing Proprietary Limited
Address: GPO Box 1185, Brisbane, Queensland 4001, Australia

Year: 1985
Pages: 272
Price: A$20.00
Geographical area: Australia.
Next edition: April 1986; A$22.00; Publishing and Marketing Australia.

Compendium of rural research and development* 18

Publisher: Department of Primary Industry
Address: Edmund Barton Building, Broughton Street, Barton, ACT 2600, Australia
Year: 1984
Pages: 836
Price: Free
ISBN: 0-644-04122-6
Geographical area: Australia.

Current plant taxonomic research on Australian flora, 1980-81 19

Editor: Robert W. Johnson
Publisher: Queensland Department of Primary Industries
Address: PO Box 46, Brisbane 4001, Queensland, Australia
Year: 1981
Pages: 125
Availability: OP
Price: Free

Green pages: directory of non-government environmental groups in Australia 20

Publisher: Australian Conservation Foundation
Address: 672B Glenferrie Road, Hawthorn, Victoria 3122, Australia
Year: 1985
Price: $23.50
ISSN: 0727-0119
Geographical area: Australia and worldwide.
Content: The directory's arrangement is: international, then national, then by each state in Australia. Approximately 818 entries, which include the names of conservation and environment groups, their address or contact point, their areas of interest (subject and geographical), and membership numbers (where available). The directory has an index of group names.

Rural industry directory 1983* 21

Editor: Graham Macafee
Publisher: Australian Government Publishing Service
Address: GPO Box 84, Canberra, ACT 2601, Australia
Year: 1985

Price: A$10.00
ISSN: 0812-1729
Geographical area: Australia.
Content: The directory provides a contact reference for government departments, agencies, and statutory authorities responsible for primary industry management, research, education and administration in Australia. It also covers associations, societies, and organizations connected with primary industry. The directory has two indexes; an index of names and an index of organizations. It contains 1200 entries.
Not available on-line.

EARTH AND SPACE SCIENCES

Australian maps 1984 22

Editor: Glenys McIver
Publisher: National Library of Australia
Address: Parkes Place, Parkes, Canberra, ACT 2600, Australia
Year: 1985
Price: $18.00
ISSN: 0045-0677
Geographical area: Australia and territories.
Content: The number of entries varies from year to year. Entries are arranged by Boggs and Lewis classification (alphanumeric representation code representing area and subject divisions). The publication contains: title/statement of responsibility, edition, scale, publication and physical description details, description and notes, subject and area classification. There are indexes to author, subject, area, title/series.
Maps included in this publication are those received on legal deposit to the National Library. Retrospective material is included where it is considered to be a valuable inclusion. The publication aims to provide a comprehensive list of maps representing areas within Australia and its territories, or other maps of Australian association. It includes atlases in sheet map format, and maps - including topographic, cadastral, resources, thematic, planning, road, and tourist maps.
It is not available on-line.

Inventory of water resources research in Australia 23

Editor: Australian Water Resources Council
Publisher: Australian Government Publishing Service
Address: GPO Box 84, Canberra, ACT 2601, Australia
Year: 1979
Pages: 207
Price: $11.80

ISBN: 0-642-04674-3
Content: Covers research into water resources and directly related matters, including research projects in progress and those completed or proposed to commence during the year. Details are included of finance and brief description of the projects.

Map collections in Australia: a directory 24

Editor: N.M. Rauchle
Publisher: National Library of Australia
Address: Parkes Place, Parkes, Canberra, ACT 2600, Australia
Year: 1980
Pages: ix + 141
Price: $4.50
ISBN: 0-642-99205-3
Content: 199 entries for map collections and map collectors; 33 pages of map reference materials and map publishers in Australia. It is arranged in five sections - principal mapping authorities in Australia, map collections (subdivided by state with entries alphabetically organized within these divisions), private map collections, map reference material, and map publishers.
Each entry in the map collections section states name of map collection, address, total printed map sheets, rare and historical maps, reference books (including gazetteers), aerial photographs, atlas volumes, landsat imagery, map specialization, major map specialization, main geographical area, special collection, and access to collection.
Not available on-line.

Water research in Australia: current projects 1983 25

Publisher: Department of Resources and Energy
Address: GPO Box 858, Canberra, ACT 2601, Australia
Year: 1983
Pages: xii + 149
Price: Free
ISSN: 0810-736X
Geographical area: Australia.
Content: This publication incorporates over 1000 summaries of water related research projects currently undertaken in Australia by individuals and research organizations. Research projects are grouped according to broad subject categories, and within these, are listed in alphabetical order by project title. Two indexes allow retrieval by the names of members of the research team and the organizations carrying out or sponsoring the research, and by subject, using terms from the AQUALINE Thesaurus and additional Australian terms. Information included in the publication is taken from the Australian computerized data base, STREAMLINE which is available on the national network, AUSINET. AUSINET can be accessed through several international networks.
Next edition: May 1985; free.

ELECTRONICS

Australian electronics directory 26

Editor: Ross Mackay
Publisher: Technical Indexes Proprietary Limited
Address: Box 98, Cheltenham, Victoria 3192, Australia
Year: 1984-85
Pages: 350
Price: A$72.00
Geographical area: Australasia.
Content: Over 600 references to local companies and their divisions who represent approximately 200 worldwide manufacturers in the electronics industry; over 1800 product categories specific to the electronics industry; over 3600 principal and trade name cross references; 84 computer generated charts which break product areas into detailed sub-groups and identify the Australian suppliers for each one; an index to Australia's only electronics data system which contains over 120 000 pages of technical product data on microfiche.
Next edition: July 1985; A$72.00.

ENGINEERING AND TRANSPORTATION

Association of Consulting Engineers New Zealand: list of members 27

Editor: R.I. Irons
Publisher: Bryce Francis Limited
Address: 11 A-17 Marion Street, PO Box 6255, Wellington, New Zealand
Year: 1984
Pages: vi + 158
Price: NZ$20.00
Geographical area: New Zealand and 25 branches throughout world.
Content: The directory consists of profiles of 138 different firms. It also lists 35 other firms at the back who have not included a profile. The index includes a geographical list of firms, also chart or matrix showing areas of activity of each firm. The introduction includes some text on selecting a consulting engineer and engagement. Also listed are the six publications available from ACENZ and their prices.

Australian engineering directory 28

Editor: Ross Mackay
Publisher: Technical Indexes Proprietary Limited
Address: Box 98, Cheltenham, Victoria 3192, Australia
Year: 1985
Pages: 480
Price: A$72.00
Geographical area: Australasia.
Content: Nearly 1300 Australian companies, giving their head office, branch, and distributor locations plus a listing of key personnel; 2260 product categories covering all facets of the engineering environment in Australia; over 7000 overseas principals and trade names showing the Australian agents; 113 computer generated charts which break product areas into detailed subgroups and identify the Australian suppliers for each one; an index to the Australian engineering data system, which contains over 120 000 pages of technical product data on microfiche.
Next edition: December 1985; A$72.00.

INDUSTRY AND MANUFACTURING

Kompass register: Australia* 29

Publisher: Peter Isaacson Publications Proprietary Limited
Address: 49-49 Porter Street, Pranhan, Victoria 3181, Australia
Year: 1985
Price: £100.00
Content: Names and addresses of individual products, arranged by Standard Industrial Classification.
Next edition: May 1986.

New Zealand manufacturers directory 30

Publisher: Braynart Group Limited
Address: Private Bag, Newmarket, Auckland, New Zealand
Year: 1985
Availability: NZ$25.00
Content: List of members of New Zealand Manufacturers' Federation, including: name; address; telex; telephone; chief executive; branches; subsidiaries; products; trade names. Not available on-line.

MEDICAL AND BIOLOGICAL SCIENCES

Guide to New Zealand Entomology 31

Editor: Graeme W. Ramsay, Pritam Singh
Publisher: Entomological Society of New Zealand
Address: Entomology Division, Department of Scientific and Industrial Research, Private Bag, Auckland 1020, New Zealand
Year: 1982
Pages: 72
Price: NZ$9.00
ISSN: 0110-4527
Series: Bulletin of the Entomological Society of New Zealand, number 7
Geographical area: New Zealand.
Content: Entries cover: 300 members of the society, their addresses and interests; lists of entomological literature; lists of institutions, associated societies, and clubs; and lists of entomological collections. Not available on-line.

Medical directory of Australia 32

Publisher: Australasian Medical Publishing Company
Address: PO Box 116, Glebe, New South Wales 2037, Australia
Year: 1980
Pages: xvi + 882
Price: A$60.00 in Australia; A$65.00 overseas
ISBN: 0-85557-022-9
Geographical area: Australia; New Zealand; Papua New Guinea.
Content: There are 36 000 entries alphabetically arranged with personal names including qualifications, professional details, and addresses; 157 page gazeteer giving geographic locations for all individual entries; 132 page directory of government and non-government organizations in the area of health and related services, university medical schools, hospitals, medical colleges and societies, and research institutes. Entries are alphabetically indexed with full list of abbreviations. Not available on-line.
Next edition: 1986.

Hospitals and health services yearbook 33

Editor: J.F. Ross
Publisher: Peter Isaacson Publications
Address: PO Box 174, Prahran 3181, Melbourne, Victoria, Australia
Year: 1985
Pages: 464
Price: A$70.00
ISSN: 0810-7513
Geographical area: Australia.

5 EUROPE INCLUDING USSR

GENERAL SCIENCES

Adressbuch des deutschsprachigen   1
buchhandels
[Directory of the German book trade]
Publisher: Buchhändler-Vereinigung GmbH
Address: Postfach 2404, Grossber Hirschgraben 17-21, D-6000 Frankfurt-am-Main, German FR
Year: 1984-85
Pages: 2154 (in 3 vols)
Price: DM118.00
ISBN: 3-7657-1266-3
Geographical area: West Germany, Switzerland, Austria.
Content: More than 28 000 addresses of the book trade in the Federal Republic of Germany and West Berlin, Austria, Switzerland, and other countries. Volume 1 covers publishing houses; volume 2 book shops, and volume 3 organizations.
Next edition: July 1985; DM118.00.

Annuaire CNRS*   2
[Yearbook of French Scientific Research]
Publisher: Centre National de la Recherche Scientifique
Address: 54 boulevard Raspail, F-75270 Paris Cedex 6, France
Year: 1985
Content: Current research projects in mathematics, physics, chemistry, earth and space science, oceanography, biology, life sciences, social sciences, and humanities sponsored by the centre. Entries include: project title, name and address of laboratory, director, staff, research in progress, publications, equipment. The publication is in French, and is available on-line from CNRS and Questel under the title 'CNRSLAB'.
Next edition: 1986.

Archive, bibliotheken und   3
dokmumentationsstellen der Schweiz/
Archives, bibliothèques et centres de
documentation en Suisse/Archivi,
biblioteche e centri di
documentazione in Svizzera
[Swiss archives, libraries, and documentation centres]
Publisher: Amt für Wissenschaft und Forschung/Office de la Science et de la Recherche
Address: PO Box 2732, CH-3001 Berne, Switzerland
Year: 1976
Pages: 805
Price: SFr32.00
Geographical area: Switzerland and Liechtenstein.
Content: There are 683 entries arranged geographically, which include the following data: name of institution, address, telephone number, telex number, telegraphic address, date established, subject area, holdings, catalogues, documentation activities, reference and loan arrangements, opening hours, Swiss union catalogue membership, information services, special equipment, reprographic services, publications, bibliography.
Indexes: alphabetical index, subject index (in German and in French).
Not available on-line.

ASLIB directory of information   4
services in the United Kingdom:
volume 1 - information sources in
science, technology and commerce
Editor: Ellen M. Codlin
Publisher: Aslib
Address: Information House, 26-27 Boswell Street, London WC1N 3JZ, UK
Year: 1982
Pages: vi + 789
Price: £55.00

ISBN: 0-85142-166-0
Content: Lists 2954 organizations alphabetically.
Next edition: 1986-87.

Association of consulting scientists: members and services 1985-86   5

Publisher: Association of Consulting Scientists
Address: Owles Hall, Buntingford, Hertfordshire SG9 9PL, UK
Year: 1985
Pages: 67
Price: £5.00
Geographical area: United Kingdom.
Content: The 66 entries are arranged alphabetically by name of member, each entry fully detailed in terms of address, number of staff, services, etc.
Index by services and facilities available.
Not available on-line.
Next edition: 1987.

Berufsschulen und sonderberufsschulen in Bayern   6
[Vocational schools in Bavaria]

Publisher: Bayerisches Landesamt für Statistik und Datenverarbeitung, München
Address: Neuhauser Strasse 51, Box 200303, 8000 München 2, German FR
Year: 1985
Price: DM7.00
Series: Statistische berichte des Bayerischen landesamtes für statistik und datenverarbeitung, BII 2
Content: There are 242 entries arranged alphabetically according to location. Entries include: name of school, address; county; type of school (ie state, county, private); number of classes; number of pupils; number of teachers; number of lessons.

Biblioteksvejviser   7
[Library directory]

Editor: Christian Götzsche
Publisher: Danish Library Association
Address: Trekronergade 15, DK 2500 Valby, Denmark
Year: 1984
Pages: 178
Price: DKr256.00
ISBN: 87-87244-14-4
Geographical area: Denmark.
Content: Danish associations and institutions; other Nordic associations and institutions; Nordic library periodicals; Danish foundations related to libraries; Danish research and special libraries; Danish public libraries.
Index in English and Danish.
Next edition: September 1985; DKr300.00.

British archives: a guide to archive resources in the United Kingdom   8

Editor: Julia Sheppard, Janet Foster
Publisher: Macmillan Press
Address: 4 Little Essex Street, London WC2R 3LF, UK
Year: 1982
Pages: 534
Price: £25.50 (hardback); £9.95 (paperback)
ISBN: 0-333-37868-7
Content: The following broad groups of repositories are covered: national record offices, central and local; national libraries and museums, central and local; university libraries and archives; societies and colleges; institutions and organizations, including religious archives; private collections. The guide itself is arranged alphabetically by city and within city and each entry contains details held, history, policy, access and services.

British calibration service: approved laboratories and their measurements   9

Editor: British Calibration Service
Publisher: Her Majesty's Stationery Office
Address: St Crispins, Norwich, Norfolk, UK
Year: 1985
Pages: 42
Price: Free
Content: Entries for 150 laboratories arranged in numerical approval number under given field of measurement (currently 13); alphabetical and numerical indexes. Entries give name of company, address, telephone number, and name of head of laboratory; an abbreviated list of measurements is given under each entry.
At the back of the book there is a dot matrix table which enables readers to select any one of around 170 measurement parameters and find the laboratory which can carry these out.

British qualifications, fifteenth edition: a comprehensive guide to educational, technical, professional and academic qualifications in Britain   10

Publisher: Kogan Page
Address: 120 Pentonville Road, London N1 9JN, UK
Year: 1985
Pages: 920
Price: £20.00 (hardback); £14.50 (paperback)
ISBN: 0-85038-909-7 (hardback); 0-85038-910-0 (paperback)
Series: Careers and jobs
Content: Listing of all the academic, educational, technical, and professional qualifications available in Britain today and all the universities, colleges, or polytechnics which run the appropriate courses leading to those qualifications, together with course entrance require-

ments. It also gives details of the relevant professional and trade bodies for a wide range of career fields, with information on how to qualify for membership and what awards they give. Some 200 career areas are covered.
Next edition: November 1985; £21.00 and £15.50.

British universities' guide to graduate study 1985-86   11

Publisher: Association of Commonwealth Universities
Address: John Foster House, 36 Gordon Square, London WC1H OPF, UK
Year: 1985
Pages: xii + 236
Price: £15.00
ISBN: 0-85143-093-7
Geographical area: United Kingdom.
Content: This guide to graduate taught courses, lasting at least six months, contains three main sections: an introduction to graduate education at universities in Britain; a short description of the content of each of 2500 courses - method of assessment, length, title and qualification awarded are also given for each course; profile of each university which, inter alia, describes its location, size, academic organization, residential accommodation and social facilities. Index to subject areas.
Next edition: January 1986.

Bulgarian Academy of Sciences 1984 reference book   12

Editor: Blagovast Hristov Sendov
Publisher: Publishing House of the Bulgarian Academy of Sciences
Address: 7 Novemvr 1, 1040 Sofia, Bulgaria
Year: 1984
Pages: x + 74
Geographical area: Bulgaria.
Content: Details of the 95 scientific organizations, academicians, corresponding members and remaining scientific staff are given up to March 31, 1984. The material is arranged in 3 sections: the General Assembly; the Presidium; the centres for research. A list of periodical publications, an alphabetical index of names and a diagram of the operative structure of the scientific units of the academy are also provided.

Bulletin of special courses   13

Publisher: London and South Eastern Regional Advisory Council for Further Education
Address: Tavistock House South, Tavistock Square, London WC1H 9LR, UK
Geographical area: Southeast England.
Content: Details of short non-examination courses primarily designed for those already possessing qualifications at degree, final professional, or HNC.

Catalogue of British official publications not published by HMSO 1985   14

Publisher: Chadwyck-Healey Limited
Address: 20 Newmarket Road, Cambridge CB5 8DT, UK
Year: 1985
Price: £130.00
ISBN: 0-85964-167-8
Content: Details of 30 000 official publications in agriculture, education, employment, food, health and medicine, health and safety, libraries and information science, transport, etc, representing over 400 organizations. Subject and author index; index to sources; keyword index.
The publication is also available in six bi-monthly issues at an annual subscription of £170.00 (ISSN 0260-5619).
Next edition: 1986.

Civic trust environmental directory, sixth edition*   15

Publisher: Civic Trust
Address: 17 Carlton House Terrace, London SW1Y 5AW, UK
Year: 1984
Price: £2.50
ISBN: 900849-43-6
Geographical area: United Kingdom.
Content: Entries for nearly 300 organizations involved with amenity, conservation, and the environment, giving address, brief description of work and aims, and an indication of facilities, publications, etc.

Compendium of advanced courses in colleges of further and higher education: full time and sandwich courses in polytechnics and other colleges outside the university sector   16

Editor: R. Eberhard
Publisher: London and South Eastern Regional Advisory Council for Further Education
Address: Tavistock House South, Tavistock Square, London WC1H 9LR, UK
Year: 1985-86
Pages: 112
Price: £2.75
ISBN: 0-85394-112-2
Geographical area: England and Wales.
Content: Entry requirements for courses; lists of subjects and options available; names, addresses, and telephone numbers of colleges.
Next edition: September 1985; £3.00.

Compendium of university entrance requirements for first degree courses in the United Kingdom 1986-87   17

Publisher: Association of Commonwealth Universities
Address: John Foster House, 36 Gordon Square, London WC1H OPF, UK
Year: 1985
Pages: 368
Price: £8.95
ISBN: 0-85143-096-1
Geographical area: United Kingdom.
Content: Detailed description of minimum requirements, in terms of the General Certificate of Education, for entry to each of the 8500 first degree courses offered by 86 universities and university colleges in the United Kingdom. Details of courses and course requirements are given in 108 tables, with new courses and 'sandwich courses' listed separately, and profiles of each admitting institution. Five appendices give information on, inter alia, attitude of universities towards candidates with certain non-GCE qualifications and towards mature candidates. Index to courses and subjects. (The publication does not cover the Open University, University of Buckingham or external degree courses).
Next edition: June 1986.

Councils, committees and boards: a handbook of advisory, consultative, executive and similar bodies in British public life   18

Editor: Lindsay Sellar
Publisher: CBD Research Limited
Address: 154 High Street, Beckenham, Kent BR3 1EA, UK
Year: 1984
Pages: xiv + 430
Price: £48.00
ISBN: 0-900246-43-X
Content: Details of government advisory committees, committees of enquiry, public boards and authorities, Royal Commissions, and various non-governmental advisory and consultative councils and committees, listed alphabetically giving address, telephone, abbreviated title, composition, chairman, secretary, terms of reference, activities, publications.
Index of abbreviated titles; index of chairmen; subject index.
Not available on-line.
Next edition: January 1987; £55.00.

Current British directories, tenth edition   19

Publisher: CBD Research Publications
Address: 154 High Street, Beckenham, Kent BR3 1EA, UK
Year: 1985
Pages: xxii + 558
Price: £60.00
ISBN: 0-900246-40-5
Geographical area: United Kingdom and Republic of Ireland.
Content: Around 400 local directories, alphabetically by town with publisher, address, telephone, content of directory, latest edition, price, number of pages, etc; around 2300 national directories, alphabetically by title with similar information; index of around 650 directory publishers, showing address, telephone and telex numbers with publications; subject index.
Not available on-line.
Next edition: February 1987; £70.00.

Current European directories, second edition   20

Publisher: CBD Research Publications
Address: 154 High Street, Beckenham, Kent BR3 1EA, UK
Year: 1981
Pages: xx + 413
Price: £45.00
ISBN: 0-900246-30-8
Geographical area: Europe, excluding United Kingdom and Republic of Ireland.
Content: International, national, city, and specialized directories and similar reference works. Bibliography of all European countries (publisher with address, telephone, data, frequency, price, etc).
General subject index; alphabetical index of titles; indexes in English, French, and German.
Not available on-line.
Next edition: October 1987; £60.00.

Data bases in Europe: a directory to machine-readable data bases and data banks in Europe   21

Publisher: Commission of the European Communities
Address: Directorate General XIII-B, Bâtiment Jean Mounet, Luxembourg, Luxembourg
Year: 1982
Pages: 232
Content: The directory is arranged in alphabetical order within subjects. It is available on-line (Diane Euronet).

Degree course guides   22

Editor: Jenny Knight
Publisher: Hobsons Limited
Address: Bateman Street, Cambridge CB2 1LZ, UK
Year: 1984-85; 1985-86
Price: Bound volumes £39.90 each; separate subject guides £2.95 each; set £34.65

ISBN: 0-86021-575-X (Bound volume 1984-85); 0-86021-741-8 (1985-86)
Geographical area: United Kingdom.
Content: There are 35 subject guides giving detailed comparisons of first-degree courses in United Kingdom universities, polytechnics and colleges, plus eight technological subjects gathered together under the title Technology. The guides are available separately, in sets or in bound volumes.

Directory for the environment: organizations in Britain and Ireland 1984-85   23

Editor: Michael J.C. Barker
Publisher: Routledge and Kegan Paul
Address: 14 Leicester Square, London WC2H 7PH, UK
Year: 1984
Pages: ix + 281
Price: £8.95
ISBN: 0-7102-0227-X
Content: Approximately 1000 entries, ordered alphabetically by name of organization, giving: address; contact name; aims; activities; status; publications. There is a subject index.
Not available on-line.
Next edition: September 1986; £12.50.

Directory of British associations and associations in Ireland, seventh edition   24

Editor: G.P. Henderson, S.P.A. Henderson
Publisher: CBD Research Limited
Address: 154 High Street, Beckenham, Kent BR3 1EA, UK
Year: 1982
Pages: xiv + 473
Price: £47.50
ISBN: 0-900246-39-1
Content: Details of interests, activities and publications of trade associations, scientific and technical societies, professional institutes, learned societies, research organizations, chambers of trade and commerce, agricultural societies, trade unions, cultural sports and welfare organizations in the United Kingdom and in the Republic of Ireland. The directory includes 6000 associations alphabetically with addresses, telephone, telex, personnel, number of branches in the United Kingdom, sphere of interest, specialist groups or sections, activities, affiliations to other organizations, number and categories of members, publications, former name or names.
Index of abbreviated names; subject index.
Not available on-line.
Next edition: September 1985; £60.00.

Directory of centres for outdoor studies in England and Wales   25

Editor: Carolyn Cocke
Publisher: Council for Environmental Education
Address: School of Education, University of Reading, London Road, Reading RG1 5AQ, UK
Year: 1981
Pages: ii + 48
Price: £1.80
ISBN: 0-906711-01-0
Geographical area: England and Wales.
Content: Entries for 500 field centres, each containing details on location, facilities, and services of each centre.

Directory of courses of further and higher education in maintained colleges in the region   26

Publisher: East Anglian Regional Advisory Council for Further Education
Address: 2 Looms Lane, Bury St Edmunds, Suffolk IP33 1HE, UK
Year: 1985
Price: £1.75
ISBN: 0-902044-10-9
Geographical area: East Anglia.
Content: Approximately 1200 entries, listing course, title, college(s) at which course is run and mode of attendance. Entries are arranged in subject order, alphabetically by subject.

Directory of directors*   27

Publisher: Skinner, Thomas, Directories
Address: Windsor Court, East Grinstead House, East Grinstead, Sussex, UK
Year: 1985
Price: £53.00
Geographical area: United Kingdom.
Content: Lists over 52 000 company directors giving name, address, telephone, and telex.
Next edition: March 1986.

Directory of environmental journals and media contacts   28

Editor: Tom Cairns
Publisher: Council for Environmental Conservation
Address: Zoological Gardens, Regent's Park, London NW1 4RY, UK
Year: 1985
Pages: iv + 49
Price: £2.00
ISBN: 0-903158-24-8
Geographical area: United Kingdom.

Content: Over 200 entries, arranged in three sections: section 1 - environmental journals, magazines and 'in-house' newsletters; section 2 - national daily and Sunday newspapers; news agencies (environmental contacts); section 3 - national radio and television (environmental contacts).

Directory of European associations, part 1: national industrial, trade and professional, third edition 29

Editor: I.G. Anderson
Publisher: CBD Research Publications
Address: 154 High Street, Beckenham, Kent BR3 1EA, UK
Year: 1981
Pages: lxix + 540
Price: £50.00
ISBN: 0-900246-35-9
Geographical area: Europe, excluding United Kingdom and Republic of Ireland.
Content: Details of industrial, trade and professional associations of all countries of Europe except the United Kingdom and Republic of Ireland, giving address, telephone, telex, abbreviated title, authorized translations of title, sphere of interest, number and categories of members, principal activities, regular publications, former name or names.
Alphabetical index of titles of organizations; index of abbreviated titles; subject indexes in English, French, and German.
Not available on-line.
Next edition: 1986.

Directory of European associations, part 2: national learned, scientific and technical societies, third edition 30

Publisher: CBD Research Publications
Address: 154 High Street, Beckenham, Kent BR3 1EA, UK
Year: 1984
Pages: lii + 331
Price: £52.50
ISBN: 0-900246-42-1
Geographical area: Europe, excluding United Kingdom and Republic of Ireland.
Content: Details of learned, scientific and technical associations of all countries of Europe except the United Kingdom and Republic of Ireland (address, telephone, telex, abbreviated title, authorized translations of title, sphere of interest, number and categories of members, principal activities, regular publications, former name or names). Alphabetical index of titles of organizations; index of abbreviated titles; subject indexes in English, French, and German.
Not available on-line.
Next edition: May 1988.

Directory of first degree and diploma of higher education courses 1984-85 31

Publisher: Council for National Academic Awards
Address: 344-354 Gray's Inn Road, London WC1X 8BP, UK
Year: 1984
Pages: vi + 129
Price: Free
ISBN: 0-902071-15-7
Geographical area: United Kingdom.
Content: General information; course information under subject headings; addresses of polytechnics and colleges; certificate and diploma of higher education index; college index; subject index.
Next edition: September 1985; free.

Directory of further education 32

Editor: Denis Curtis
Publisher: Hobsons Limited
Address: Bateman Street, Cambridge CB2 1LZ, UK
Year: 1985-86
Pages: 862
Price: £43.00 (hardback); $37.50 (paperback)
ISBN: 0-86021-701-9 (hardback); 0-86021-702-7 (paperback)
Geographical area: United Kingdom.
Content: Guide to over 6500 courses leading to formal qualifications in almost 750 United Kingdom polytechnics and colleges of further and higher education.

Directory of institutions and individuals active in environmentally-sound and appropriate technologies 33

Editor: United Nations Environmental Programme
Publisher: Pergamon Press Limited
Address: Headington Hill Hall, Oxford OX3 0BW, UK
Year: 1979
Pages: 172
Price: £16.25
ISBN: 0-08-025658-9
Geographical area: United Kingdom.

Directory of lectures in natural history and environmental issues 34

Editor: Edwina Milesi
Publisher: Council for Environmental Conservation
Address: Zoological Gardens, Regent's Park, London NW1 4RY, UK
Year: 1985
Geographical area: United Kingdom.
Content: The directory contains over 1000 entries (details not available at time of publication).

Directory of libraries in Ireland 35

Editor: Margaret Barry, Alun Bevan
Publisher: Library Association of Ireland and Library Association (Northern Ireland Branch)
Address: c/o 53-54 Upper Mount Street, Dublin 2, Ireland
Year: 1983
Pages: iv + 100
Price: £5.50
ISBN: 0-946037-01-9
Geographical area: Northern Ireland and Republic of Ireland
Content: Lists 189 libraries/information units plus library schools. Each entry contains: address; name of librarian; telephone and telex; opening hours; departmental libraries; book stock; number of serials; types of catalogue; classification scheme in use; special collections; subject specializations and other relevant details (eg lending restrictions). Indexes: name and special collections; subject specialization.

Directory of natural history and 36
related societies in Great Britain and
Ireland*

Editor: A. Meenan
Publisher: British Museum (Natural History)
Address: Cromwell Road, London SW7 5BD, UK
Year: 1983
Price: £15.00
ISBN: 0-565-00859-5
Geographical area: United Kingdom.
Content: Entries cover national societies and associations, local societies and groups, university and school societies. Approximately 750 entries.

Directory of operational computer 37
applications in United Kingdom
libraries and information units

Editor: C.W.J. Wilson
Publisher: Aslib
Address: Information House, 26-27 Boswell Street, London WC1N 3JZ, UK
Year: 1977
Pages: iii + 196
Price: 15.00
ISBN: 0-85142-092-3

Directory of postgraduate and post- 38
experience courses 1984-85

Publisher: Council for National Academic Awards
Address: 344-354 Gray's Inn Road, London WC1X 8BP, UK
Year: 1984
Pages: iv + 84
Price: Free
ISBN: 0-902071-16-5
Geographical area: United Kingdom.
Content: Introduction; course information under subject headings; addresses of polytechnics and colleges; college index; subject index.
Next edition: September 1985; free.

Directory of rare book and special 39
collections in the United Kingdom and
the Republic of Ireland

Editor: Moelwyn Williams
Publisher: Library Association Publishing
Address: 7 Ridgemount Street, London WC1E 7AE, UK
Year: 1985
Pages: xiii + 664
Price: £75.00
ISBN: 0-85365-646-0
Content: The directory lists all collections of rare books (ie printed before 1851) and special collections which are, or can be made, available to bona fide scholars and researchers. The collections listed are those in colleges, schools, churches and cathedrals, societies and institutes, National Trust properties and private houses, and large national, university, and public libraries. The names of libraries and institutions have been arranged in alphabetical order within the respective regions. Details of address, telephone number, hours of opening, conditions of admission, research facilities and a brief history are given for each library, followed by a description of each collection, and relevant catalogues an published references.

Directory of technical and further 40
education 1986

Publisher: Longman Group Limited
Address: Sixth Floor, Westgate House, The High, Harlow, Essex CM20 1NE, UK
ISBN: 0-582-90061-1
Content: A comprehensive guide to colleges in the UK offering vocationally oriented courses. Information is given for polytechnics, technical colleges and colleges of further education. Details of the more vocationally directed aspects of university courses are also given.

Directory of the Commission of the 41
European communities*

Publisher: EC Publications Office
Address: 5 Rue du Commerce, 2985 Luxembourg
Year: 1985
Price: £1.70
Content: List of individuals with specific responsibility within European Community institutions.

Directory of university-industry liaison services   42

Publisher: Brunel University Industrial Services Bureau
Address: Kingston Lane, Uxbridge, Middlesex UB8 3PH, UK
Year: 1984
Pages: vi + 82
Price: £4.00
ISSN: 1043-5035
Geographical area: Great Britain.
Content: Entries for 49 universities, arranged alphabetically.

DOC Italia: annuario degli enti di studio, ricerca, cultura, e informazione   43
[DOC Italy: yearbook of research, cultural and information organizations]
Publisher: Editoriale Italiana
Address: Via Vigliena 10, Roma 00192, Italy
Year: 1985
Pages: 1500
Price: L100 000
Content: Over 5000 scientific, cultural, and educational institutions and centres in Italy with 50 000 names of chief officials operating in them. Entries are arranged alphabetically and each gives history, address, key personnel, activities, projects, publications. The book has three indexes: alphabetical, analytical, and of abbreviations.
Next edition: 1986; L150 000.

Education year book 1985   44

Publisher: Longman Group Limited
Address: Sixth Floor, Westgate House, The High, Harlow, Essex CM20 1NE, UK
Pages: 954
Price: £27.00
ISBN: 0-582-90404-8
Geographical area: United Kingdom.
Content: Provides full details of local education authorities; a complete listing of state and independent secondary, middle and special schools; and extensive coverage of higher and further education. The directory also contains information on a wide range of education related organizations, including careers services, educational travel, voluntary service, teachers' associations, educational equipment and teaching aids.

ENEX directory*   45

Publisher: Peter Peregrinus Limited
Address: PO Box 8, Southgate House, Stevenage, Hertfordshire SG1 1HQ, UK
Year: 1981
Price: £14.00
ISBN: 0-906048-58-3
Geographical area: Europe.
Content: Listing of individuals and organizations with expertise in areas of environmental significance.

ENREP directory*   46

Publisher: Peter Peregrinus Limited
Address: PO Box 8, Southgate House, Stevenage, Hertfordshire SG1 1HQ, UK
Year: 1982
Price: £25.00
ISBN: 0-906048-90-7
Geographical area: Europe.
Content: Entries for some 22 000 research projects on environmental subjects.

Environmental directory: national and regional organizations of interest to those concerned with amenity and environment   47

Editor: Saskia Hallam
Publisher: Civic Trust
Address: 17 Carlton House Terrace, London SW1Y 5AW, UK
Year: 1984
Pages: 68
Price: £2.50
ISBN: 900849-43-6
Geographical area: United Kingdom.
Content: There are 290 entries covering organizations concerned with amenity and conservation, especially in the built environment, government departments and agencies, voluntary societies, professional institutions, educational bodies, trade associations, etc. Entries give address, telephone number and a brief description giving aims, membership, and activities, also symbols to indicate available facilities and services. They are arranged alphabetically and there is a subject index.
Not available on-line.
Next edition: 1986; £2.50.

European communities and other European organisations yearbook, sixth edition 1985*   48

Editor: G. Seingry
Publisher: Editions Delta
Address: 92-94 Square E. Platsky, B-1040 Bruxelles, Belgium
Year: 1984
Price: BF1800
ISBN: 2-8029-0047-1

Content: Information on the structure, operation, and activities of the European Communities and private or public bodies which contribute to European integration. Fully indexed.
Next edition: 1985; BF1950.

European Community: the practical guide and directory for business, industry and trade, second edition, 1985*   49

Editor: B. Morris, K. Boehm
Publisher: Macmillan Press
Address: Houndmills, Basingstoke, Hampshire, UK
Year: 1985
Price: £45.00
ISBN: 0-333-37069-4
Content: Covers twelve main areas of interest: agriculture, fisheries, and forestry; consumer protection; energy; environment; external trade and customs; finance; health and social welfare; industry and trade; information and media; standards institutions; transport; work and employment.

European companies: a guide to sources of information fourth edition   50

Editor: G.P. Henderson
Publisher: CBD Research Limited
Address: 154 High Street, Beckenham, Kent BR3 1EA, UK
Year: 1985
Price: £45.00
Content: The guide gives for each country in Europe: official registry of companies; legal forms of company; stock exchanges; descriptions of publications containing information about companies, with translations into English of principal balance sheet and other headings; credit reporting and commercial information services; financial newspapers and periodicals.
Not available on-line.

European research centres: a directory of organizations in science, technology, agriculture and medicine, sixth edition   51

Publisher: Longman Group Limited
Address: Sixth Floor, Westgate House, The High, Harlow, Essex CM20 1NE, UK
Year: 1985
Pages: 2100
Price: £195.00
ISBN: 0-582-90027-1
Series: Reference on research
Content: Provides detailed information on approximately 19 000 research establishments within Europe including major industrial research laboratories in private and public corporations, government laboratories, research funding organizations, university research institutes and university departments conducting research. In two volumes, the directory is indexed by establishment and by subject.

European sources of scientific and technical information, sixth edition   52

Editor: Anthony P. Harvey
Publisher: Longman Group Limited
Address: Sixth Floor, Westgate House, The High, Harlow, Essex CM20 1NE, UK
Year: 1984
Pages: x + 368
Price: £105.00
ISBN: 0-582-90152-9
Series: Reference on research
Geographical area: Europe, including the USSR.
Content: A detailed guide to 1500 key information sources on science and technology in Europe. The guide is arranged under twenty-five subject headings, and information centres are listed by country under each subject area. It includes details of national offices of information, patents and standards offices, and organizations active in identified scientific fields with library facilities available to the public.
The book has been completely revised for this new edition, and is fully indexed by title and by subject.
Next edition: December 1986

First destinations of polytechnic students   53

Editor: Patricia Pearce
Publisher: Committee of Directors of Polytechnics
Address: 309 Regent Street, London W1R 7PE, UK
Year: 1984
Pages: xi + 238
Price: £7.50
Geographical area: England and Wales.
Content: Annual statistical analysis of student output from the 30 polytechnics in England and Wales. Four tables covering: first degrees; higher national diplomas; diplomas of higher education; postgraduate certificate in education. Each table is sub-divided to show by subject discipline first destinations, employer categories, and type of work.
Not available on-line.
Next edition: September 1985; £8.50.

Graduate studies 1985-86   54

Editor: Jenny Knight
Publisher: Hobsons Limited
Address: Bateman Street, Cambridge CB2 1LZ, UK
Year: 1985-86

Pages: 960
Price: £63.00
ISBN: 0-86021-703-5
Geographical area: United Kingdom.
Content: Information on all areas of study and research open to graduates in the United Kingdom. The material is divided into humanities and social sciences; biological, health and agricultural sciences; physical sciences; and engineering and applied sciences.

Guide de l'Ingénierie* 55
[Engineering Guide, 20th edition]
Publisher: DIT
Address: 11 rue de Madrid, 75008 Paris, France
Year: 1984-85
Price: F520.00
Geographical area: France.
Content: Alphabetical list of 600 societies of engineering, also of leading figures in the engineering world in France.

Guide to government department and other libraries, 26th edition 56
Publisher: British Library Science Reference Library
Address: 25 Southampton Buildings, London WC2A 1AW, UK
Year: 1984
Pages: 104
Price: £10.00
ISBN: 0-712307-09-5
Geographical area: United Kingdom.
Content: There are 630 entries, arranged in a subject sequence and supplemented by an alphabetical index of institutions. Entries include title of institution, address, telephone number, departments, staff contacts, hours of opening, publications produced and stock.
Not available on-line.
Next edition: 1986; £12.00.

Guide to specialist facilities and courses for handicapped people in post-school educational institutions in the region 57
Publisher: London and South Eastern Regional Advisory Council for Further Education
Address: Tavistock House South, Tavistock Square, London WC1H 9LR, UK

Handbuch der bibliotheken: Bundesrepublik Deutschland, Österreich, Schweiz 58
[Handbook of libraries: German Federal Republic, Austria, Switzerland]
Editor: Helga Lengenfelder
Publisher: Saur Verlag, K.G.
Address: Pössenbacherstrasse 2B, PO Box 71 10 09, 8000 München 71, German FR
Year: 1984
Pages: xiv + 329
Price: DM78.00
ISBN: 3-598-10522-3

Handbuch der Deutschen wissenschaftlichen akademien und gesellschaften 59
[German scientific academies and societies handbook]
Editor: Friedrich Domay
Publisher: Franz Steiner Verlag Wiesbaden GmbH
Address: Box 347, Birkenwaldstrasse 44, D-7000 Stuttgart 1, German FR
Year: 1977
Pages: xviii + 1209
Price: DM240.00
ISBN: 3-515-02172-8
Geographical area: West Germany.

Handbuch der universitäten und fachhochschulen Bundesrepublik Deutschland, Österreich, Schweiz 60
[Handbook of universities and faculties in the German Federal Republic, Austria, and Switzerland]
Editor: Helga Lengenfelder
Publisher: Saur Verlag, K.G.
Address: Pössenbacherstrasse 2B, PO Box 71 10 09, 8000 München 71, German FR
Year: 1985
Pages: xxii + 329
Price: DM198.00
ISBN: 3-598-10534-7
Content: Listing of around 280 universities and colleges with around 8500 departments. Included are the names of directors, and heads of departments.

Hatrics directory of resources, eighth edition 61
Publisher: Hampshire Technical Research Industrial Commercial Service
Address: c/o Hampshire County Library, 81 North Walls, Winchester SO23 8BY, UK
Year: 1985
ISBN: 0-90103-11-6
Geographical area: Hampshire and adjoining areas.
Content: The 350 organizations of Hatrics cooperative library and information service are listed alphabetically by organization. Entries give name and address, contact

name and title, main interests, telephone and telex numbers, etc.
Indexes of report holdings and standards holdings; subject index by interest. Supplement of fax numbers.

Higher education in the United Kingdom, 1984-86 — 62

Publisher: Longman Group Limited
Address: Longman House, Burnt Mill, Harlow, Essex CM20 2JE, UK
Year: 1984
Price: £7.50
ISBN: 0-582-94718-3
Series: British Council Publications
Content: A handbook for overseas students containing details of university and college courses in the United Kingdom. One section contains addresses of colleges, universities, professional bodies, British Council offices overseas, examining boards and an index to the directory of subjects.

Industrial research in the United Kingdom: a guide to organizations and programmes, eleventh edition — 63

Publisher: Longman Group Limited
Address: Sixth Floor, Westgate House, The High, Harlow, Essex CM20 1NE, UK
Year: 1985
Pages: 655
Price: £105.00
ISBN: 0-582-90029-8
Series: Reference on research
Content: With approximately 4000 entries the guide gives profiles of research laboratories carrying out or funding research in the United Kingdom, and provides contact details of trade associations and professional societies. There is a personal name index, a titles of establishments index, and a subject index.

Industrial technology: a guide to sources of information in the United Kingdom available to developing countries — 64

Publisher: Tropical Development and Research Institute
Address: 127 Clerkenwell Road, London EC1R 4DB, UK
Year: 1979
Pages: vi + 21
Price: £0.60
ISBN: 0-85954-096-0
Content: There are 49 entries arranged in two parts, covering: organizations directly concerned with the needs of developing countries; other public or corporate bodies which can help developing countries. Entries include: name; address; telephone, telegram and telex; general information; subjects.

Information professionals directory: who's who in librarianship and information science — 65

Editor: Janet Shuter
Publisher: Elm Publications
Address: Seaton House, Kings Ripton, Cambridgeshire PE17 2NJ, UK
Year: 1984
Pages: iv + 108
Price: £15.75
ISBN: 0-946139-20-2
Geographical area: United Kingdom.
Content: Approximately 750 entries, arranged alphabetically within subject area, and giving name of person, qualifications, job title, address and telephone number, professional and related activities, publications, areas of expertise, overseas work experience.
Next edition: 1986; £17.00.

Information technology in the UK* — 66

Editor: J. Howlett
Publisher: Blackwell Scientific Publications
Address: 8 John Street, London WC1N 2ES, UK
Year: 1985
Price: £3.95
ISBN: 0-63201-287-0
Content: Alphabetical listing of councils, societies, and associations

Inventaire des centres Belges de recherche disposant d'une bibliotheque ou d'un service de documentation — 67

[Directory of Belgium research centres having library or documentation services]
Editor: Janine Verougstraete
Publisher: Centre National de Documentation Scientifique et Technique
Address: Bibliothèque Royale Albert 1er, Service de Vente des Publications, Boulevard de l'Empereur 4, B-1000 Bruxelles Belgium
Year: 1982
Pages: 456
Price: BF1000
ISBN: 2-87093-001-1
Geographical area: Belgium
Content: The fourth edition of this directory discribes 1090 research units classified alphabetically by location. The book is fully indexed by title subject and contains an index of research and development directors and information officers. Text is in French and Dutch.

Jahrbuch der Deutschen bibliotheken 68
[Yearbook of German Libraries, 51st edition]
Editor: Verein Deutscher Bibliothekare
Publisher: Verlag Otto Harrassowitz
Address: PO Box 2929, Taunusstrasse 14, D-6200 Wiesbaden, German FR
Year: 1985
Pages: 650
Price: DM88.00
ISBN: 3-447-02544-1
Content: Detailed information of about 600 scientific libraries and of about 110 institutions for librarianship as well as the names and personal dates of the members of the Verein Deutscher Bibliothekare.
Next edition: September 1987; DM98.00.

Kurschners deutscher gelehrten- 69
kalender 1983
[German scientists directory 1983]
Editor: Werner Schnider
Publisher: Walter de Gruyter
Address: Genthiner Strasse 13, 1 Berlin 30, German FR
Pages: xiv + 5265
ISBN: 3-110-08558-5
Geographical area: Germany, Austria, Switzerland.
Content: The directory, now in its fourteenth edition, contains almost 43 000 entries of German-speaking scholars and scientists giving biographical and bibliographical details. The entries are listed both alphabetically by the individual's name, and also by discipline, thereby facilitating the locating of scholars in a given field.
Next edition: January 1987; DM500.00.

Library and information networks in 70
the United Kingdom
Editor: Jack Burkett
Publisher: Aslib
Address: Information House, 26-27 Boswell Street, London WC1N 3JZ, UK
Year: 1979
Pages: vi + 260
Price: £17.00
ISBN: 0-85142-117-2
Content: Covers government and other organizations concerned with research and development in scientific and technical fields, including the British Library.

Library and information networks in 71
Western Europe
Editor: Jack Burkett
Publisher: Aslib
Address: Information House, 26-27 Boswell Street, London WC1N 3JZ, UK
Year: 1983
Pages: iv + 139
Price: £13.00
ISBN: 0-85142-168-7

Museji Jugoslavije 72
[Yugoslavian museums]
Editor: Branka Šulc
Publisher: Muzejski Dokumentacioni Centar
Address: Mesnička 5, 41000 Zagreb, Yugoslavia
Year: 1984
Pages: 12
ISSN: 0350-2325
Next edition: 1986.

Museums in Great Britain with 73
scientific and technological collections
Editor: J.E. Smart
Publisher: Science Museum
Address: Exhibition Road, London SW7 2DD, UK
Year: 1978
Pages: 88
Availability: OP
ISBN: 0-901805-19-X
Content: Sixteen entries on national museums; twelve entries on larger provincial museums; four entries on developing museums. Gazetteer listing 335 museums alphabetically by location.
Not available on-line.

Museums yearbook 74
Editor: Steve Caplin
Publisher: Museums Association
Address: 34 Bloomsbury Way, London WC1A 25F, UK
Year: 1985
Pages: 216
Price: £22.00
ISBN: 0-902102-60-5
Geographical area: Great Britain and Ireland.
Content: The yearbook includes museums and art galleries; administering authorities; affiliated institutional members; overseas, and other types of members; related organizations and specialist groups (eg Commonwealth Association of Museums, Regional Arts Association). There are many relevant addresses.
Next edition: February 1986; £22.00

Nederlandse museum gids 75
[Netherlands museums guide]
Editor: Rudi Molegraaf
Publisher: Staatsuitgeverij (J. Nijland)
Address: PO Box 20014, Christoffel Plantijnstrasse 2, 2500 EA 's-Gravenhague, Netherlands

Year: 1982
Pages: ii + 329
Price: DFl25.00
ISBN: 90-12-03665-8
Content: The publication lists 728 museums of the Netherlands.

Norges Teknisk-Naturvitenskapelige Forskningsråd årsberentning 76
[Royal Norwegian Council for Scientific and Industrial Research, annual report 1984]
Publisher: Royal Norwegian Council for Scientific and Industrial Research
Address: PO Box 70 Tåsen, N-0801 Oslo 8, Norway
Pages: 44
Price: Free
Geographical area: Norway.
Content: Key information on seventeen Norwegian institutions for scientific and technological research and development. Each entry contains the following information: name, acronym; address; telephone, telex, telefax; number of employees/number of scientists; updated list of activities; name of chairman of the board, executive director, information officer.
Next edition: June 1986; free.

Norske vitenskapelige og faglige biblioteker: en håndbok 77
[Handbook of Norwegian professional libraries]
Editor: Libena Vokac
Publisher: Riksbibliotektjenesten
Address: POB 2439 Solli, 0202 Oslo 2, Norway
Year: 1984
Pages: xii + 300
Availability: OP
ISBN: 82-7195-046-0
Content: There are 341 entries of libraries arranged by subject and covering humanities, social science, science, economics, technology, biomedicine, and agriculture, as well as academic libraries and twenty county (public) libraries. Entries include: name, address, etc; operating since (year); office hours; collections; special subjects covered; subject catalogues; equipment; services for users; guides for users.
Indexes: alphabetical library index; geographical library index; subject index (Norwegian and English).
Not available on-line.
Next edition: 1987; NKr160.00.

NSCA reference book 78
Editor: Jane Dunmore
Publisher: National Society for Clean Air
Address: 136 North Street, Brighton, Sussex BN1 1RG, UK
Year: 1985
Pages: viii + 294
Price: £10.95
ISBN: 0-903474-28-X
Geographical area: Mainly United Kingdom, includes Europe.
Content: Deals with air pollution; noise; water; pollution of the land; conservation and current research programmes and directories. Has a trade directory and a buyer's guide.

On-line bibliographic databases 79
Editor: James L. Hall, Marjorie J. Brown
Publisher: Aslib
Address: Information House, 26-27 Boswell Street, London WC1N 3JZ, UK
Year: 1983
Pages: xiv + 383
Price: £38.00
ISBN: 0-85142-167-9
Geographical area: United Kingdom.
Content: Entries for 179 bibliographic data bases.

On-line information retrieval 1965-76 80
Editor: James L. Hall
Publisher: Aslib
Address: Information House, 26-27 Boswell Street, London WC1N 3JZ, UK
Year: 1977
Pages: vii + 125
Price: £13.00
ISBN: 0-85142-094-X
Geographical area: United Kingdom.
Content: Over 900 references from the monograph, periodical and report literature are listed. There are, in addition, tabulations of generally available on-line data bases, with a guide to typical host systems/installations and access costs, and also on-line systems, developments and installations, with addresses, data bases, key references and notes. Author, report number, and subject/name indexes are provided.

Pan European associations 81
Editor: C.A.P. Henderson
Publisher: CBD Research Limited
Address: 154 High Street, Beckenham, Kent BR3 1EA, UK
Year: 1983
Pages: xii + 426
Price: £40.00
ISBN: 0-900246-37-5
Content: Details of about 2000 organizations alphabetically by names in English, with around 3000 cross-references from names in twelve other European languages (names in other languages, abbreviated title, address, telephone, telex, objects, activities, number and

categories of members, countries represented among members, affiliations to world organizations, official languages, publications, former name or names).
Index of abbreviated titles; subject index.
Not available on-line.
Next edition: July 1986; £55.00.

Patent information and documentation in Western Europe 82

Editor: H. Bank, M. Fenat-Haessig, M. Roland
Publisher: Saur Verlag, K.G.
Address: Pössenbacherstrasse 2B, PO Box 71 10 09, 8000 München 71, German FR
Year: 1980
Pages: 268
Price: DM80.00
ISBN: 3-598-10518-9
Content: Patent literature of the Western European countries: Austria, Belgium, Denmark, Finland, France, West Germany, Greece, Ireland, Italy, Luxembourg, Netherlands, Norway, Portugal, Spain, Sweden, Switzerland and United Kingdom and the non-European countries Japan, Soviet Union and USA. In addition, the book covers the European Patent Office, the International Patent Documentation Center (INPADOC) and the World Intellectual Property Organization (WIPO).
Information is given on the patent office of each country as well as on other official and private patent information services. The user learns among other things how patents are published, where the documents can be obtained and at what price. Periodical and other publications on patents are listed. A survey of the stock of patent literature at each patent office facilitates the search for information.
The handbook is made complete by a list by country of periodicals on patent information and on legal protection of industrial property as well as a chapter on abstracting journals in English and French.

Polish Academy of Sciences: directory 83

Editor: Edward Haron
Publisher: Ossolineum, Publishing House of the Polish Academy of Sciences
Address: Rynek 9, 50-106 Wroclaw, Poland
Year: 1983
Pages: xxi + 206
Availability: OP
ISBN: 83-04-00872-6
Content: Organization, committees, and members of the academy.

Pollution research and the research councils 84

Publisher: Natural Environment Research Council
Address: Polaris House, North Star Avenue, Swindon, Wiltshire SN2 1EU, UK
Year: 1985
Pages: 109
ISSN: 0141-4674
Content: Entries give details of research projects, including subject, funding body, and laboratory involved. There is a subject index.
Next edition: 1986.

Polytechnic courses handbook 1985-86 85

Editor: Committee of Directors of Polytechnics
Publisher: Pitman Publishing Limited
Address: 128 Long Acre, London WC2E 9AN, UK
Year: 1984
Pages: xiii + 442
Price: £7.30
ISBN: 0-273-02204-0
Geographical area: England and Wales.
Content: Approximately 1550 entries (plus cross references) describing all full-time and sandwich advanced courses in the 30 polytechnics in England and Wales (annually updated), giving two-page profiles of each polytechnic, first degree courses, higher national diploma courses, advanced non-degree courses, second-stage advanced courses. Appendices give lists of taught higher degrees and diplomas and of other postgraduate courses. Each course section is subdivided into broad subject areas which in turn are sub-divided into subject disciplines. Comprehensive index of courses and principal subjects.
Not available on-line.
Next edition: September 1985; £7.70.

Polytechnics directory 86

Publisher: Council of Polytechnic Librarians
Address: c/o Oxford Polytechnic Library, Headington, Oxford OX3 OBP, UK
Year: 1983
Pages: 62
Price: £1.00
Geographical area: England and Wales.
Content: There are 30 entries, arranged alphabetically by name of polytechnic. Entries contain names of senior library staff, and names of senior members of the polytechnic.
Not available on-line.
Next edition: 1985; £3.00.

Public reference services in the UK: a   87
directory of information and specialist
staff, second edition

Editor: Mary Toase
Publisher: Libarary Association - Reference, Special and Information Section
Address: Oakwood, Hexham, Northumberland NE46 4LJ, UK
Year: 1984
Price: £7.20
ISBN: 0-946347-02-6
Content: There are 172 entries arranged geographically by county. Entries include name of library, address, name of senior librarians eg. music librarian, reference librarian, etc. Alphabetical index of libraries.

Pyttersen's Nederlandse almanak: 88
handbook van personen en
instellingen in Nederland de
Nederlandse Antillen en Suriname

[Pyttersen's Netherlands almanac: handbook of persons and organizations in the Netherlands, the Netherlands Antilles and Surinam]
Editor: J.J. Honssen
Publisher: Van Loghum Slaterus
Address: Postbus 23, 7400 GA Deventer bv, Overijssel, Netherlands
Year: 1985
Pages: 862
Price: DFl121.00
ISBN: 9060019547
Content: Associations, institutions, etc are listed according to subjects, which include: commerce and industry; economics and commercial science; finance, credit and insurance; engineering and technology; building and housing; agriculture, forestry and horticulture; cattle-breeding, apiculture, dairying and fishing industry; horse-breeding and equestrian sports; traffic and transport; shipping; information; medicine and public health; mental retardation; old people; physically handicapped people; partially and totally blind people; partially and totally deaf people; energy; environmental control; municipal affairs and public utility services; public libraries. Not available on-line.
Next edition: June 1986; DFl125.00.

Register of consulting scientists,   89
contract research organizations, and
other scientific and technical services

Editor: D.J.B. Copp
Publisher: Hilger, Adam, Limited
Address: Techno House, Radcliffe Way, Bristol BS1 6NX, UK
Year: 1984
Pages: x + 82
Price: £15.00
ISBN: 0-85274-751-9
Geographical area: United Kingdom.
Content: This directory identifies over 350 individuals and organizations offering scientific consultancy or contract research services within the United Kingdom, ranging from full-time professional consultants and testhouses, through independent organizations carrying out contract research and university organizations, to part-time consultants and others. All entries are classified according to type of organization, type of service and range of technical subjects offered, and are fully indexed. Not available on-line.

Repertoire des universités   90

[Guide to universities]
Publisher: Office National d'Information sur les Enseignements et les Professions
Address: 75635 Paris Cedex 13, France
Year: 1983
Pages: 380
Availability: OP
Geographical area: France.
Content: Entries give addresses, level of degree obtainable, and subjects.
Next edition: December 1985; Fr120.00.

Research establishments*   91

Publisher: Data Research Group
Address: Bridge House, Great Missenden, Buckinghamshire, UK
Year: 1984
Price: £33.00
Geographical area: United Kingdom.
Content: Listing of 570 establishments divided by town and county.

Research in British universities,   92
polytechnics and colleges

Publisher: British Library
Address: Boston Spa, Wetherby, West Yorkshire LS23 7BQ, UK
Year: 1984
Pages: 3 vols
Price: Vol 1 £44.00; vol 2 £44.00; vol 3 £33.00; set £109.00
ISBN: 0-7123-2021-0
Content: Over 3000 research units at educational institutions in the United Kingdom; also includes research by other institutions and by government departments. Entries include: institution and department names, names of researchers, project descriptions, sponsoring bodies, dates of the work. Addresses are provided in 'Department Index'. Arrangement: Volume 1, physical

sciences; Volume 2, biological sciences; Volume 3, social sciences including government and other research outside educational institutions; classified by subject within volumes. Indexes: keyword/subject, researcher name (indexes in each volume for that volume only).
Next edition: Autumn 1985.

Research libraries and collections in the UK: a selective inventory and guide   93

Editor: Stephen Roberts, Alan Cooper, Lesley Gilder
Publisher: Clive Bingley Limited
Address: 7 Ridgemount Street, London WC1E 7AE, UK
Year: 1978
Pages: 288
Price: £19.50
ISBN: 0-85157-258-8
Geographical area: United Kingdom.

Research supported by the Economic and Social Research Council   94

Editor: David Wainwright, Ian Miller
Publisher: Economic and Social Research Council
Address: 1 Temple Avenue, London EC4Y OBD, UK
Year: 1984
Pages: iv + 214
Price: £6.50
ISBN: 0-86226-139-2
ISSN: 0266-2159
Geographical area: United Kingdom.
Content: Details of research funded in the social sciences in British universities, polytechnics and colleges, and research institutes, listing title, institution, investigator, period of award and amount of award, with abstract.
Not available on-line.
Next edition: October 1985.

Schweizer museumsfuhrer   95
[Guide to the Swiss museums]
Editor: Martin R. Schärer
Publisher: Paul Haupt Berne Publishers (Switzerland)
Address: Flakenplatz 14, 3001 Berne, Switzerland
Year: 1985
Pages: 404
Price: SFr29.00
ISBN: 3-258-03409-5
Geographical area: Switzerland and the Principality of Liechtenstein.
Content: Entries for 590 museums in Switzerland and Principality of Liechtenstein arranged by alphabetic order of towns. Each entry states address, opening hours, directors, exhibition and literature.
Not available on-line.

Schweizerische Naturforschende Gesellschaft: verhandlungen, administrativer teil   96
[Swiss Natural Science Society, Swiss Academy of Sciences: translations, administration volume]
Publisher: Swiss Academy of Sciences
Address: Box 2535, Hirschengraben 11, 3001 Berne, Switzerland
Year: 1985
Price: SFr18.00

Science and technology report*   97
Editor: John Turney
Publisher: Pluto Press
Address: The Works, 105A Torriano Avenue, London NW5 2RX, UK
Year: 1984
Price: £8.95
ISBN: 0-86104-761-3
Geographical area: United Kingdom.
Content: Includes bibliography and directory of useful contacts and organizations.

Scottish conservation directory*   98
Editor: N. Allen
Publisher: Scottish Development Agency
Address: Conservation Bureau, Roseberry House, Haymarket Terrace, Edinburgh EH12 5EZ, UK
Year: 1984
Price: £2.95
ISBN: 0-905574-05-2

Słownik Polskich towarzystw naukowych: volume I towarzystwa naukowe działajace obecnie w Polsce   99
[Dictionary of Polish scientific societies: volume I: scientific societies at present active in Poland]
Editor: Leon Łos
Publisher: Ossolineum, Publishing House of the Polish Academy of Sciences
Address: Rynek 9, 50-106 Wrocław, Poland
Year: 1978
Pages: iv + 550
Availability: OP
Content: There are 172 entries in alphabetical order with general and specialized division; index of persons and index of the names of societies.

Social services year book 1985-86   100
Publisher: Longman Group Limited
Address: Sixth Floor, Westgate House, The High, Harlow, Essex CM20 1NE, UK
Year: 1985

Pages: 1000
Price: £27.00
ISBN: 0-582-90407-2
Geographical area: United Kingdom.
Content: The complete directory of the social services in the UK, with extensive listings of organizations and institutions active at all levels. The directory covers government departments, local authority social services departments including special schools, health authorities, the prison service, advice and counselling services, and a wide range of other welfare organizations - all with full address, telephone number and names of key personnel. Information is also given on education, training, journals and suppliers of specialist equipment and services.

Someone to talk to directory 1985: a directory of self-help and community support agencies in the United Kingdom and Republic of Ireland 101

Editor: Dick Thompson, Penny Webb, Matthew Pudney
Publisher: Mental Health Foundation
Address: 8 Hallam Street, London W1N 6DH, UK
Year: 1985
Pages: vii + 682
Price: £20.00
ISBN: 0-901944-08-4
Content: The directory includes some 10 000 entries divided by subject matter into 22 sections containing national and local organizations. The subjects aim to be as comprehensive as possible and include addiction, ageing, disability, education, family matters, medical disorders and voluntary organizations. Each agency has an address, telephone number and a brief description of the services it offers. There is an index providing a detailed listing of the types of problems and subjects contained in the directory and in which section the relevant agencies can be found.

Sources of information in environmental pollution 102

Editor: C.M. Lambert
Publisher: Department of the Environment and Department of Transport Library Services
Address: 2 Marsham Street, London SW1P 3EB, UK
Year: 1983
Pages: ii + 127
Price: £3.50
ISBN: 0-7184-0193-X
Series: Library information series
Geographical area: United Kingdom.
Content: There are 561 entries, covering government departments, parliamentary bodies, local authorities, official and advisory bodies, national and voluntary organizations, research and higher education, sources of further information.
Not available on-line.

Sources of scientific and technical information in Ireland 103

Publisher: Institute for Industrial Research and Standards
Address: Ballymun Road, Dublin 9, Ireland
Year: 1983
Pages: 44
Price: £5.00
ISBN: 0-900450-72-X
Geographical area: Republic of Ireland.

Special collections in German libraries 104

Editor: Walther Gebhardt
Publisher: Walter de Gruyter
Address: Genthiner Strasse 13, 1 Berlin 30, German FR
Year: 1977
Pages: xxiv + 722
Availability: OP
ISBN: 3-11-0058-39-1
Geographical area: Germany and West Berlin.
Content: The publication covers about 9000 libraries, and includes information on all special collections. It is indexed alphabetically and by subject matter.

Statistics Europe: sources for social, economic, and market research, fourth edition 105

Editor: Joan M. Harvey
Publisher: CBD Research Limited
Address: 154 High Street, Beckenham, Kent BR3 1EA, UK
Year: 1981
Pages: xiv + 508
Price: £42.50
ISBN: 0-900246-36-7
Content: Bibliography of sources of statistical information for each country, showing publisher, frequency, price, contents, etc; central statistical offices (address, telephone, scope of activities); other organizations collecting statistics.
Alphabetical index of organizations; alphabetical index of titles; subject index.
Not available on-line.
Next edition: 1986; £55.00.

Subject collections in European libraries 106

Editor: Richard C. Lewanski
Publisher: Bowker Publishing Company
Address: 58/62 High Street, Epping, Essex CM16 4BU, UK
Year: 1978
Pages: 900
Availability: OP
ISBN: 0-85935-011-8
Content: Information about special collections of books located in European libraries. The descriptions for some 12 000 Dewey-arranged entries include library name and location, director's name, date of foundation, brief history, present holdings, special collections, etc.

Suomen tieteellisten kirjastojen opas vetenskapliga bibliotek i Finland 107

[Guide to the research and special libraries of Finland, sixth edition]
Editor: Matti Liinamaa, Marjatta Heikkilä
Publisher: Finnish Research Library Association
Address: POB 217, SF-00171 Helsinki, Finland
Year: 1981
Pages: 174
Availability: OP
ISBN: 951-95382-1-6
Content: Information about 365 Finnish libraries and institutions.
Next edition: Summer 1986; FMk220.00.

Technical and scientific writers' register 108

Editor: John Dawes
Publisher: Society of Authors
Address: 84 Drayton Gardens, London SW10 9SB, UK
Year: 1984
Pages: 48
Availability: OP
ISBN: 0-9503856-7-0
Geographical area: United Kingdom.
Content: Index to subjects; entries for authors arranged alphabetically, giving name, address, telephone, specialization, qualifications, biography, and publications.

TNO: a key to research facilities 109

Editor: Peter M. Baven
Publisher: TNO Corporate Communication Department
Address: PO Box 297, 2501 BD The Hague, Netherlands
Year: 1984
Pages: xix + 100
Availability: OP

ISBN: 90-6743-026-9
Geographical area: Netherlands.
Content: Alphabetical list of keywords. Entries arranged according to: technology for society; building and construction research; industrial products and services; technical-scientific research; nutrition and food research; health research; national defence research; policy research and information. Interinstitutional research and addresses of r&d institutes.
Next edition: June 1986; free.

TNO: research applied 110

Publisher: TNO Corporate Communication Department
Address: PO Box 297, 2501 BD The Hague, Netherlands
Year: 1984
Pages: vii + 64
Availability: OP
Geographical area: Netherlands.
Content: Alphabetical list of keywords. Subjects covered/areas covered: industrial technology; concern for quality; energy; building and living; the environment; health; nutrition and food; national defence research; policy studies and information; information centres; development cooperation, addresses of r&d institutes.
Next edition: June 1986; free.

UK on-line search services 111

Editor: J.B. Deunette
Publisher: Aslib
Address: Information House, 26-27 Boswell Street, London WC1N 3JZ, UK
Year: 1982
Pages: vi + 106
Price: £14.00
ISBN: 0-85142-165-2
Content: Covers 97 existing services plus six planned services. Arranged alphabetically by name of service with one entry to each page. Each entry gives name and address, availability, coverage, charges, speed of service, participation by enquirers, date service began, number of searches per year, staffing and backup services.
Index to coverage of services and geographical index.

Verzeichnis der spezialbibliotheken in der Bundesrepublik Deutschland einschliesslich West-Berlin 112

[Directory of special libraries in the German Federal Republic and West Berlin]
Editor: Arbeitsgemeinschaft der Spezialbibliotheken eV
Publisher: Vieweg, Friedr., und Sohn Verlagsgesellschaft mbH

EUROPE INCLUDING USSR

Address: PO Box 58 29, Faulbrunnenstrasse 13, 6200 Wiesbaden, German FR
Year: 1970
Pages: viii + 208
Price: DM98.00
ISBN: 3-528-08289-5
Content: Specialist libraries with subject, geographical and name indexes. English summary.

Vodič kroz muzeje, galerije i zbirke u SR Hrvatskoj   113
[Guide through museums and galleries of the Republic of Croatia]
Publisher: Muzejski Dokumentacioni Centar
Address: Mesnička 5, 41000 Zagreb, Yugoslavia
Year: 1981
Pages: 124
Geographical area: Yugoslavia.
Content: Published in English and German.

Waste management research 1978   114
Editor: Patricia A. Ross
Publisher: Department of the Environment and Department of Transport Library Services
Address: 2 Marsham Street, London SW1P 3EB, UK
Year: 1979
Pages: iii + 44
Price: £3.10
Series: Library information series
Geographical area: United Kingdom.
Content: The publication contains 176 entries with sections on: industrial, commercial, and domestic waste handling; disposal and treatment; reclamation; food, drink, and fish industry waste; farm wastes.

Water services yearbook 1985   115
Editor: Victor H. French
Publisher: Industrial and Marine Publications Limited
Address: Fuels and Metallurgical Journals Limited, Queensway House, 2 Queensway, Redhill, Surrey RH1 1QS, UK
Year: 1985
Pages: xxii + 233
Price: £25.00
ISBN: 0-86108-172-2
Geographical area: United Kingdom.
Content: Details of ten water authorities; details of water companies; details of government-controlled bodies; statistics; associations and institutions; buyers' guide to 'who supplies what'; classified index to products.
Next edition: January 1986; £27.50.

Who's who in science in Europe: a biographical guide in science, technology, agriculture and medicine, fourth edition   116
Publisher: Longman Group Limited
Address: Sixth Floor, Westgate House, The High, Harlow, Essex CM20 1NE, UK
Year: 1984
Pages: 2500
Price: £295.00
ISBN: 0-582-90109-X
Series: Reference on research
Content: The guide provides professional and biographical details of over 25 000 senior scientists in government, industrial, academic and independent organizations. Each entry includes, where available; full name; year of birth; higher education and degrees obtained; present job, employer and year appointed; previous professional experience; directorships held; membership of societies and national committees; major publications; main professional and research interests; telephone number and full postal address.
In three volumes the guide also gives a personal name index and a section classifying entries by country according to main research interest.

Who's who in technology, second edition   117
Publisher: Who's Who International Red Series
Address: PO Box 1150, D8031 Worthsee, München, German FR
Year: 1984
Pages: xxxvi + 3626 (in 3 vols)
Price: £106.75 (for the set)
ISBN: 3-92122-036-X
Series: Who's Who: the international red series.
Geographical area: Europe.
Content: Detailed biographical listings of individuals, including names, place and date of birth, marital status, current address, education, speciality, career history, etc; index of names - leading figures in science and technology, listed alphabetically within each speciality; appendices of libraries, universities, schools, information centres, and organizations; company profiles.

Who's who of British scientists 1980-81, third edition   118
Publisher: Simon Books Limited
Address: Executive House, 136 South Street, Dorking, Surrey, UK
Year: 1980
Pages: xv + 589
Price: £24.75
ISBN: 0-86229-001-5

Content: An alphabetical list of approximately 6000 entries with an addenda of 73 names. Entries give full names, date of birth, position, qualifications, past appointments, selected publications, professional interests and address. There is a list of British winners of Nobel prizes for science, abbreviations (including those of qualifications and awards), checklists of scientific research establishments and scientific societies and professional institutes. Does not include in detail medicine, psychology, psychiatry and veterinary sciences.

AGRICULTURE AND FOOD SCIENCE

Adressbuch der Deutschen tierarzteschaft 119
[Veterinary surgeons directory of the Federal Republic of Germany]
Publisher: Schlütersche Verlagsanstalt und Druckerei GmbH und Co
Address: POB 54 40, Georgswall 4, 3000 Hannover 1, German FR
Year: 1983
Pages: 576
Price: DM78.00
ISBN: 3-87706-025-0
Content: Section A lists official and professional organizations of the veterinary profession in the Federal Republic of Germany such as government ministries and research centres, training establishments, and scientific and press bodies. Section B contains the names, addresses and specialities of veterinary surgeons listed by 'Land' (state) of the Federal Republic; there is also an entry for foreign veterinarians. Section C gives the list of names alphabetically, and Section D an alphabetical list of place names. Both of these sections have a cross reference to Section B. Section E contains advertisements.
Next edition: December 1985.

Agricultural education: full time and sandwich courses serving England and Wales* 120
Publisher: National Consultative Committee for Agricultural Education
Address: Shire Hall, Bury St Edmunds, Suffolk IP33 2AN, UK

Agricultural research service and institutes and units of the agricultural research service 121
Publisher: Agricultural and Food Research Council
Address: 160 Great Portland Street, London W1N 6DT, UK
Year: 1983
Pages: 55
Price: Free
Geographical area: United Kingdom.
Content: There are four main sections: AFRC institutes; AFRC units and groups; state-aided institutes in England and Wales; state-aided institutes in Scotland. All four sections are further sub-divided alphabetically into the relevant research establishments.

Agricultural science in the Netherlands (including the former guide, Wageningen, centre of agricultural science) 1985-87 122
Publisher: International Agricultural Centre
Address: Postbox 88, 6700 AB Wageningen, Netherlands
Year: 1985
Pages: 352
Price: DFl15.00
ISBN: 90-70785-03-X
Geographical area: The Netherlands.
Content: The directory contains 158 entries in 22 chapters, as well as the following: subject matter index; index of names; index in Dutch for Dutch users; list of abbreviations.
Next edition: January 1988.

Courses and training for careers in horticulture 123
Editor: Barry Maxim
Publisher: Horticultural Education Association
Address: c/o Canterbury Printers Limited, 11 Best Lane, Canterbury, Kent CT1 2JD, UK
Year: 1984
Pages: iii + 45
Price: £1.00
Geographical area: United Kingdom and Republic of Ireland.
Content: A summary of the careers, qualifications, courses, and training schemes available in horticulture and related subjects.

Directory of agricultural, horticultural and fishery co-operatives in the United Kingdom, sixth edition* 124
Publisher: Plunkett Foundation for Co-operative Studies

Address: 31 St Giles, Oxford OX1 3LF, UK
Year: 1984
Price: £8.50
ISBN: 0-85042-0636-6
Content: Information on approximately 600 cooperatives giving name, address, telephone and telex, names of officers, society or company registration number, type of activities, turnover, and number of members. Also contains details of central cooperative organizations, marketing boards, and federal bodies.

Directory of European agricultural organizations 125

Publisher: Kogan Page Limited
Address: 120 Pentonville Road, London N1 9JN, UK
Year: 1984
Pages: 720
Price: £39.00
ISBN: 0-85038-929-1
Geographical area: Western Europe.
Content: Description of the agricultural organizations in the European Economic Community, the applicant states, and the European Free Trade Association countries. It presents a general picture of the European and national agricultural organizations, decision-making procedures within the organizations, the channels they use to exert influence, their priority policies, and activities. Full details of the following are given for each organization: size and location; contact names and addresses; internal structure and hierarchies.

Directory of marine science and fisheries related degree and diploma courses available at colleges and universities in the United Kingdom 126

Editor: S.J. Lockwood
Publisher: Ministry of Agriculture, Fisheries and Food, Directorate of Fisheries Research
Address: Pakefield Road, Lowestoft, Suffolk NR33 0HT, UK
Year: 1983
Pages: iv + 52
Price: Free
ISSN: 0309-3670
Series: Library information leaflet.
Content: There are 50 entries, with a table of contents. Entries are under three sections: post-graduate degree courses; diploma courses; first degree courses. Within each section entries are under universities and colleges.

Food industry directory 1985-86* 127

Editor: K. Rasmussen
Publisher: Newman Books Limited
Address: 48 Poland Street, London W1V 4PP, UK

Pages: 174
Price: £25.00 (paperback)
ISBN: 0-7079-6935-2

Food trades directory and food buyer's yearbook 1985-86* 128

Editor: K. Rasmussen
Publisher: Newman Books Limited
Address: 48 Poland Street, London W1V 4PP, UK
Pages: 1098
Price: £57.00
ISBN: 0-7079-6934-4

Forschungsstätten der landbauwissenschaften, ernährungswissenschaften, forstwissenschaften, holzwirtschaftswissenschaften, des naturschutzes, der landschaftspflege, der veterinärmedizin in der Bundesrepublik Deutschland 1984 129

[Research establishments in agriculture, nutrition, veterinary medicine, forestry and timber in the German Federal Republic, 1984]
Publisher: Zentralstelle für Agrardokumentation und -information
Address: Villichgasse 17, Postfach 200569, D-5300 Bonn 2, German FR
Pages: xiv + 611
Availability: OP
Content: There are 1511 entries which give: name and address of institutions; telephone; name and title of the head of the institution; name of the assistant; number of scientists; descriptions of main activities; correspondence languages.
Personal index; subject index; place index.
Not available on-line.

Index of agricultural and food research 130

Publisher: Agricultural and Food Research Council
Address: 160 Great Portland Street, London W1N 6DT, UK
Year: 1982
Pages: x + 231
Price: £3.00
ISBN: 0-7084-0255-0
Geographical area: United Kingdom.
Content: An alphabetical listing of the 33 institutes and units of the research service with the names of those responsible for scientific liaison. Details of research in progress at the institutes and units, arranged alphabetically by institute and department. Heads of departments are given together with an alphabetical listing of

senior staff and their research responsibilities. Research organization of the United Kingdom agricultural departments, and available research grants. There is a subject index and a name index.
Next edition: 1985.

Made in Ireland food directory 131

Publisher: Institute for Industrial Research and Standards
Address: Ballymun Road, Dublin 9, Ireland
Year: 1978
Pages: 140
Price: IR£7.00
ISBN: 0-900450-50-3
Geographical area: Republic of Ireland.

Poultry world disease directory 132

Publisher: Farmers Publishing Group
Address: Surrey House, Sutton, Surrey SM1 4QQ, UK
Year: 1985
Pages: viii + 30
Price: £1.00
Geographical area: United Kingdom.
Content: Index of common poultry diseases; descriptions of common poultry diseases and treatment; medicaments, insecticides, disinfectants, equipment, services, etc; names and addresses of suppliers.

Register of farm surveys in the United 133
Kingdom and Republic of Ireland 2

Editor: J.A. Cousins, D.J. Fitzgerald, W.J.K. Thomas
Publisher: University of Exeter, Agricultural Economics Unit
Address: St German's Road, Exeter EX4 6TL, UK
Year: 1979
Pages: iv + 94
Price: £1.50
ISSN: 0141-9900
Content: The publication lists 154 entries classified by subjects, and indexed as follows: by name of researcher; by name of organization; by location of farm; of surveys with national or widespread samples of farms.
The publication is not available on-line.
Next edition: Not planned.

Research in forestry and wood science 134
in Finland

Editor: Viljo Holopainen
Publisher: Society of Forestry in Finland
Address: Unioninkatu 40 B, SF-00170 Helsinki, Finland
Year: 1984
Pages: 60
Price: FMk30.00

ISBN: 951-651-060-4
Geographical area: Finland.
Content: The publication gives details of six state and five private institutions, with an index of personnel listed by departments.

Royal College of Veterinary Surgeons 135
registers and directory

Publisher: Royal College of Veterinary Surgeons
Address: 32 Belgrave Square, London SW1X 8QP, UK
Year: 1983-84
Price: £10.50
Content: Alphabetical list of all registered veterinary surgeons.
Next edition: August 1985; £11.50.

Survey of potato research in the UK 136
1983

Editor: Julian R. Carpenter
Publisher: Potato Marketing Board
Address: PO Box 55, Cowley, Oxford OX4 3NA, UK
Pages: 88
Price: £0.65
Content: Approximately 370 entries comprise a classified list of research projects in progress. Each entry gives a brief description of the work with the name of the organization and the individuals carrying out the work. There is an index of research institutes and a name index.
Not available on-line.
Next edition: 1986.

Verzeichnis der land-und 137
ernährungswissenschaflichen verbände
zusammengeschlossen im Rahmen der
EG

[Directory of non-governmental agricultural organization set up at European Community level]
Editor: Generaldirektion Informationsmarkt und Innovation der EG
Publisher: Saur Verlag, K.G.
Address: Pössenbacherstrasse 2B, PO Box 71 10 09, 8000 München 71, German FR
Year: 1985
Pages: 420
Price: DM96.00
ISBN: 3-598-10504-5
Geographical area: Western Europe.

Veterinary register of Ireland* 138

Publisher: Veterinary Council
Address: 53 Lansdowne Road, Ballsbridge, Dublin 4, Ireland

Year: 1985
Pages: 125
Price: IR£5.00
Geographical area: Republic of Ireland.
Content: The register contains 1663 entries, listing: names and addresses of registered veterinary surgeons; names of veterinary surgeons in the Department of Agriculture and in local authorities, in the National University, and in the Agricultural Institute.

WWOOF directory of organic organisations in the UK and other relevant bodies — 139

Editor: Susan Coppard, Chris Mager
Publisher: Working Weekends on Organic Farms
Address: 19 Bradford Road, Lewes, Sussex BN7 1RB, UK
Year: 1984
Pages: i + 21
Price: £1.00
ISBN: 0-9509502-0-3
Content: About 120 entries giving details of organizations, journals, and establishments advocating or giving training in organic methods of farming, horticulture, and gardening.
Next edition: 1985; £1.00.

Yearbook of wines, alcohol and spirits of the Common Market — 140

Editor: G. Francis Seingry
Publisher: Editions Delta
Address: 92-94 Square E. Plasky, B-1040 Bruxelles, Belgium
Year: 1979
Price: BFr1500
ISBN: 2-8029-0012-9
Series: EEC directories

CHEMICAL AND MATERIALS SCIENCE

Adressbuch deutscher chemiker — 141

[Directory of German chemists]
Publisher: Verlag Chemie GmbH
Address: PO Box 1260/1280, D-6940 Weinheim, German FR
Year: 1983-84
Pages: 410
Price: DM128.00
ISBN: 3-527-26028-5
Content: Not available on-line.

Annuaire CNRS chimie* — 142

[CNRS Chemical yearbook]
Publisher: Centre National de la Recherche Scientifique
Address: 15 quai Anatole France, 75700 Paris, France
Year: 1983
Price: F150.00
ISBN: 2-222-03401-9
Geographical area: France.
Content: Research activities of CNRS and associated laboratories.

Chemfacts: Belgium, second edition — 143

Publisher: Chemical Data Services
Address: Quadrant House, The Quadrant, Sutton, Surrey SM2 5AS, UK
Year: 1981
Pages: 92
Price: £45.00
Content: Survey of 67 essential chemicals and some 50 major Belgian producers on a national scale. Three products are included for the first time - fatty acids, polyether polyols, and potassium sulphate. Information on each chemical consists of a product description; market trends with details of production, imports and exports spanning 10 years in general; a plant data section listing major producers, plant locations, present and planned capacities, processes feedstocks, licensors and contractors wherever possible; a map which pinpoints plant sites for each chemical; and a trade breakdown for 1978 and 1979, which details the foreign flow of trade with volumes.
Detailed company information on the major producers surveys their overall activities, interests, directorate, subsidiaries, ownership and financial data where available.

Chemfacts: Federal Republic of Germany, third edition — 144

Publisher: Chemical Data Services
Address: Quadrant House, The Quadrant, Sutton, Surrey SM2 5AS, UK
Year: 1982
Pages: 146
Price: £60.00
Content: The publication covers 95 products. Almost one third, including epoxy resins, fatty alcohols, MTBE, methyl methacrylate, polyacetal resins, polyisobutenes and zeolites, appear for the first time. Company profiles survey 80 chemical manufacturers.

Chemfacts: France, first edition — 145

Publisher: Chemical Data Services
Address: Quadrant House, The Quadrant, Sutton, Surrey SM2 5AS, UK

Content: Survey of 100 industrial chemicals and 100 major French producers on a national scale. Information on each chemical consists of a product description; market trends in production, imports and exports spanning 1966 to 1977 in general; the trade breakdowns for 1976 and 1977 detailing the foreign flow of trade with volume and value totals; a plant data section listing major producers, plant locations and where available present and planned capacities, processes, feedstocks, licensors, and contractors.

Chemfacts: Italy, first edition 146

Publisher: Chemical Data Services
Address: Quadrant House, The Quadrant, Sutton, Surrey SM2 5AS, UK
Year: 1979
Pages: 136
Price: £40.00
Content: Survey of 94 industrial chemicals and 91 major Italian producers on a national scale. Information on each chemical consists of a product description; market trends in production, imports and exports spanning 1967 to 1977; trade breakdowns for 1976 and 1977 detailing the foreign flow of trade with volume and value totals; a plant data section listing major producers, plant locations and where available present and planned capacities, processes, feedstocks, licensors and contractors; a map pinpoints plant locations for each chemical together with major neighbouring cities. The company information in Section 2 of the book contains addresses, telephone and telex numbers, manufacturing activities and where available the directorate, history and present structure of the company, financial data, number of employees, foreign and domestic subsidiaries, their undertakings and details of ownership.

Chemfacts: Netherlands, second edition 147

Publisher: Chemical Data Services
Address: Quadrant House, The Quadrant, Sutton, Surrey SM2 5AS, UK
Year: 1981
Pages: 112
Price: £45.00
Content: Each of the 83 profiles of chemical products contains a brief description of the product followed by detailed plant data arranged in clear tabular form, with a map of plant locations. All available production figures since 1971, total trade figures from 1971 to 1980, and country-by-country trade breakdowns for 1979 and 1980 with percentages for exports and imports, are also tabulated.
The 47 chemical producers mentioned in the plant data section are surveyed in detailed company profiles.

Chemfacts: Portugal, first edition 148

Publisher: Chemical Data Services
Address: Quadrant House, The Quadrant, Sutton, Surrey SM2 5AS, UK
Year: 1983
Pages: 75
Price: £55.00
Content: The publication surveys 54 chemical products made in Portugal including petrochemicals, plastics, synthetic fibres, fertilizers, inorganics, and industrial gases. Plant data and plant location maps are presented in tabular form together with production, import and export figures from 1972 to 1981 and trade breakdowns for 1979, 1980 and 1981. In addition the book contains profiles of the 30 companies producing the chemicals surveyed.

Chemfacts: Scandinavia, second edition 149

Publisher: Chemical Data Services
Address: Quadrant House, The Quadrant, Sutton, Surrey SM2 5AS, UK
Year: 1981
Pages: 145
Price: £45.00
Content: New plants for producing MTBE and benzene in Finland together with a re-organization of petrochemical companies, start-up of new polyolefin plants in southern Norway, production of linear LD polyethylene in Sweden - these are some of the developments since the first edition. The range of products covered includes for the first time acetone, acetylene, carboxymethylcellulose, copper sulphate, cumene, ethanol, ethylene dichloride, fatty acids, MTBE, SBR latex and sodium chlorate. Each profile contains a brief description of the product followed by details of the major producers, plant locations, capacities (present and planned), feedstocks, processes, licensors and contractors, with a map.
All available production figures since 1975, total trade figures from 1975 to 1980, and country-by-country trade breakdowns for 1978 and 1979 with percentages for the volume, are presented.
76 major chemical producers are surveyed.

Chemfacts: Spain, second edition 150

Publisher: Chemical Data Services
Address: Quadrant House, The Quadrant, Sutton, Surrey SM2 5AS, UK
Year: 1982
Pages: 128
Price: £60.00
Content: Profiles of 93 products, including polymeric products such as epoxy resins, industrial gases such as oxygen and nitrogen, inorganics such as sodium sulph-

ate, and organics such as fumaric acid, all of which are new to this edition. Profiles of 108 manufacturers appear in the company information section.

Chemfacts: United Kingdom, third edition — 151

Publisher: Chemical Data Services
Address: Quadrant House, The Quadrant, Sutton, Surrey SM2 5AS, UK
Content: Profiles of some 96 products and 69 companies.

Chemical company profiles: Western Europe, third edition — 152

Publisher: Chemical Data Services
Address: Quadrant House, The Quadrant, Sutton, Surrey SM2 5AS, UK
Year: 1982
Pages: 384
Price: £60.00
Content: Up-to-date profiles of some 1750 chemical manufacturers in 19 countries. Arranged by country, each profile gives the address details, manufacturing activities, and wherever possible a full picture of the company's history, present structure, future plans, directors, recent financial data and a list of associated companies.
A full index of company names completes the book.

Chemical directory of Northern Europe: manufacturers and traders of chemicals and allied products in Scandinavia and Finland and Iceland — 153

Publisher: Stamex bv
Address: PO Box 400, 3770 AK Barneveld, Netherlands
Year: 1977
Availability: OP

Chemicalien adresboek — 154

[Directory of the Dutch chemical and allied industry and trade]
Publisher: Stamex bv
Address: PO Box 400, 3770 AK Barneveld, Netherlands
Year: 1985
Pages: 540
Price: Dfl160.00
Content: Alphabetical list of chemical and pharmaceutical companies with details of address, telephone number, directors' names. Manufacturers included concern the following: cosmetics, pharmaceutical specialities, paints, and laquers.

Chemicals 1985: chemicals on the UK market* — 155

Editor: E. Wilson
Publisher: Chemical Industries Association
Address: Alembic House, 93 Albert Embankment, London SE1 7TU, UK
Year: 1985
Pages: 200

Chemische jaarboek der Koninklijke Nederlandse Chemische Vereniging — 156

[Chemical yearbook of the Royal Netherlands Chemical Society]
Publisher: Koninklijke Nederlandse Chemische Vereniging
Address: PO Box 90613, 2509 LP, 's-Gravenhage, Netherlands
Year: 1985
Pages: 160
Price: Dfl50.00
Next edition: April 1987.

Composite materials: a directory of European research* — 157

Editor: A.R. Bunsell, A. Kelly
Publisher: Butterworth and Company (Publishers) Limited
Address: 88 Kingsway, London WC2B 6AB, UK
Year: 1985
Price: £60.00

Corrosion prevention directory — 158

Publisher: National Corrosion Service
Address: National Physical Laboratory, Teddington, Middlesex TW11 0LW, UK
Year: 1978
Pages: viii + 137
Geographical area: United Kingdom.

Directory of consulting practices in chemistry and related subjects — 159

Publisher: Royal Society of Chemistry
Address: Burlington House, Piccadilly, London W1V 0BN, UK
Year: 1985
Pages: 120
Price: £10.00
Geographical area: United Kingdom.
Content: There are 200 entries comprising a guide to services, and a subject guide, with an alphabetical index of consulting practices.
Not available on-line.

Directory of municipal wastewater treatment plants 160

Publisher: Institute of Water Pollution Control
Address: 53 London Road, Maidstone, Kent ME16 8JH, UK
Year: 1972
Price: £10.00
Geographical area: England, Scotland and Wales.

ENREP directory of solid waste and chemical waste* 161

Publisher: Peter Peregrinus Limited
Address: PO Box 8, Southgate House, Stevenage, Hertfordshire SG1 1HQ, UK
Year: 1980
Price: £25.00 (2 vols)
ISBN: 0-906048-28-1
Geographical area: Europe.
Content: Details of more than 1000 research projects. Subject indexes in: Danish and English (volume two); Dutch and French (volume three); and German and Italian (volume four).

Entoma Europe* 162

Publisher: Stamex BV
Address: POB 400, 3770 AK Barneveld, Netherlands
Content: Lists European manufacturers' associations and institutions classified by country and alphabetically. Index of trade names of chemicals for the agriculture sector. Published every three years in English.

Europäische Föderation Korrosion: jahrsbericht 1978-84 163

[European Federation of Corrosion: annual report]
Publisher: DECHEMA
Address: PO Box 97 01 46, Theodor-Fluiss-Allee 25, 6000 Frankfurt am Main 97, German FR
Year: 1985
Pages: 200-250
Content: Part one covers the years 1978-1982; Part two lists member societies; Part three covers 1983-84.
Next edition: 1987.

European paint manufacturers: manufacturers of paints, varnishes, enamels, lacquers, printing inks, solvents and paint removers 164

Publisher: Stamex bv
Address: POB 400, 3770 AK Barneveld, Netherlands
Year: 1977
Availability: OP

European plastics: manufacturers of raw material, semi-finished products and auxiliaries for the plastics industry 165

Publisher: Stamex bv
Address: POB 400, 3770 AK Barneveld, Netherlands
Year: 1978
Availability: OP

Firmenhandbuch chemische industrie 1985-87, Bundesrepublik Deutschland und Berlin (West) 166

[Directory of the German chemical industry 1985-87, Federal Republic of Germany and West Berlin]
Publisher: Econ Verlag GmbH
Address: Grupellostrasse 28, Postfach 92 29, D-4000 Düsseldorf 1, German FR
Year: 1985
Pages: 500
Price: DM148.00
Content: Information on 18000 manufacturers and wholesalers of chemical products. Listing of 11 000 chemicals and chemical products and approximately 2000 trade names, with an alphabetical index in the English language.
Next edition: June 1988; DM148.00.

Guide to rubber and plastics test equipment 167

Editor: R.P. Brown
Publisher: Rapra Technology Limited
Address: Shawbury, Shrewsbury, Shropshire SY4 4NR, UK
Year: 1979
Pages: iv + 261
Price: £10.00
ISBN: 0-902348-19-1
Geographical area: United Kingdom.

Handbook of industrial materials, first edition 168

Publisher: Trade and Technical Press Limited
Address: Crown House, London Road, Morden, Surrey SM4 5EW, UK
Year: 1977
Pages: 700
Price: £48.00 (UK); £54.00 (elsewhere)
ISBN: 85461-060-X
Geographical area: United Kingdom.
Content: Comprehensive coverage of types and applications of materials. Buyers' guide section, editorial and advertisers indexes.

EUROPE INCLUDING USSR

Institute of Water Pollution Control yearbook 169

Publisher: Institute of Water Pollution Control
Address: 53 London Road, Maidstone, Kent ME16 8JH, UK
Year: 1984
Pages: lii + 184
Price: £10.00
Geographical area: United Kingdom.
Content: Fifty four company profiles; details of IWPC structure, officers, committees, and publications; list of 2625 members in three sections with addresses; buyers' and specifiers' guide.
Next edition: September 1985; £10.00.

Irish chemicals directory 170

Publisher: Institute for Industrial Research and Standards, Industrial Research Centre
Address: Ballymun Road, Dublin 9 Ireland
Year: 1984
Pages: 158
Price: IR£7.00
ISBN: 0-900450-77-0
Geographical area: Republic of Ireland.

Irish plastics and rubber directory 171

Publisher: Institute for Industrial Research and Standards
Address: Ballymun Road, Dublin 9, Ireland
Year: 1983
Pages: 152
Price: IR£7.00
ISBN: 0-800450-75-4
Geographical area: Republic of Ireland.

Paint and resin directory* 172

Publisher: Turret-Wheatland Limited
Address: Penn House, Penn Place, Rickmansworth, Hertfordshire WD3 1SN, UK
Year: 1984
Price: £15.00
Content: Listing of suppliers, product guide, and brand names.
Next edition: December 1985.

Plastics in building: index of applications and suppliers* 173

Publisher: British Plastics Federation
Address: 5 Belgrave Square, London SW1X 8PH, UK
Year: 1977
Geographical area: United Kingdom.
Content: A classified index to applications and suppliers; includes an alphabetical list of suppliers with their address, a trademark index and a list of British Standards and Agrément Certificates relevant to the use of plastics in building.

Review of research activities in polymer science and technology, sixth edition 174

Publisher: Plastics and Rubber Institute
Address: 11 Hobart Place, London SW1W 0HL, UK
Year: 1979
Pages: ii + 28
Price: £5.00
Geographical area: United Kingdom.
Content: There are 53 entries divided into two sections: universities; and research associations. Each entry lists: current research projects; names of personnel conducting research; experimental facilities available.
Next edition: September 1985; £8.00.

EARTH AND SPACE SCIENCES

Annuaire CNRS sciences de la terre, de l'océan, de l'atmosphère et de l'espace* 175

[CNRS yearbook of earth, oceanic, atmospheric, and space sciences]
Publisher: Centre National de la Recherche Scientifique
Address: 15 quai Anatole France, 75700 Paris, France
Year: 1983
Price: F150.00
ISBN: 2-222-03400-0
Geographical area: France.
Content: Research activities of CNRS and associated laboratories.

Directory of British caving clubs 1984 176

Editor: Tony Oldham
Publisher: Oldham, Anne
Address: Rhychydwr, Crymych, Dyfed SA41 3RB, UK
Year: 1985
Pages: ii + 22
Price: £2.00
Geographical area: United Kingdom and Republic of Ireland.
Content: The directory contains 340 entries, listing, caving clubs in alphabetical order, and giving the name of club, name of secretary or permanant address, date club was formed, number of members, publications issued, any other information, eg details of club hut, control of which caves etc.
Not available on-line.

Directory of marine technology research 177

Publisher: Science and Engineering Research Council
Address: Marine Technology Directorate, Garrick House, 3-5 Charing Cross Road, PO Box 271, London WC2Y 0HW, UK
Year: 1983
Pages: xxi + 269
Price: Free
Geographical area: United Kingdom.
Content: The directory contains lists of project titles, project summary sheets, recently completed projects, index of principal investigators, and contact points.
Next edition: September 1985; free.

Directory of mines and quarries* 178

Editor: P.M. Harris
Publisher: British Geological Survey
Address: Keyworth, Nottinghamshire NG12 5GG, UK
Year: 1984
Price: £21.50
ISBN: 0-85272-082-3
Geographical area: United Kingdom.
Content: Details of mineral workings in the UK giving location, name, map reference, commodity, geology, and operating company; names and details of companies.
Next edition: 1986.

Directory of research and development activities in the United Kingdom in land survey and related fields 179

Editor: I.J. Dowman
Publisher: Surveyors Publications
Address: 12 Great George Street, London SW1P 3AD, UK
Year: 1982
Pages: 65
Price: £3.00
ISBN: 0-85406-152-5
Geographical area: United Kingdom.
Content: The directory lists information on 154 organizations under the following main headings: organizations engaged primarily in research; organizations which undertake research from time to time; other useful contacts - manufacturers, professional bodies, and periodicals and publications. There is an index to organizations.
Not available on-line.

Directory of UK research in climatology 180

Publisher: Science and Engineering Research Council
Address: Polaris House, North Star Avenue, Swindon, Wiltshire SN2 1EU, UK
Year: 1981
Pages: ii + 105
Price: Free
Content: The directory lists climatic research projects being carried out in UK universities and research centres. Entries list name of organization, project leader, address and telephone, personnel involved, aims of project, details of funding, etc.

Geological directory of the British Isles 181

Editor: Judith A. Diment
Publisher: Geological Society
Address: Burlington House, Piccadilly, London W1V 0JU, UK
Year: 1978
Pages: 109
Price: £6.00
Content: A guide to information sources, listed geographically and by name of organization.

Geologists directory, third edition 1985* 182

Editor: M.P. Henton
Publisher: Institute of Geologists
Address: 2nd Floor, Geological Society Apartments, Burlington House, Piccadilly, London W1V 9HG, UK
Price: £12.50
ISBN: 0-9506-9061-9
Geographical area: United Kingdom.
Content: Sources of geological information including government departments, local authorities, oil, mining and quarrying companies, universities and colleges, consultants, libraries and museums, buyers' guide for geological equipment.

Geologists yearbook 1977 183

Publisher: Blandford Press Limited
Address: Link House, West Street, Poole, Dorset BH15 1LL, UK
Year: 1977
Pages: 274
Availability: OP
ISBN: 0856420484
Series: Dolphin yearbooks
Geographical area: United Kingdom.

Geology in museums - a bibliography and index 184

Editor: T. Sharpe
Publisher: National Museum of Wales
Address: Cathays Park, Cardiff CF1 3NP, UK
Year: 1983

Pages: 128
Price: £2.50 (£3.70 by post)
ISBN: 0-7200-0281-8
Series: Geological series
Geographical area: United Kingdom.
Content: There are 1000 entries on literature relevant to geological curation, with an author index with full bibliographical citation and keywords and a subject (or keyword) index.

Guide to the coalfields 185

Editor: Bob Sansom
Publisher: Colliery Guardian
Address: Queensway House, 2 Queensway, Redhill, Surrey RH1 1QS, UK
Year: 1984
Pages: 576
Price: £23.00
ISBN: 0-86108-158-7
Geographical area: United Kingdom.
Content: The guide list National Coal Board mines by area, and also contains sections on relevant learned societies and institutions, universities and technical colleges, trade unions, and small mines federation, as well as indexes to personnel and mines, and a buyers' guide.

Hydrological research in the United Kingdom 1965-70; 1970-75; 1975-80 186

Editor: C. Kirby
Publisher: Institute of Hydrology
Address: Maclean Building, Crowmarsh Gifford, Wallingford, Oxfordshire OX10 8BB, UK
Year: 1980
Pages: vi + 150
Price: Free
Geographical area: United Kingdom.
Content: Details of: development and organization of hydrological research in UK; a tabular statement of research; hydrological research by government and other agencies; hydrological research in UK universities. The publication also contains appendices on selected publications, and societies and institutions, and an index to institutions involved in hydrology.
Next edition: 1985/1986; free.

Jahrbuch fur bergbau, energie, mineralöl und chemie* 187

[Mining, energy, mineral oil and chemistry yearbook]
Publisher: Verlag Glückauf GmbH
Address: Postfach 10 39 45, 4300 Essen 1, German FR
Year: 1984
Price: DM78.00
ISBN: 3-7739-0426-6
Next edition: September 1985; DM78.00

Matter of degree: directory of geography courses 188

Editor: Molly Weber
Publisher: Geo Books
Address: Regency House, 34 Duke Street, Norwich NR3 3AP, UK
Year: 1984
Pages: 39
Price: £1.50
ISBN: 0-86094-161-2
Series: A matter of degree
Geographical area: United Kingdom.
Content: There are 122 entries listed in alphabetical order. Each entry is a course (or courses) provided by a centre of higher education, and consists of the following information: name of institution; postal address; address of department if different from main postal address; courses; requirements for undergraduate entry; to whom applications should be made; brief description of the course; department publications; course content of geography, computing, field work; how course is assessed; postgraduate courses; teaching staff, their titles and research interests; statistical information.
The publication is not available on-line.
Next edition: June 1985; £1.75.

Remote sensing of earth resources: list of UK groups and individuals engaged in remote sensing with a brief account of their activities and facilities, fifth edition 189

Editor: E.J. Lindsay
Publisher: Department of Trade and Industry
Address: Ashdown House, 23 Victoria Street, London SW1, UK
Year: 1981
Pages: v + 386
Price: £15.00
Content: The list is arranged in five sections: government departments; education and training; manufacturers; index of non-UK countries; and also indexes of subjects and organizations.

United Kingdom research on the history of geological sciences directory 190

Editor: H.S. Torrens
Publisher: Royal Society
Address: 6 Carlton House Terrace, London SW1Y 5AG, UK
Year: 1983
Pages: 32
Price: £3.00
ISBN: 0-85403-218-5

Content: There are 87 entries listed alphabetically, giving name, address, field of interest, literature references. There are no indexes.
Not available on-line.

Volcanological research in the United Kingdom 1971-75: a survey of research activities   191

Editor: M.J. Le Bas
Publisher: Royal Society
Address: 6 Carlton House Terrace, London SW1Y 5AG, UK
Year: 1979
Pages: 93
Availability: OP
ISBN: 0-85403-125-1
Content: Details of 67 research projects listed alphabetically by research institute/university, giving names of research workers, research interest, and key literature references; lists of research workers by subject, and alphabetically, and of research institutions, approximately alphabetically.
Not available on-line.

ELECTRONICS

British robot association members' handbook 1984-85*   192

Publisher: British Robot Association
Address: 28-30 High Street, Kempston, Bedford MK42 7AJ, UK
Year: 1984
Price: £50.00
Geographical area: United Kingdom.
Content: The publication gives address and contact names of all robot suppliers in the United Kingdom, with details of robotics consultants and academic institutions involved in robotic research.

Codata directory of data sources of science and technology*   193

Editor: D.G. Watson
Publisher: Codata Bulletin, 24
Address: Headington Hill Hall, Oxford OX3 0BW, UK
Year: 1977
Pages: 42

Computer companies in the UK, 1984, first edition*   194

Publisher: Eurolec/David Raynor
Address: 6 Woodbury Lane, Clifton, Bristol BS8 2SD, UK
Year: 1983
Price: £42.00
ISBN: 0-900614-59-5
Content: Entries for about 760 computer companies, giving name, address, type of company, employees, directors, sales contacts, related companies, and industrial classification.

Directory of electronics, instruments and computers   195

Editor: James Robertson
Publisher: Morgan-Grampian Book Publishing Company Limited
Address: 30 Calderwood Street, London SE18 6QH, UK
Year: 1985
Pages: vi + 490
Price: £18.00
ISBN: 0-86213-061-1
Geographical area: United Kingdom.
Content: Names, addresses, telephone and telex, etc of over 3500 manufacturers and suppliers of equipment; list of over 4000 trade names; buyers guide giving listings for over 2500 products supplied by 3000 leading UK companies; list of 2000 overseas companies with UK agents' addresses; details of nearly 150 trade associations; distributors - geographical guide and map, alphabetical list of manufacturers with distributors, names and addresses of distributors.
Not available on-line.
Next edition: February 1986; £20.00.

Directory of research lasers and expertise in universities, polytechnics and SRC establishments in the UK   196

Publisher: Rutherford Appleton Laboratory
Address: Chilton, Didcot, Oxfordshire OX11 0QX, UK
Year: 1978
Pages: 29
Availability: OP

Electrical and electronics trades directory 1985   197

Editor: A.E. Newill
Publisher: Peter Peregrinus Limited
Address: PO Box 8, Southgate House, Stevenage, Hertfordshire SG1 1HQ, UK
Pages: 608
Price: £40.00

ISBN: 0-86341-031-6
Geographical area: United Kingdom.
Content: The directory lists 3500 manufacturers, 3850 product headings, 3700 manufacturers' representatives, 1150 wholesalers, 250 trade associations, and 2100 trade names.
Not available on-line.
Next edition: February 1986; £45.00.

EURASIP directory: directory of European signal processing research institutions — 198

Editor: Jan J. Gerbrands
Publisher: D. Reidel Publishing Company
Address: PO Box 17, 3300 AA Dordrecht, Netherlands
Year: 1984
Pages: viii + 430
Price: Dfl155.00
ISBN: 90-277-1824-5
Geographical area: Europe, including Bulgaria and Poland.
Content: There are 379 entries from seventeen countries, listed by country, institutions arranged alphabetically by city. Each entry gives name of organization, location, name of contact person, address, thesaurus-controlled interest statement, and free-text research-group description.
The directory is not available on-line. Enquiries concerning leasing or loan of the directory on tape should be addressed to the Editor, Jan J. Gerbrands, Department of Electrical Engineering, Delft University, PO Box 5031, 26000 GA Delft.
There are no supplements as of 1st July 1985.

EUSIDIC database guide* — 199

Publisher: Learned Information
Address: Besselsleigh, Abingdon, Oxfordshire, UK
Year: 1983
Price: £25.00
ISBN: 0-904933-37-7
Content: Listing of over 1800 data bases publicly available in Europe.
Next edition: 1985.

Irish electronics and electrical directory — 200

Publisher: Institute for Industrial Research and Standards
Address: Ballymun Road, Dublin 9, Ireland
Year: 1982
Pages: 141
Price: IR£8.00
ISBN: 0-900450-68-1
Geographical area: Republic of Ireland.

Microcomputer companies in the UK* — 201

Publisher: Eurolec/David Rayner
Address: 6 Woodbury Lane, Clifton, Bristol BS8 2SD, UK
Year: 1983
Price: £32.00
ISBN: 0-900614-58-7
Content: Details of 1400 companies giving name, address, type of company, products or services offered, employees, year of formation, directors, sales contacts, etc.

Optical sensor component directory — 202

Editor: E.W. Lecznar, P.A. Shepherd
Publisher: ERA Technology Limited
Address: Cleeve Road, Leatherhead, Surrey KT22 7SA, UK
Year: 1985
Geographical area: United Kingdom.
Content: The directory has three main sections:
Product tables, based on the following component categories - detectors, optical filters, gratings, integrated optic components and subsystems, emitters, materials relevant to optical transducers, lenses, mirrors, optical fibres, modulators, polarizers, prisms, optical receivers and transmitters, splitters, wavelength shifters, Bragg frequency shifters and miscellaneous components.
Manufacturers/agents - a list of manufacturers and/or UK agents included in the directory with an indication of the products they supply in quick reference form.
Manufacturers' addresses - an alphabetical listing of manufacturers included in the first two sections, with addresses, telephone, telex, and telefax numbers and personal contacts where available.
The directory is available on-line via the ERA Information Services Department.

Product directory 1984-85 — 203

Publisher: Electronic Engineering Association
Address: Leicester House, 8 Leicester Street, London WC2H 7BN, UK
Pages: 40
Price: Free
Geographical area: United Kingdom.
Content: The association's product directory gives broad details of the products, equipment, and services offered by its members within the electronic capital equipment industry.
Next edition: September 1985; free.

Studien- und Forschungsführer Informatik — 204

[Studies and Research in Informatics]
Editor: Wilfried Brauer, Wolfhart Haacke, Siegfried Münch

Publisher: Springer-Verlag
Address: Postfach 10 52 80, Tiergartenstrasse 17, D-6900 Heidelberg, German FR
Year: 1984
Pages: xii + 177
Price: DM12.80
ISBN: 3-540-13810-2
Geographical area: Germany.

ENERGY SCIENCES

ANEP 85: European petroleum yearbook   205

Editor: Thomas Vieth
Publisher: Otto Vieth Verlag
Address: Postfach 701606, 2000 Hamburg 70, German FR
Year: 1984
Pages: 460
Price: DM158.00
ISSN: 0342-6947
Content: Details of oil and gas fields in Western Europe, North Sea oil and gas, crude oil, product and natural gas pipelines, refineries survey of Western Europe, oil and gas in Eastern Europe, oil and gas statistics. Information about over 2000 companies in all European countries: oil and gas companies, trading and distribution, storage and transport, associations, government and administration offices, institutes, etc. Suppliers' directory and buyers' guide.
Next edition: November 1985; DM158.00.

Annuario nazionale dell'energia*   206
[National energy yearbook, third edition]
Editor: L.A. Salvi
Publisher: Inter-ed srl
Address: Via Cassia 1134/A, 00189 Roma, Italy
Year: 1985
Price: $100.00
ISBN: 0392-8403
Geographical area: Italy.
Content: Entries on 400 research centres and 4500 firms operating in the energy sector; who's who in energy in Italy. Mainly in Italian, but with abstracts in English.
Next edition: December 1985.

Directory of current UK r&d relevant to underwater and offshore instrumentation and measurement, 1976   207

Publisher: Construction Industry Research and Information Association
Address: 6 Storey's Gate, London SW1P 3AU, UK
Pages: 57
Price: £2.00 (to members); £10.00 (to non-members)
Series: UEG publications

Directory of UK renewable energy suppliers and services   208

Editor: Dr C.M.A. Johansson
Publisher: Solar Energy Unit, University College
Address: Newport Road, Cardiff CF2 1TA, UK
Year: 1984
Pages: i + 22
Price: £5.00
ISSN: 0266-8041
Content: List of products, services, and suppliers; A-Z company information; list of trade names; location of companies; trade associations, comprising 53 entries. Not available on-line.
Next edition: July 1985; £6.00.

Electricity supply handbook 1985   209

Editor: G.A. Jack
Publisher: Business Press International
Address: Quadrant House, The Quadrant, Sutton, Surrey SM2 5AS, UK
Pages: ix + 304
ISBN: 0-617-00396-3
Geographical area: United Kingdom.
Content: There are over 2000 entries, covering: generating boards; tariffs; government departments relevant to the electricity industry; negotiating bodies; training boards; power stations; electricity boards.
Not available on-line.
Next edition: January 1986; £10.00.

Energy: a register of research development and demonstration in the United Kingdom: part one - energy conservation; part three - renewable energy   210

Editor: J. Furnival
Publisher: Her Majesty's Stationery Office
Address: PO Box 569, London SE1 9NH, UK
Year: 1979
Pages: Part One, 102; Part Three, 440
Availability: OP
ISBN: Part One, 0-70-580821-1; Part Three, 0-70-580772-X

EUROPE INCLUDING USSR

Geographical area: United Kingdom.
Content: Entries give details of Department of Energy r&d projects. Part One contains 98 entries, and Part Three 422 (Part Two was never published), indexed to investigators, participating organizations, and sponsors.
Next edition: Not planned.

European biomass directory 211

Editor: J. Coombs
Publisher: Macmillan Press Limited
Address: 4 Little Essex Street, London WC2R 3LF, UK
Year: 1985
Pages: 306
Price: £40.00
ISBN: 0-333-39289-2
Geographical area: Western Europe.
Content: Biomass means the production and utilization of useful energy products from biological materials. It includes: production of methanol from wood, biomass systems for electric power production; the use of vegetable oils for diesel fuel substitutes, biogas production on modern farms; agricultural wastes as energy potential and alcohol production from biomass.
The book consists of two parts.
Part 1 taken country by country in West Europe, will give a state of art in each, followed by list of all non-commercial and commercial organizations describing what they offer. Part 2 is a product guide.

European offshore oil and gas directory* 212

Publisher: Telford, Thomas, Limited
Address: 1-7 Great George Street, London SW1P 3AA, UK
Year: 1985
Price: £40.00
Content: Guide to every aspect of the European offshore scene; company directory.

Gas directory and who's who, 88th edition* 213

Editor: J. Hedges, D. Steadman
Publisher: Benn Business Information Services
Address: Union House, Eridge Road, Tunbridge Wells, Kent TN9 1RW, UK
Year: 1985
Price: £34.00
ISBN: 0-86382-021-2
Geographical area: United Kingdom.
Content: Guide to the British Gas Corporation; suppliers to the industry; classified buyers' guide; British gas statistics; trade names; trade associations.
Next edition: February 1986; £36.00.

Handbook of energy data and calculations 214

Editor: Peter Osborn
Publisher: Butterworths Scientific Limited
Address: PO Box 63, Westbury House, Bury Street, Guildford, Surrey GU2 5BH, UK
Year: 1985
Price: £37.50
ISBN: 0-408-01327-3
Geographical area: United Kingdom.
Content: The handbook gives data charts and tables, calculation and analysis procedures, directory of products and services, bibliography and sources. 660 entries contain: product, name, address, telephone, and product details.
Not available on-line.

Institute of Energy directory of qualified energy consultants, third edition* 215

Publisher: Energy Publications
Address: PO Box 147, Cambridge CB1 1NY, UK
Year: 1982
Price: £13.50
ISBN: 0-905332-18-0
Geographical area: United Kingdom.
Content: Guide to 369 consultants; subject specialization index.

Local area networks: a European directory of suppliers and systems 216

Editor: Yvonne Collier
Publisher: Online Publications
Address: Pinner Green House, Ash Hill Drive, Pinner, Middlesex HA5 2AE, UK
Year: 1985
Pages: 232
Price: £50.00: $75.00 (overseas)
ISBN: 0-68353-043-5
Content: The directory provides a source of information on the suppliers of local area networks in Europe, together with a brief description of the systems they offer. Presented in a standard format, brief technical data provided by the manufacturers may be accessed by the sixty-five manufacturers or by the system name. The largest section of the directory covers suppliers, cross-referenced by system and by country. Twenty European countries are included, and entries give contacts, addresses, telephone and telex numbers.

Major energy supply companies of Western Europe* 217

Editor: R. Whiteside
Publisher: Graham and Trotman Limited

Address: Sterling House, 66 Wilton Road, London SW1V 1DE, UK
Pages: 180
Availability: £45.00; $72.00
ISBN: 0-86010-712-4
Content: A new directory providing information on over 600 Western European companies involved in the energy supply business. This covers gas and electricity supply, oil and gas exploration and production, oil refining, nuclear power generation and engineering, coal mining, and fuel distribution. Information includes: company name and address, telephone, telex, names of directors and senior management, business activities, brand names, subsidiaries, bankers, financial information for last 2 years, and numbers of employees.
Next edition: 1985

North Sea oil and gas directory, thirteenth edition 1985* 218

Publisher: Spearhead Publications Limited
Address: Rowe House, 55-59 Fife Road, Kingston upon Thames, Surrey KT1 1TA, UK
Year: 1985
Price: £32.95
ISSN: 0265-5039
Content: Names, addresses, status, etc of over 15 000 key personnel in over 3500 companies involved in the offshore industry.
Next edition: April 1986.

World directory of energy information - volume 1 219

Publisher: Gower Publishing Company Limited
Address: Gower House, Croft Road, Aldershot, Hampshire GU11 3HR, UK
Year: 1981
Pages: x + 326
Price: £35.00
ISBN: 0-566-02198-6
Geographical area: Western Europe.
Content: Part One provides an overview of the main features of energy in Western Europe, with ten maps and numerous tables.
Part Two looks in detail at 17 individual countries in a standard sequence: the general development of the economy and energy consumption; the oil, coal, gas, and electricity sectors; imports and exports of energy; energy policies.
Part Three contains information on the activity of more than 400 organizations in the energy sector. These include government departments, official agencies, state undertakings, private companies, professional institutions, and trade associations.
Part Four consists of a wide-ranging bibliography of over 900 publications relevant to the supply and consumption of energy in Western Europe. The volume concludes with a section defining the terms and units used in the discussion of energy issues, an index to publishing bodies and addresses, and an appendix giving energy units and terms.
Not available on-line.

ENGINEERING AND TRANSPORTATION

CIRIA guide to sources of information in the construction industry 220

Publisher: Construction Industry Research and Information Association
Address: 6 Storey's Gate, London SW1P 3AU, UK
Year: 1984
Pages: 126
Price: £8 (to members); £10 (to non-members)
ISBN: 0-86017-217-1
Geographical area: United Kingdom.

Consulting engineers* 221

Publisher: Data Research Group
Address: Bridge House, Great Missenden, Buckinghamshire, UK
Year: 1983
Pages: £33.00
Geographical area: United Kingdom.
Content: List over 1500 engineers divided geographically, with telephone numbers.

Consulting engineers who's who and yearbook* 222

Editor: John Austin
Publisher: Municipal Publications
Address: 178-202 Great Portland Street, London W1N 6NH, UK
Year: 1985
Price: £19.50
ISBN: 0-90055-239-5
Geographical area: United Kingdom.
Content: The book identifies consulting engineering practices in accordance with their various disciplines and applications of major engineering projects, giving biographical details of members of the association and firms to which they belong.
Next edition: December 1985.

EUROPE INCLUDING USSR

Directory of current United Kingdom research and development relating to offshore structures and pipelines — 223

Publisher: Construction Industry Research and Information Association
Address: 6 Storey's Gate, London SW1P 3AU, UK
Year: 1977
Pages: vi + 287
Price: £6 (to members); £25 (to non-members)
ISBN: 0-86017-028-4
Series: UEG publications
Geographical area: United Kingdom.

Directory of current United Kingdom research and development relevant to underwater inspection and repair — 224

Publisher: Construction Industry Research and Information Association
Address: 6 Storey's Gate, London SW1P 3AU, UK
Year: 1979
Pages: v + 246
Price: £7 (to members); £28 (to non-members)
ISBN: 0-86017-134-5
Series: UEG publications
Geographical area: United Kingdom.

Directory of official architecture and planning 1985-86 — 225

Publisher: Longman Group Limited
Address: Sixth Floor, Westgate House, The High, Harlow, Essex CM20 1NE, UK
Year: 1985
Pages: 400
Price: £25
ISBN: 0-860-95048-4
Geographical area: United Kingdom.
Content: The established reference work for the construction industry, this directory is a comprehensive guide to officials and organizations involved with planning, architecture and related work. In this new edition sections have been expanded to include more coverage of industrial development, and the names and locations of architects in many large companies. Key features include: coverage of local authority officials in architecture, planning, surveying and related departments; a new section with a list of training and professional bodies; and an address book of organizations and associations relevant to construction together with brief details on each.

Engineering companies 1984-85* — 226

Publisher: Data Research Group
Address: Bridge House, Great Missenden, Buckinghamshire, UK
Year: 1985
Price: £52.00
Geographical area: United Kingdom.
Content: Entries for engineering companies listed by towns and counties with telephone numbers. Part one - electrical and electronics has 3660 entries; part two - hydraulics, 540; part three - mechanical handling, 589; part four - precision engineering, 4440.

Guide to information services in marine technology — 227

Editor: Arnold Myers
Publisher: Institute of Offshore Engineering
Address: Heriot-Watt University, Riccarton, Edinburgh EH14 4AS, UK
Year: 1979 with subsequent amendments to date
Pages: xi + 125
Price: £3.00
ISBN: 0-904046-06-0
Geographical area: Emphasis on UK/North Sea although coverage is worldwide.
Content: There are 115 entries arranged alphabetically, giving name, address, contact persons, subject interest, library facilities, information services, publications. The guide concludes with a bibliography and a subject index. Not available on-line.
Next edition: 1986; £4.00.

Guide to sources of information in the construction industry, fourth edition* — 228

Publisher: Construction Industry Research and Information Association
Address: 6 Storey's Gate, London SW1P 3AU, UK
Year: 1984
Price: £10.00
ISBN: 0-86017-217-1
Geographical area: United Kingdom.
Content: Alphabetical list of advisory services; list of organizations classified according to type and area; list of organizations alphabetically by name.

Heating, ventilating, refrigeration and air conditioning year book — 229

Editor: Donald Edwards
Publisher: Heating and Ventilating Contractors' Association
Address: 34 Palace Court, London W2 4JG, UK
Year: 1985
Pages: 420
Price: £21.00 (United Kingdom); £28.00 (elsewhere)
ISBN: 0-903783-02-9
ISSN: 0140-1947
Geographical area: United Kingdom and Europe.
Next edition: July 1986; £25.00 and £32.00.

HVAC redbook　　230

Publisher: Heating and Ventilating Publications (Developments) Limited
Address: 111 St James's Road, Croydon, Surrey CR9 2TH, UK
Year: 1984
Pages: 455
Price: £20.00
Geographical area: United Kingdom.
Content: Description of air-conditioning, heating, and ventilation products manufactured and distributed in the UK. Lists trade names, government departments, manufactures, and manufacturers.
Next edition: April 1986.

Irish construction directory　　231

Publisher: Institute for Industrial Research and Standards
Address: Ballymun Road, Dublin 9, Ireland
Year: 1980
Pages: 608
Price: IR£12.00
ISBN: 0-900450-51-7
Geographical area: Republic of Ireland.

Irish engineering directory 1985　　232

Publisher: Institute for Industrial Research and Standards
Address: Ballymun Road, Dublin 9, Ireland
Pages: 196
Price: IR£10.00
Geographical area: Republic of Ireland.

Professional services to the construction industry　　233

Publisher: Institute for Industrial Research and Standards
Address: Ballymun Road, Dublin 9, Ireland
Year: 1977
Pages: 204
Availability: OP
ISBN: 0-900450-56-8
Geographical area: Republic of Ireland.

Refrigeration and air conditioning year book　　234

Publisher: Maclaren Publishers Limited
Address: PO Box 109, Maclaren House, Scarbrook Road, Croydon, Surrey CR9 1QH, UK
Year: 1985
Pages: 266
Price: £16.00
Geographical area: United Kingdom.
Next edition: February 1986; £17.00.

Register of research on machine tools and related production engineering　　235

Publisher: Machine Tool Industry Research Association
Address: Hulley Road, Macclesfield, Cheshire SK10 2NE, UK
Year: 1976
Pages: x + 166
Availability: OP
Content: Information on 638 research projects; 99 British university and polytechnic research departments and industrial research organizations; project descriptions; effort, stage reached, source of support; published references; named contact and address. Arrangement and index by subject title.

United Kingdom government departments and other agencies concerned with the offshore industry　　236

Publisher: Construction Industry Research and Information Association
Address: 6 Storey's Gate, London SW1P 3AU, UK
Year: 1985
Price: £6 (to members); £25 (to non-members)
Series: UEG publications

Vendor profiles of suppliers to the offshore industry　　237

Publisher: Industrial Press
Address: Business Press International, Quadrant House, The Quadrant, Sutton, Surrey SM2 5AS, UK
Year: 1985
Series: Supplement to Petroleum Times
Geographical area: Europe.
Content: Profiles give company details such as activities, products, performance and plans, to assist purchasers of equipment and services for the offshore industry. Products covered range from abrasives, hydraulics, power generation to well-head equipment, platform modules, and production units.
Next edition: March 1986

Verkfraedingtal: aeviagrip islenzkra verkfraedinga og annarra felagsmanna verkfraedinafélags Islands　　238

[Dictionary of Icelandic engineers and members of the Association of Chartered Engineers in Iceland]
Editor: Jón E. Vestdal
Publisher: Verkfraeoingafélag Islands

Address: Sigtúni 7, 105 R, Postbox 645, 121 R, Reykjavik, Iceland
Year: 1981
Pages: xvi + 623

Who's who in West European automotive components* 239

Editor: F. Firth
Publisher: John Martin Publishing
Address: 24 Old Bond Street, London W1, UK
Year: 1985
Price: £60.00
Content: Entries for several thousand companies, giving products and services, location, inter-company relationships and distributors, key personnel, financial information, etc.

INDUSTRY AND MANUFACTURING

Adhesives handbook 240

Editor: J. Shields
Publisher: Butterworths Scientific Limited
Address: PO Box 63, Westbury House, Bury Street, Guildford, Surrey GU2 5BH, UK
Year: 1983
Pages: x + 350
Price: £40.00
ISBN: 0-408-01356-7
Geographical area: United Kingdom.
Content: The handbook contains about 450 entries, giving product, trade names, and chemical composition, also plant processing equipment and manufacturers, and testing services.
Not available on-line.
Next edition: 1988.

Adhesives Euro-guide, second edition* 241

Publisher: Industrial Aids Limited
Address: 14 Buckingham Palace Road, London SW1W 0QP, UK
Year: 1984
Price: £25.00
Geographical area: Western Europe.
Content: Lists over 200 adhesive companies in Belgium, France, West Germany, Netherlands, and United Kingdom, and includes information on adhesives, types used, and end use industries served.

Adresboek van de oostvlaamse industrie 242

[Directory of East Flemish industry]
Publisher: Regional Development Authority for East-Flanders (GOMOV)
Address: ICC - Floraliapaleis - bus 6, B - 9000 Gent, Belgium
Year: 1982
Pages: xxx | 540
Price: BF1000
Content: The directory consists of three volumes, each giving a different arrangement of entries: Volume One - companies listed alphabetically; Volume Two - companies listed by branch of activity; Volume Three - companies are listed geographically. A 1984 supplement is available free of charge.
Next edition: December 1985; BF1000.

Aerosol review 243

Editor: Neil Eisberg
Publisher: Morgan-Grampian (Process Press) Limited
Address: 30 Calderwood Street, Woolwich, London SE18 6QH, UK
Year: 1984
Pages: 134
ISBN: $16.00
Geographical area: United Kingdom.
Content: Listing of over 2000 products under 25 main headings, containing 80 individual product groups. Also included are details of products imported into the UK and exported, whilst a buyers' guide lists components, machinery and services worldwide. A review looks at the last 12 months in the industry and future trends.
Not available on-line.
Next edition: August 1985; £16.00.

Association of Bronze and Brass Founders buyers guide* 244

Publisher: Association of Bronze and Brass Founders
Address: 136 Hagley Road, Edgbaston, Birmingham B16 9PN, UK
Year: 1985
Pages: 15
Price: Free
Geographical area: United Kingdom.
Content: Details of members and production capacities.
Next edition: 1985 may

Association of Hydraulic Equipment Manufacturers: directory of members 245

Publisher: Association of Hydraulic Equipment Manufacturers
Address: 192-198 Vauxhall Bridge Road, London SW1V 1DX, UK

Year: 1985
Pages: xv + 85
Price: £10.00
Geographical area: United Kingdom.
Content: Details of member companies and association services and publications.

Automated manufacturing directory 1985   246

Editor: Richard Weston
Publisher: Morgan-Grampian Book Publishing Company Limited
Address: 30 Calderwood Street, Woolwich, London SE18 6QH, UK
Year: 1985
Pages: xvi + 314
Price: £25.00
ISBN: 0-86213-059-X
Geographical area: United Kingdom.
Content: Guide to computer-assisted manufacturing systems and services available from over 500 UK companies.
Not available on-line.
Next edition: December 1985; £25.00

Autotrade directory   247

Editor: Pat Brown
Publisher: Morgan-Grampian Book Publishing Company Limited
Address: 30 Calderwood Street, Woolwich, London SE18 6QH, UK
Year: 1984
Pages: viii + 248
Price: £15.00
ISBN: 0-86213-051-4
Geographical area: United Kingdom.
Content: The full contents includes a complete list of all companies appearing in the buyers' guide, an index to trade names used by companies in the motor-trade, a comprehensive buyers' guide containing over 1350 product headings relating to the motor-trade, an alphabetical list of factors together with a geographical location guide under UK town names, a list of cash-and-carry establishments in geographical order, a guide to car manufacturers and concessionaires, a list of useful motor trade addresses, and a diary of major motor trade events.
Not available on-line.

Belgian Exports   248

Publisher: ABC Belge pour le Commerce et l'Industrie BV
Address: Passage International 6 B 10, Bruxelles B-1000, Belgium
Year: 1985
Pages: 432

Geographical area: Belgium.
Content: Commercial data on 4200 Belgian exporting companies, sub-divided in a product guide and a company guide.
Next edition: February 1986.

Belgisch ABC voor handel en industrie; ABC Belge pour le commerce et l'industrie   249

[Belgian ABC for commerce and industry]
Publisher: ABC Belge pour le Commerce et l'Industrie BV
Address: Passage International 6 B 10, Bruxelles B-1000, Belgium
Year: 1984
Pages: 1700
Price: BF2800
Geographical area: Belgium.
Content: Commercial data on 16 500 Belgian producing and importing companies, subdivided in a product guide and a company guide.
Next edition: October 1985; BF3000.

Binsted's directory of food trade marks and brand names   250

Editor: Adrian M. Binsted
Publisher: Food Trade Press Limited
Address: 29 High Street, Green Street Green, Orpington, Kent BR6 6LS, UK
Year: 1983
Pages: vi + 187
Price: $26.50
ISBN: 0-90037-931-6
ISSN: 0067-8651
Geographical area: United Kingdom.
Content: Over 3750 new trade marks and brand names; over 1400 new company addresses; listing including manufactured food products, ingredient materials, catering supplies, and imported foods.
Next edition: September 1985; £36.00.

British Aerosol Manufacturers' Association: 24th annual report 1984   251

Publisher: British Aerosol Manufacturers' Association Limited
Address: Alembic House, 93 Albert Embankment, London SE1 7TU, UK
Pages: 47
Price: Free
Geographical area: United Kingdom.
Content: Includes list of member companies, and representatives on association committees.
Next edition: January 1986; free.

British machine tools and equipment 252

Publisher: Machine Tool Trades Association
Address: 62 Bayswater Road, London W2 6JU, UK
Year: 1984
Pages: 125
Price: £10.00
Content: Details of products manufactured by UK machine tool (and associated equipment) companies.
Next edition: June 1984; £10.00.

Buyers' guide to north east industry 253

Publisher: North of England Development Council
Address: Bank House, Carliol Square, Newcastle upon Tyne NE1 6XE, UK
Year: 1980-81
Pages: 204
Price: Free
Geographical area: Northumberland, Tyne and Wear, Durham, Cleveland.
Content: First half by Standard Industrial Classification, second half by alphabetical classification. Name, address and telephone number and product of each firm. Not available on-line.

Buyers' guide to pumps 1985 254

Publisher: British Pump Manufacturers' Association
Address: 3 Pannells Court, Chertsey Street, Guildford, Surrey GU1 4EU, UK
Year: 1985
Price: $5.00
Geographical area: United Kingdom.

Chemische industrie der Schweiz und ihre nebenprodukte 255

[Swiss chemical industry and related products]
Editor: C. Laemmel
Publisher: Verlag für Wirtschaftsliteratur GmbH
Address: Birmendorferstrasse 421, 8055 Zürich, Switzerland
Year: Vol 1, 1983; Vol 2, 1985
Pages: Vol 1, 306; vol 2, 444
Availability: Vol 1 OP
Price: SF70.00 (Vol 2)
ISBN: 3-909214-02 (Vol 1)
Content: Information about Swiss firms according to table of contents; volume one contains index of companies; volume two contains index of products.
Next edition: Vol I 1987, Vol II 1989; SF70.00 each.

Computer users' year book 1985 256

Editor: Alison Murdoch
Publisher: VNU Business Publications bv
Address: VNU House, 32-34 Broadwick Street, London W1A 2HG, UK
Year: 1985
Pages: 1700 (in 2 vols)
Price: £57.25
Geographical area: United Kingdom.
Content: Volume one covers technical information, hardware and suppliers, including: a guide to computers, peripherals and communications; computer peripherals and data preparation equipment; training companies and colleges; names and addresses of equipment and service suppliers.
Volume two is the services volume, including: computer service bureaux; data preparation services; computer consultants and systems houses; leasing and broking; recruitment; salary survey, and the directory of computer installations.
Not yet available on-line.
Next edition: January 1986.

Computing Services Association: directory of members and services 257

Publisher: Computing Services Association
Address: Hanover House, 73/74 High Holborn, London WC1V 6LE, UK
Year: 1984
Pages: 30
Price: Free
Geographical area: United Kingdom.
Content: Information included on 200 companies' products and services.
Next edition: September 1985; free.

Computing Services Association: members survey 258

Publisher: Computing Services Association
Address: Hanover House, 73/74 High Holborn, London WC1V 6LE, UK
Year: 1984
Pages: 155
Price: £95.00
Geographical area: United Kingdom.
Content: Statistical survey of member companies.

Computing Services Association: official reference book 1985 259

Publisher: Computing Services Association
Address: Hanover House, 73/74 High Holborn, London WC1V 6LE, UK
Pages: x + 350
Price: £25.00
ISSN: 0266-7916
Geographical area: United Kingdom.
Content: The publication includes: list of members, associates, and affiliates; members' services index.
Next edition: September 1986.

Construction plant and equipment annual 260

Editor: Christine Casley
Publisher: Morgan-Grampian Book Publishing Company Limited
Address: 30 Calderwood Street, Woolwich, London SE18 6QH, UK
Year: 1985
Pages: xii + 564
Price: $30.00
ISBN: 0-86213-060-3
Geographical area: United Kingdom.
Content: This guide to earthmoving and construction equipment provides detailed specifications for over 600 models. Data covers dozers, graders, excavators, haulers, cranes, compactors, site, concrete and access equipment. A buyers' guide of components and accessories is also included. Details more than 320 companies.
Not available on-line.

Deutsche branchen-fernsprechbuch 261

[German trade register of firms classified according to trade]
Publisher: Deutscher Adressbuch-Verlag GmbH
Address: Holzhofallee 38, POB 11 03 20, D-6100 Darmstadt, German FR
Year: 1985
Pages: 1700
Price: DM210.00
ISBN: 3-87148-114-9
Content: Addresses of 300 000 firms from industry and trade, transport and organizations. Detailed trade classification outline and German trade and goods register with approximately 25 000 catchwords. Directions for use in English and French. Index of goods and services.
Next edition: December 1985; DM210.00.

Deutsche firmen-alphabet: industrie, handel, verkehr, organisationen 262

[German trade register of firms from A-Z: industry, trade, commerce, and associations]
Publisher: Deutscher Adressbuch-Verlag GmbH
Address: Holzhofallee 38, POB 11 03 20, D-6100 Darmstadt, German FR
Year: 1985
Pages: 1600
Price: DM95.00
ISBN: 3-87148-116-5
Content: There are 300 000 entries, alphabetically classified according to firm's names, with addresses and telephone numbers. 9000 trade marks are also listed.
Next edition: January 1986; DM101.00.

Deutsches bundes-adressbuch 263

[German trade register of firms according to places]
Publisher: Deutscher Adressbuch-Verlag GmbH
Address: Holzhofallee 38, POB 11 03 20, D-6100 Darmstadt, German FR
Year: 1984
Pages: 1980
Price: DM100.00
ISBN: 3-87148-113-0
Content: Lists 300 000 addresses of firms from industry, trade and transport, central governments and authorities, municipal and local governments. Towns are listed alphabetically with postal codes, and firms within the towns are also listed alphabetically. There is a trade section index.
Next edition: July 1985; DM100.00.

Directory of computer training/ directory of management training 264

Editor: Colin Steed
Publisher: Badgemore Park Enterprises Limited
Address: 71A High Street, Maidenhead, Berkshire SL6 1JX, UK
Year: 1985
Pages: 1280 (in 2 vols)
Price: £55.00 (for the set)
ISBN: 0-947586-02-4
Geographical area: United Kingdom.
Content: Information on more than 3500 courses from over 100 training organizations.
Next edition: December 1985; £65.00.

Directory of European Community trade and professional associations 265

Editor: G. Francis Seingry
Publisher: Editions Delta
Address: 92-94 Square E. Plasky, B-1040 Bruxelles, Belgium
Year: 1985
Pages: 500
Price: BFr2500
ISBN: 2-8029-6051-X
Series: EEC directories
Content: Over 500 European organizations from the sectors of industry, crafts, small and medium-sized enterprises, trade unions, consumers. Entries give name, address, telephone and telex, names of officers, list of national member organizations (totally 5000). Indexes in English, French and German.

EUROPE INCLUDING USSR

Directory of manufacturers of vacuum plant, components and associated equipment in the UK, 1982 266

Editor: J.S. Colligon
Publisher: Pergamon Press Limited
Address: Headington Hill Hall, Oxford OX3 0BW, UK
Pages: 56
Price: £7.95
ISBN: 0-08-029323-9
Geographical area: United Kingdom.

Dutch companies with their UK agents, representatives 267

Publisher: Netherlands British Chamber of Commerce
Address: 307 High Holborn, London WC1V 7LS, UK
Year: 1984
Pages: 124
Price: £15.00
Geographical area: United Kingdom.
Content: The directory gives details on over 700 British companies which are representing or importing Dutch products into the UK. Full particulars are given on the names and addresses of the Dutch suppliers and their UK contacts, together with the products carried.
Next edition: 1986; £15.00.

Einkaufs der deutschen industrie 268
[Buyers' guide to German industry]

Publisher: Deutscher Adressbuch-Verlag GmbH
Address: Holzhofallee 38, POB 11 03 20, D-6100 Darmstadt, German FR
Year: 1984
Pages: 1400
Price: DM100.00
ISBN: 3-87148-112-2
Content: Lists 220 000 sources of supply for 75 000 products, and 9000 trademarks with explanations. Product headings and trademarks are in alphabetical order, with name, address, and telephone of manufacturer under each. There are indexes in English and French.
Next edition: June 1985; DM100.00

Engineer buyers guide 269

Editor: James Robertson
Publisher: Morgan-Grampian Book Publishing Company Limited
Address: 30 Calderwood Street, Woolwich, London SE18 6QH, UK
Year: 1985
Pages: viii + 700
Price: £18.00
ISBN: 0-86213-062-X
Geographical area: United Kingdom.
Content: Published for all sections of the engineering industry, this guide lists under more than 3500 headings, the products and services provided by 2900 leading manufacturers and distributors throughout the United Kingdom. The guide is divided into sections giving addresses, UK agents for foreign firms, trade names, buyers' guide, and associations.
Not available on-line.
Next edition: January 1986; £20.00

European coil coating directory 270

Publisher: Fuel and Metallurgical Journals Limited
Address: Queensway House, 2 Queensway, Redhill, Surrey RH1 1QS, UK
Price: £24.95
Content: Coil coating A-Z of coaters and suppliers giving addresses, telephone and telex numbers, cable addresses, year of establishment, number of employees, parent companies and subsidiaries, works, sales outlets, product ranges, capacities, widths, thicknesses, line speeds, coating systems, brand names, and key contact personnel; bookmark glossary. The information is given in English, French and German.

European confectionery and bakery products: manufacturers of bakery products, cocoa, chocolate, and their products, confectionery titbits, seasoned crackers, etc. 271

Publisher: Stamex BV
Address: PO Box 400, 3770 AK Barneveld, Netherlands
Year: 1977
Availability: OP

European glass directory and buyers' guide 272

Editor: A. Laverick
Publisher: Fuel and Metallurgical Journals Limited
Address: Queensway House, 2 Queensway, Redhill, Surrey RH1 1QS, UK
Year: 1985
Pages: 200
Price: £25.50
ISBN: 0-86108-170-6
Geographical area: Europe.
Content: Information on the manufacturers, processors and users of glass, and the companies who supply them with plant, equipment and services.
One master index provides addresses of all companies, in strict alphabetical sequence, showing category and nationality. This section also lists telephone and telex numbers, company directors, sales contacts and other essential commercial information.

Comprehensive classified entries are arranged in three separate product guides: products of the glass industry covering glass containers, flat glass, domestic glassware, illuminating glass, industrial and technical glassware, laboratory, medical, surgical and scientific glassware (which are in turn sub-divided into specific product types); plant, equipment and services with more than 200 product headings; raw materials covering in excess of 80 constituents. There is a numerical index system included for easy reference, and also a section giving a country-by-country breakdown of firms listed in the master index. Brand names of products are alphabetically listed, together with a brief description and the names of users. List of European associations giving names, addresses and aims of glass industrial associations, institutes and societies, educational institutions and schools, research laboratories, etc.
Next edition: December 1985; £28.00

Europe's 15 000 largest companies, tenth edition 1985   273

Publisher: ELC International
Address: 227-229 Chiswick High Road, London W4 2DW, UK
Price: £90.00
Series: Largest companies
ISBN: 09-4805-800-5
Content: Financial and statistical business information for companies in sixteen European countries. Four major sections identify the five hundred largest companies, largest industrials, largest trading companies, and largest service companies.

Fab guide: a buyer's directory for welding fabrication engineers   274

Publisher: Fuel and Metallurgical Journals Limited
Address: Queensway House, 2 Queensway, Redhill, Surrey RH1 1QS, UK
Year: 1985
Pages: 60
Price: £12.00
Geographical area: United Kingdom.
Next edition: January 1986.

Filters and filtration handbook, first edition   275

Editor: R.H. Warring
Publisher: Trade and Technical Press Limited
Address: Crown House, London Road, Morden, Surrey SM4 5EW, UK
Year: 1981
Pages: 520
Price: £50.00 (UK); £57.00 (elsewhere)
ISBN: 0-85461-086-3

Geographical area: United Kingdom.
Content: Types and applications with classified index of equipment and components, together with lists of manufacturers and suppliers.

Food manufacture: ingredient and machinery survey   276

Editor: Hugh Darrington
Publisher: Morgan-Grampian (Process Press) Limited
Address: 30 Calderwood Street, Woolwich, London SE18 6QH, UK
Year: 1985
Pages: 184
Price: £16.00
Geographical area: United Kingdom.
Content: List of ingredient suppliers, their products and applications; a guide to packaging materials; a list of trade and professional organizations; and finally a list of food and drink manufacturers.
Not available on-line.
Next edition: June 1986; £18.00.

Food trade research register: volume 1 - food technology, tenth edition*   277

Publisher: Institute of Grocery Distribution
Address: Grange Lane, Letchmore Heath, Watford WD2 8, UK
Year: 1984
Price: £30.00
Geographical area: United Kingdom.
Content: Information on fourteen research centres in four sections: food manufacturing; food microbiology; food biochemistry; food processing.

Food trade research register: volume 2 - food business research, tenth edition*   278

Publisher: Institute of Grocery Distribution
Address: Grange Lane, Letchmore Heath, Watford WD2 8, UK
Year: 1984
Price: £30.00
Geographical area: United Kingdom.
Content: Information on 203 projects at 64 centres.

Foundry directory and register of forges   279

Publisher: Metal Bulletin Books Limited
Address: Park House, Park Terrace, Worcester Park, Surrey KT4 7HY, UK
Year: 1983
Pages: 240
Price: £14.00
ISBN: 0-900542-90-X

Geographical area: United Kingdom.
Content: Information covers the UK foundry industry: regional list of iron founders (capacities and brands); buyers' guide (by types of castings produced); steel and non-ferrous founders (address capacity, brands); classified list of companies providing equipment supplies and services for the foundry directory. There are approximately 1000 entries.
Next edition: November 1985.

Gas directory 1985 and who's who 280

Editor: Dorothy Steadman
Publisher: Benn Information Services Limited
Address: PO Box 20, Sovereign Way, Tonbridge, Kent TN9 1RQ, UK
Pages: 224
Price: £34.00 (United Kingdom); £38.00 (elsewhere)
ISBN: 0-86382-021-2
Geographical area: United Kingdom and Europe.
Content: Details of organization and personnel in British Gas Corporation; who's who in the gas industry, 2500 entries; British gas statistics; suppliers to the gas industry, 1250 entries; buyers' guide to machinery, equipment, materials, and services; entries for 300 overseas gas supply companies; list of trade names; list of approved domestic appliances; buyers' guide to liquid petroleum gas.
Next edition: February 1986; £36.00 (United Kingdom); £40.00 (elsewhere).

Giesserei- und metallindustrie der Schweiz 281
[Foundry and metal industry of Switzerland]
Editor: C. Laemmel
Publisher: Verlag für Wirtschaftsliteratur GmbH
Address: Birmensdorferstrasse 421, 8055 Zürich, Switzerland
Year: 1985
Pages: 224
Price: SFr50.00
ISBN: 3-909-214
Content: Information about 600 Swiss firms according to table of contents.
Next edition: 1987; SF50.00.

Handbok för nordisk träindustri 282
[Handbook of the northern wood industries, thirteenth edition]
Editor: Rune Lindqvist
Publisher: AB Svensk Trävarutidning
Address: Observatoriegatan 17, 113 29 Stockholm, Sweden
Year: 1983-84
Pages: 960
Price: SKr225.00
Geographical area: Mainly Sweden, Norway, Denmark, Finland.
Content: Details of: trade associations and organizations; sawmills and manufacturers of wooden houses, plywood, chipboard, etc; manufacturers of woodpulp, paper and board; agents, exporters, sales organizations etc. Entries include names, addresses, telephone and telex, personnel, activities, bankers, etc. There are indexes to shipping marks (alphabetically arranged), shipping districts with lists of sawmills in each, and names of companies, firms and personnel.
Next edition: June 1986; Skr250.00

Handbook of hose, pipes, couplings and fittings, first edition 283

Editor: M.J. Barber
Publisher: Trade and Technical Press Limited
Address: Crown House, London Road, Morden, Surrey SM4 5EW, UK
Year: 1985
Pages: 450
Price: £45.00 (UK); £51.00 (elsewhere)
ISBN: 0-85461-091-X
Geographical area: United Kingdom.
Content: Types and applications together with glossary, trade names index, classified index, alphabetical listing of manufacturers, buyers' guide section, editorial, and advertisers indexes.

Handbook of industrial fasteners, third edition 284

Publisher: Trade and Technical Press Limited
Address: Crown House, London Road, Morden, Surrey SM4 5EW, UK
Year: 1985
Pages: 700
Price: £52.00 (UK); £60.00 (elsewhere)
ISBN: 0-85461-097-9
Geographical area: United Kingdom.
Content: Types and applications of various fastenings, together with a buyers' guide section, editorial and advertisers indexes.

Handbook of instruments and instrumentation, first edition 285

Publisher: Trade and Technical Press Limited
Address: Crown House, London Road, Morden, Surrey SM4 5EW, UK
Year: 1976
Pages: 650
Price: £45.00 (UK); £51.00 (elsewhere)
ISBN: 0-85461-064-2
Geographical area: United Kingdom.
Content: Measurement techniques, specialized fields, data section, together with buyers' guide section, editorial and advertisers indexes.

Handbook of mechanical power drives, third edition 286

Publisher: Trade and Technical Press Limited
Address: Crown House, London Road, Morden, Surrey SM4 5EW, UK
Year: 1981
Pages: 700
Price: £52.00 (UK); £60.00 (elsewhere)
ISBN: 0-85461-085-5
Geographical area: United Kingdom.
Content: Systems, fundamentals, equipment, related/assorted mechanisms. Buyers' guide section, editorial and advertisers indexes.

Handbook of noise and vibration control 287

Editor: R.H. Warring
Publisher: Trade and Technical Press Limited
Address: Crown House, London Road, Morden, Surrey SM4 5EW, UK
Year: 1983
Pages: 600
Price: £52.00 (UK); £60.00 (elsewhere)
ISBN: 0-85461-093-6
Geographical area: United Kingdom.
Content: Measurement instruments and techniques, prevention and insulation, legislation. Buyers' guide section, editorial and advertisers indexes.

Handbuch der Schweizerischen textil-, bekleidungs- und lederwirtschaft 288

[Swiss textile, clothing and leather directory]
Editor: C. Laemmel
Publisher: Verlag für Wirtschaftsliteratur GmbH
Address: Birmendorferstrasse 421, 8055 Zürich, Switzerland
Year: 1982
Pages: 220
Availability: OP
Content: Information about 1600 Swiss firms according to table of contents.
Next edition: 1986; SF60.00.

Holland exports/commercial gardening and farming 289

Publisher: ABC voor Handel en Industrie CV
Address: PO Box 190, 2000 AD Haarlem, Netherlands
Year: 1985
Pages: 236
Price: Free of charge (controlled circulation)
ISBN: 90-70729-14-8
Series: Holland exports
Content: Information, in English, about 1100 Dutch manufacturers and merchants/exporters of agriculture, horticulture, floriculture, livestock breeding and everything in this connection. There is an alphabetical list of products in English, French, German, and Spanish; a product index; a company index (alphabetically by company name, each entry giving: name, address, telephone, telex, bankers, export manager(s), number of employees and product range; an alphabetical list of companies.
Not available on-line.
See separate entries for: Holland Exports/Industrial Products; Holland Exports/Consumer Goods (Non Food); Holland Exports/Consumer Goods (Food)
Next edition: February 1986; free of charge (controlled circulation).

Holland exports/consumer goods (food) 290

Publisher: ABC voor Handel en Industrie CV
Address: PO Box 190, 2000 AD Haarlem, Netherlands
Year: 1985
Pages: 184
Price: Free of charge (controlled circulation)
ISBN: 90-70729-17-2
Series: Holland exports
Content: Information, in English, about 950 Dutch manufacturers and merchants/exporters of foodstuffs and stimulants and related products. There is an alphabetical list of products in English, French, German, and Spanish; a product index; a company index (alphabetically by company name, each entry giving name, address, telephone, telex, bankers, export manager(s), number of employees and product range; an alphabetical list of companies.
Not available on-line.
See separate entries for: Holland Exports/Industrial Products; Holland Exports/Consumer Goods (Non Food); Holland Exports/Commercial Gardening and Farming.
Next edition: February 1986; free of charge (controlled circulation).

Holland exports/consumer goods (non food) 291

Publisher: ABC voor Handel en Industrie CV
Address: PO Box 190, 2000 AD Haarlem, Netherlands
Year: 1985
Pages: 136
Availability: OP
ISBN: 90-70729-16-4
Series: Holland exports
Content: Information is given, in English, about 830 Dutch manufacturers and merchants/exporters of durable and non-durable consumer goods and related products. There is an alphabetical list of products in English, French, German, and Spanish; a product index; a com-

pany index (alphabetically by company name. Each entry giving name, address, telephone, telex, bankers, export manager(s), number of employees and product range); an alphabetical list of companies.
Not available on-line.
See separate entries for: Holland Exports/Industrial Products; Holland Exports/Consumer Goods (Food); Holland Exports/Commercial Gardening and Farming.
Next edition: February 1986; free of charge (controlled circulation).

Holland exports/industrial products 292

Publisher: ABC voor Handel en Industrie CV
Address: PO Box 190, 2000 AD Haarlem, Netherlands
Year: 1985
Pages: 610
Availability: OP
ISBN: 90-70729-15-6
Series: Holland exports
Content: Information is given, in English, about 2070 Dutch manufacturers and merchants/exporters of finished products, semi-finished articles, and raw materials. There is an alphabetical list of products in English, French, German, and Spanish; a product index; a company index (alphabetically by company name), each entry giving name, address, telephone, telex, bankers, export manager(s), number of employees and product range); and an alphabetical list of companies.
Not available on-line.
See separate entries for: Holland Exports/Consumer Goods (non Food); Holland Exports/Consumer Goods (Food); Holland Exports/Commercial Gardening and Farming.
Next edition: February 1986; free of charge (controlled circulation).

Hydraulic handbook, eighth edition 293

Editor: R.H. Warring
Publisher: Trade and Technical Press Limited
Address: Crown House, London Road, Morden, Surrey SM4 5EW, UK
Year: 1983
Pages: 600
Price: $52.00 (UK); £60.00 (elsewhere)
ISBN: 0-85461-094-4
Geographical area: United Kingdom.
Content: Principles, components, systems, instruments, applications, and surveys of hydraulic equipment, together with a buyers' guide section, editorial and advertisers indexes.

Iceland - yearbook of trade and industry* 294

Publisher: Iceland Review
Address: Hofdabakki 9, 121 Reykjavik, Posthof 93, Iceland
Year: 1985
Price: $12.00
Content: Details of fishing industry, power-intensive industries, traditional exports, and industrial products
Next edition: 1986.

Index of manufacturers of artificial sports surfaces 295

Editor: A.L. Cox, B.J. Wain
Publisher: Rapra Technology Limited
Address: Shawbury, Shrewsbury, Shropshire SY4 4NR, UK
Year: 1984
Pages: v + 82
Price: £10.00
Geographical area: United Kingdom.
Content: Around 300 surfaces from 64 manufacturers listed under sports for which each surface is recommended. Names and addresses of manufacturers and distributors and an index of trade names are also included.

Industrial development guide 296

Editor: Cambridge Information and Research Services
Publisher: Longman Group Limited
Address: Sixth Floor, Westgate House, The High, Harlow, Essex CM20 1NE, UK
Year: 1985
Pages: xii + 300
Price: £28.00
ISBN: 0-582-90335-1
Geographical area: United Kingdom, Ireland, Isle of Man, Channel Islands.
Content: The guide brings together over 300 pages of information on national, regional, and local initiatives and programmes concerning industrial development opportunities. Key features of the 1985 edition are: detailed surveys and profiles of the counties and regions in the United Kingdom; a review of national industrial development prospects; and a comprehensive directory of people and organizations dealing with industrial development.
The national review section provides statistics and a glossary of development and related agencies which explains the background of each agency and their areas of activity. The county and regional surveys section provides names and addresses of Industrial Development Officers, CoSIRA organizers, the regional office of the Small Firms Service, skill centres and enterprise groups.
Also included is a comprehensive alphabetical directory giving contacts and addresses for national agencies, enterprise groups, local authority Industrial Development Officers and departments, and property agents.
Next edition: July 1986; £30.00.

Industrie-Kompass Deutschland, thirteenth edition   297

[Kompass to West Germany's industry and commerce]
Publisher: Kompass Deutschland Verlags- und Vertriebsges mbH
Address: PO Box 964, Wilhelmstrasse 1, D-7800 Freiburg, German FR
Year: 1985
Pages: 3300 (in 2 vols)
Price: DM385.00
ISSN: 0173-721X
Series: Kompass registers
Content: Volume One contains: alphabetical index of products and services (in German, English, and French); supply sources section by 20 main industrial groups and 420 subdivisions - 30 000 entries; trade names section. Volume Two contains: alphabetical index by company names; company information section (37 000 entries) giving postal address, telephone, telex, banks, names of directors and senior executives, share capital, annual turnover, number of employees, year of foundation, range of activities.
Available on-line.
Next edition: June 1986; DM415.00.

Irish Industrial Laboratory Directory   298

Publisher: Institute for Industrial Research and Standards
Address: Ballymun Road, Dublin 9, Ireland
Year: 1983
Pages: 112
Price: IR£5.00
ISBN: 0-900450-76-2
Geographical area: Republic of Ireland.

Kelly's manufacturers and merchants directory   299

Publisher: Kelly's Directories
Address: Windsor Court, East Grinstead House, East Grinstead, Sussex RH19 1XB, UK
Year: 1985
Pages: 2000
Geographical area: United Kingdom.
Next edition: October 1985; £60.00.

Kelly's UK Exports   300

Publisher: Kelly's Directories
Address: Windsor Court, East Grinstead House, East Grinstead, Sussex RH19 1XB, UK
Year: 1985
Pages: 2000
Price: £30.00
Geographical area: United Kingdom.
Next edition: November 1985; £35.00.

Kingdom of Denmark trade directory*   301

Publisher: Kon Danmarks Handels-Kalendar
Address: 8 Nygade 5, DK-1164 Kobenhavn K, Denmark
Year: 1985
Price: $60.00
Content: List of more than 10 000 companies; index of industry in English, French, German, and Spanish.

Kompass: Belgium   302

Publisher: Kompass Belgium SA
Address: Avenue Molière 256, 1060 Bruxelles, Belgium
Year: 1985-86
Pages: 3000 (in 2 vols)
Price: BF5.700
Series: Kompass registers
Geographical area: Belgium and Luxembourg.
Content: The guide contains 23 000 entries, giving records of suppliers and buyers. Volume 1 gives alphabetical listing under geographical location by industry or business of companies' address, telephone, telex, name of director, activities, bank, etc. Volume 2 gives industrial and commercial activities within country classified by type of activity, brands, representatives, etc.
The information is available on-line via Ekol European Kompass on-line.
Next edition: May 1986; BF63.00.

Kompass: Denmark*   303

Publisher: A/S Forlaget Kompass
Address: Hovedgade 4, Lyngby DK-2800, Denmark
Year: 1984
Price: £70.00
Series: Kompass registers
Content: Individual products and company services, classified according to Standard Industrial Classification.
Next edition: October 1985; £80.00.

Kompass: France*   304

Publisher: Kompass France
Address: 91 rue du Fauberge St Honore, 75008 Paris, France
Year: 1984
Price: £95.00
Series: Kompass registers
Content: Names and addresses of individual products and services of companies, classified according to Standard Industrial Classification.
Next edition: December 1985; £105.00.

Kompass: Holland   305

Publisher: Kompass Nederland BV

Address: Hogehilweg 15, Geb. 'California', 1101 CB Amsterdam ZO, Netherlands
Year: 1985
Pages: 3000
Price: Dfl225.00
Series: Kompass registers
Content: The publication covers about 24 500 industrial and service companies with names, addresses, names of directors and staff personnel, type of activities, arranged alphabetically by city and company names. Further information includes: names of the companies with the numbers of branches in which they are active, listed alphabetically; alphabetical list of products and services; classification of the products in which you find which company is producing or selling a product, if that product is exported and/or imported; alphabetical list of foreign companies with their Dutch representative; list of trademarks with the name and address of the Dutch company.
Next edition: August 1985; Dfl225.00.

Kompass: Italy* 306

Publisher: Etas Kompass
Address: Via Rivoltana 95, 20090 Limito, Milano, Italy
Year: 1985
Price: £95.00
Series: Kompass registers
Content: Names and addresses of individual products and company services, classified according to Standard Industrial Classification.
Next edition: March 1986.

Kompass: Norway* 307

Publisher: Kompass Norge A/S
Address: Kompass Huset, Lojkeveien 87, Stavanger N4000, Norway
Year: 1985
Price: £75.00
Series: Kompass registers
Content: Names and addresses on individual products and company services, classified according to Standard Industrial Classification.
Next edition: May 1986.

Kompass: Spain* 308

Publisher: Kompass España SA
Address: Avenida General Peron 26, Madrid 20, Spain
Year: 1984
Price: £70.00
Series: Kompass registers
Content: Names and addresses for individual products and company services, classified according to Standard Industrial Classification.
Next edition: October 1985; £75.00.

Kompass: Sweden* 309

Publisher: Kompass Sverige AB
Address: PO Box 3303, 510366 Stockholm 3, Sweden
Year: 1985
Price: £75.00
Series: Kompass registers
Content: Names and addresses for individual products and company services, classified according to Standard Industrial Classification.

Kompass: Switzerland* 310

Publisher: Kompass Schweiz Verlag
Address: AG IN Grosswiesen 14, CH-8044 Zürich-Gockhausen, Switzerland
Year: 1985
Price: £75.00
Series: Kompass registers
Content: Names and addresses for individual products and company services, classified according to Standard Industrial Classification.

Kompass: United Kingdom 311

Editor: H.M. Thomson
Publisher: Kompass Publishers Limited
Address: Windsor Court, East Grinstead House, East Grinstead, West Sussex RH19 1XD, UK
Year: 1984
Pages: xxxii + 5000 (in 2 vols)
Price: £90
ISBN: 0-86268-052-2
Series: Kompass registers
Geographical area: United Kingdom.
Content: There are 30 000 entries arranged by product or service details in Volume I and geographically in Volume 2. Content includes information on business activities, location, number of employees, financial status.
Industrial supplements (Buyers' Guides) are produced as follows: textiles and clothing industry; plastics and chemical products; electronic and electrical products; scientific and industrial instruments; industrial machinery and equipment; business services.
The information is available on-line.
Next edition: May 1985; £98.00.

Kunststoff-industrie der Schweiz 312

[Swiss plastics industry]
Editor: C. Laemmel
Publisher: Verlag für Wirtschaftsliteratur GmbH
Address: Birmendorferstrasse 421, 8055 Zürich, Switzerland
Year: Vol I, 1984; vol II, 1981
Pages: Vol I, 332; vol II, 532
Availability: OP

Content: Information about Swiss firms according to table of contents; volume one contains index of firms; volume two contains index of products.
Next edition: Vol I 1988, vol II 1986; SF65.00 each.

Laboratory equipment directory 1984   313

Publisher: Morgan-Grampian Book Publishing Company Limited
Address: 30 Calderwood Street, Woolwich, London SE18 6QH, UK
Year: 1985
Pages: vi + 272
Price: £18.00
ISBN: 0-86213-064-6
Geographical area: United Kingdom.
Content: Listing of more than 3000 products and services provided by 1500 UK companies. Entries give company name, address, telephone and telex, with key contact names. Buyers' guide listing individual products alphabetically followed by a list of suppliers. Trade names section with brief product description and name of manufacturer; overseas companies list; trade associations list; diary of forthcoming events.
Next edition: April 1986; £20.00.

Lebensmittel und- getranke-industrie   314
der Schweiz

[Swiss beverage and food industry]
Editor: C. Laemmel
Publisher: Verlag für Wirtschaftsliteratur GmbH
Address: Birmendorferstrasse 421, 8055 Zürich, Switzerland
Year: 1983
Pages: 352
Availability: OP
ISBN: 3-909214-01
Content: Information about Swiss firms according to table of contents.
Next edition: 1987; SF65.00.

London directory of industry and   315
commerce 1985*

Publisher: Bowker Publishing Company
Address: Erasmus House, 58-62 High Street, Epping, Essex CM16 4BU, UK
Year: 1984
Pages: 500
Price: £18.00
ISBN: 0-86259-064-7
Content: Information is organized in sections as follows: introductory section detailing the facilities, services and sources of information and assistance available to companies in the London area and those moving to the city; alphabetical listing of industrial, commercial and service companies within the area; classified section, alphabetically listing companies by products and services.
Next edition: 1985.

Machinery buyers' guide   316

Editor: Fred Browne
Publisher: Findlay Publications Limited
Address: Franks Hall, Horton Kirby, Dartford, Kent DA4 9LL, UK
Year: 1985
Pages: xxxii + 1440
Price: £17.00
Geographical area: United Kingdom.
Content: Lists of trade names and agencies, and products and services, arranged alphabetically, followed by list of addresses and telephone numbers. Also includes sections on trade fairs, professional and scientific institutions, research organizations, chambers of commerce, machine tool makers associations, and trade associations.
Next edition: March 1986.

Major companies in the Netherlands*   317

Publisher: Netherlands - British Chamber of Commerce
Address: 307 High Holborn, London WC1V 7LS, UK
Year: 1985
Price: £10.00
Content: Details of 300 companies listed by turnover and alphabetically, giving addresses, employees, telephone, telex, and contact person.

Major companies of Europe, fifth   318
edition 1985*

Editor: R.M. Whiteside
Publisher: Graham and Trotman Limited
Address: 66 Wilton Road, London SW1V 1DE, UK
Year: 1985
Pages: 1300 (2 vols)
Price: vol 1: £120.00, $192.00 (hard cover); £110.00, $176.00 (paperback); vol 2: £65.00, $104.00 (hard cover); £55.00, $88.00 (paperback)
ISBN: vol 1: 0-86010-602-0 (hard cover); 0-86010-601-2 (paperback); vol 2: 0-86010-604-7 (hard cover); 0-86010-603-9 (paperback)
ISSN: vol 1: 0266-934X; vol 2: 0266-9358
Series: Major companies
Content: This two-volume directory provides directory details of Europe's major companies, including company name, address, telephone, telex and cable numbers; names of directors and senior executives by job title; business activities; brand names; agencies; branches and subsidiaries; bankers; financial details; date of establishment.

A quarterly updating supplement service for both volumes is now available. The directory is also available on microfiche.
Next edition: February 1986.

Major companies of Europe 1979-85 *   319

Publisher: Graham and Trotman Limited
Address: 66 Wilton Road, London SW1V 1DE, UK
Pages: 4950; 8 vols on microfiche
Price: £365.00; $584.00
Series: Major companies
Content: This is a compendium of the volumes of the reference work Major Companies of Europe published between 1979 and 1985. Records for all these volumes are now available on microfiche.

Major companies of Scandinavia 1985 *   320

Editor: R.M. Whiteside
Publisher: Graham and Trotman Limited
Address: 66 Wilton Road, London SW1V 1DE, UK
Year: 1985
Pages: 280
Price: £45.00; $72.00
ISBN: 0-86010-635-7
Series: Major companies
Content: Details of 1200 companies including address, telephone, telex, key personnel, activities, branches, subsidiaries, bankers, financial details.

Microcomputer users year book   321

Editor: Chris Long
Publisher: VNU Business Publications bv
Address: VNU House, 32-34 Broadwick Street, London W1 2HG, UK
Year: 1985
Pages: 430
Price: £35.00
ISBN: 0-86271-048-0
Geographical area: United Kingdom.
Content: Sections, in tabular and descriptive form, on microcomputer systems; operating systems and programming languages; peripherals; consummables and accessories; communications; maintenance and care; training; consultants and systems houses; financial services; installations and user groups; glossary; supplies. Over 300 systems, both general and dedicated are listed and 1000 suppliers.
Not available on-line.

Nederlands ABC voor handel en industrie   322

[Netherlands ABC for commerce and industry]
Publisher: ABC voor Handel en Industrie CV
Address: PO Box 190, 2000 AD Haarlem, Netherlands
Year: 1984
Pages: 3000 (in 2 vols)
Price: Dfl155.00
ISBN: 90-70729-13-X
Content: Volume I gives: alphabetical list of products; list of related products; product-index; alphabetical list of foreign companies with their Dutch representatives; alphabetical list of companies. Volume II is divided into 17 sections, in each of which the companies (total of 22 000) are listed alphabetically by city and by city alphabetically by company name. The company listing gives: name, full address, telephone, telex, bankers, names of the staff, number of employees, chamber of commerce registration number, product line, import/export record.
Not yet available on-line.
Next edition: August 1985; Dfl160.00.

Norges Handels-Kalender   323

[Norwegian directory of commerce]
Publisher: S.M. Bryde A/S
Address: Sognsveien 70, 0855 Oslo 8, Norway
Year: 1985
Pages: 2968
Price: NKr400.00
ISBN: 82-7030-097-7
Series: Commercial directories
Content: The trades section comprises about 2500 trade headings, covering about 145 000 firms.
Is not yet available on-line but it will be during 1986.
Next edition: May 1986; NKr400.00.

Personnel and training databook, second edition   324

Editor: Michael Armstrong
Publisher: Kogan Page Limited
Address: 120 Pentonville Road, London N1 9JN, UK
Year: 1985
Pages: 390
Price: £18.50
ISBN: 0-85038-908-9
Geographical area: United Kingdom.
Content: Sections covering: computer-based training; computer-managed learning; payroll programs; computerized manpower planning; recruitment programs; authoring systems; interactive video.
Next edition: January 1986.

Pneumatic handbook, sixth edition 325
Editor: R.H. Warring
Publisher: Trade and Technical Press Limited
Address: Crown House, London Road, Morden, Surrey SM4 5EW, UK
Year: 1982
Pages: 550
Price: £50.00 (UK); £60.00 (elsewhere)
ISBN: 0-85461-090-1
Geographical area: United Kingdom.
Content: This is an accepted standard reference work on modern pneumatic and compressed air engineering. Buyers' guide section, editorial and advertisers indexes.

Polymers, paint and colour year book 1985 326
Editor: Derek Eddowes
Publisher: Fuel and Metallurgical Journals Limited
Address: Queensway House, 2 Queensway, Redhill, Surrey RH1 1QS, UK
Pages: 400
Price: £25.00
Geographical area: United Kingdom.
Content: Information on companies who supply the paint, ink, and polymer processing industries with essential raw materials, plant, and equipment, including tables listing characteristics of organic pigments and of extenders and fillers. A classified section lists suppliers of pigments; extenders; resins, media; solvents; plasticisers; chemicals; additives; equipment; containers; polymers (and polymer processors); plus mould, die; and patternmakers. An alphabetical section lists: consultants; associations; trade names; manufacturers of paints, varnishes, adhesives, and printing inks; and overseas paint manufacturers.
Next edition: October, 1985; £27.50.

Process engineering directory, 1985 327
Editor: James Robertson
Publisher: Morgan-Grampian Book Publishing Company Limited
Address: 30 Calderwood Street, Woolwich, London SE18 6QH, UK
Year: 1985
Pages: vi + 420
Price: £25.00
ISBN: 0-86213-065-4
Geographical area: United Kingdom.
Content: Guide to the products and services provided by over 2500 companies, indexed under 4000 product headings, produced to meet the requirements of engineers, buyers, and designers throughout the continuous process industries.
Not available on-line.
Next edition: April 1986; £28.00.

PRODEI 328
[Directory of Spanish industry, export and import]
Publisher: Capel Editorial Distribuidora SA
Address: Almirante 21, 28004 Madrid, Spain
Year: 1984-85
Pages: 1584
Price: £46.00 (surface mail); £50.00 (air-mail to Europe)
ISBN: 84-85002-05-9
Geographical area: Spain.
Next edition: June 1986; £50.00.

Pump selection systems and applications, second edition 329
Editor: R.H. Warring
Publisher: Trade and Technical Press Limited
Address: Crown House, London Road, Morden, Surrey SM4 5EW, UK
Year: 1984
Pages: 250
Price: £17.00 (UK); £19.50 (elsewhere)
ISBN: 0-85461-096-0
Geographical area: United Kingdom.
Content: A guide to the selection of pumps and pumping systems to meet the specifications of a given application. Buyers' guide section, editorial and advertisers indexes.

Pump users' handbook, second edition 330
Editor: F. Pollak
Publisher: Trade and Technical Press Limited
Address: Crown House, London Road, Morden, Surrey SM4 5EW, UK
Year: 1980
Pages: 250
Price: £16.00 (UK); £18.00 (elsewhere)
ISBN: 0-85461-070-7
Geographical area: United Kingdom.
Content: Technology, standards, materials, installation, operation and maintenance of pumps. Buyers' guide section, editorial and advertisers indexes.

Pumping manual, seventh edition 331
Editor: R.H. Warring
Publisher: Trade and Technical Press Limited
Address: Crown House, London Road, Morden, Surrey SM4 5EW, UK
Year: 1984
Pages: 600
Price: £52.00 (UK); £60.00 (elsewhere)
ISBN: 0-85461-095-2
Geographical area: United Kingdom.

Content: Comprehensive and authoritative update including the latest developments in the field of pumping, with buyers' guide section, editorial and advertisers indexes.

Quality technology handbook 332

Editor: R.S. Sharpe, J. West, D.S. Dean, D.A. Tyler, H.A. Cole
Publisher: Butterworths Scientific Limited
Address: PO Box 63, Westbury House, Bury Street, Guildford, Surrey GU2 5BH, UK
Year: 1984
Pages: vi + 486
Price: £45.00
ISBN: 0-408-01331-1
Geographical area: Great Britain and Western Europe.
Content: There are 920 entries arranged geographically, covering UK companies, societies, institutes, and organizations, as well as companies and organizations in Western Europe. Entries give name, address, overseas officers, products or services, and publications. There are indexes to subject matter and trade names.
Not available on-line.
Next edition: 1987.

Rubbicana Europe: a directory of suppliers to the rubber industry 333

Editor: David Reed
Publisher: Crain Communications Limited
Address: 20/22 Bedford Row, London WC1R 4EW, UK
Year: 1985
Pages: 344
Price: £30.00
Content: Over 1200 companies in 31 European countries are listed in the directory. It gives the full names and addresses of all the companies including telex and telephone numbers and the names of key commercial, sales and technical personnel. It also gives full details of the relevant trade and technical associations.
Over half of the directory is devoted to product indexes - sources of over 50 individual rubber types are listed, well over 200 compounding ingredients and their sources are identified, along with nearly 200 types of machines and items of processing equipment. Test equipment, too, is covered and sources of raw materials and processing equipment for polyurethanes are cross-indexed. A list of trade names competes the directory. This section gives details of the nature of the products referred to and the name of the company supplying or offering the item in question.
Not available on-line.

Rylands directory of the engineering industry* 334

Publisher: Guardian Communications Limited
Address: Albany House, Hurst Street, Birmingham B5 4BD, UK
Year: 1984-85
Pages: 724
Price: £38.00
ISBN: 0-86108-171-4
Geographical area: United Kingdom.
Content: There are approximately 20 000 entries, in two main sections. The alphabetical section comprises name, address, telephone, telex, list of directors, company description, capital authorized/issued, date of registration, brand names and trade marks. The geographical section lists company name, address, and description. There is an index to advertising.
Next edition: September 1985; £42.00.

Scandinavia's 5000 largest companies 335

Publisher: ELC International
Address: 227-229 Chiswick High Road, London W4 2DW, UK
Year: 1985
Price: £75.00
Series: Largest companies
Content: This new directory provides structured financial analysis of the 5000 largest Nordic companies, and consolidates and compares financial information of companies in one volume for Denmark, Norway, Finland and Sweden.

Seals and sealing handbook, first edition 336

Editor: R.H. Warring
Publisher: Trade and Technical Press Limited
Address: Crown House, London Road, Morden, Surrey SM4 5EW, UK
Year: 1981
Pages: 500
Price: £46.00 (UK); £52.00 (elsewhere)
ISBN: 0-85461-082-0
Geographical area: United Kingdom.
Content: Fundamentals and principles for all types of sealing for hydraulic, pneumatic, and mechanical applications, together with buyers' guide section and editorial and advertisers indexes.

Sheet metal industries yearbook 1985 337

Publisher: Fuel and Metallurgical Journals Limited
Address: Queensway House, 2 Queensway, Redhill, Surrey RH1 1QS, UK
Price: £25.50
Geographical area: United Kingdom.

Content: Technical data, standard specifications, principles of sheet metal operations, safety regulations and measures and testing methods; fully classified buyers' guide; alphabetically and geographically arranged list of stockholders.

Software users year book　　　　　338

Editor:　Grace Bays
Publisher:　VNU Business Publications bv
Address:　VNU House, 32-34 Broadwick Street, London W1 2HG, UK
Year:　1985
Pages:　2100 (in 2 vols)
Price:　£89.00
ISBN:　0-86271-049-9
Geographical area:　United Kingdom.
Content:　Volume one contains details of independent software packages, and suppliers for mainframe and mini computers, plus details of software services ie, software consultants and systems house services, recruitment companies, training companies.
Volume two contains details of independent software packages for microcomputers: 6500 independent software packages available and 1500 suppliers are listed. Index is by category, manufacturer, operating system, industry, and A-Z.
Not yet available on-line.
Next edition:　September 1986.

Stainless steel directory 1985-86　　　　　339

Editor:　K.T. Rowland
Publisher:　Modern Metals Publications Limited
Address:　14 Knoll Road, Dorking, Surrey RH4 3EW, UK
Year:　1983
Pages:　ii + 134
Geographical area:　United Kingdom.
Content:　Approximately 12 000 entries under some 800 headings including suppliers, manufacturers, processing services, alloy and scrap suppliers, trade names, and location of stockholders.
Next edition:　October 1985; £10.00.

Swiss export directory: export products and services of Switzerland　　　　　340

Publisher:　Swiss Office for the Development of Trade
Address:　PO Box, CH-8035 Zürich, Switzerland
Year:　1983
Pages:　900
Price:　£16.00
Next edition:　1986; £10.00.

Timber trades directory, 25th edition　　　　　341

Editor:　Harriet Martin
Publisher:　Benn Business Information Services Limited
Address:　PO Box 20, Sovereign Way, Tonbridge, Kent TN9 1RQ, UK
Year:　1984
Pages:　184
Price:　£25.00 (United Kingdom); £29.00 (elsewhere)
ISBN:　0-86382-000-X
Geographical area:　United Kingdom and Europe.
Content:　Alphabetical index of UK firms; three-section buyers' guide (UK) - timber and timber products; plant, equipment, and materials; services; buyers' guide to Europe; UK and European trade names.

Trade associations and professional bodies of the United Kingdom, seventh edition　　　　　342

Editor:　Patricia Millard
Publisher:　Pergamon Press Limited
Address:　Headington Hill Hall, Oxford OX3 0BW, UK
Year:　1985
Pages:　vii + 597
Price:　£29.00
ISBN:　0-08-023024-5
Geographical area:　United Kingdom.
Content:　There are 4800 entries, arranged as follows: alphabetical list of trade associations and professional bodies; geographical index by town (excluding London postal district); chambers of commerce, trade, industry, and shipping; United Kingdom offices of overseas chambers of commerce.
Not available on-line.
Next edition:　March 1986; £32.50.

Trade directory information in journals, fifth edition*　　　　　343

Publisher:　Science Reference Library
Address:　25 Southampton Buildings, London WC2A 1AW, UK
Year:　1985

Trade monopolies in Eastern Europe: the state foreign trade organizations, the chambers of commerce and the state organizations in the field of transport, banking, tourism, trade fairs etc.　　　　　344

Publisher:　Stamex BV
Address:　PO Box 400, 3770 AK Barneveld, Netherlands
Year:　1978
Availability:　OP

EUROPE INCLUDING USSR

UK's 7500 largest companies 1985-86 345
Publisher: ELC International
Address: 227-229 Chiswick High Road, London W4 2DW, UK
Price: £60.00
Series: Largest companies
Content: A source of comparative financial information on quoted and unquoted companies, the directory provides data on organizations operating in all sectors of manufacturing and service industries in the United Kingdom. Data is presented in tables in which companies are ranked by turnover, profit/loss and trading margin. In addition to global listings separate analyses are provided for major industrial sectors and the standard geographical regions. Information also includes addresses of head offices, who owns whom, dates of company formation, and Stock Exchange listing dates.

Valves, piping and pipelines, first edition 346
Editor: R.H. Warring
Publisher: Trade and Technical Press Limited
Address: Crown House, London Road, Morden, Surrey SM4 5EW, UK
Year: 1982
Pages: 530
Price: £52.00 (UK); £60.00 (elsewhere)
ISBN: 0-85461-087-1
Geographical area: United Kingdom.
Content: Types, design, construction, application. Materials and ancillaries relating to valves, pipes, and pipelines, together with buyers' guide section and editorial and advertisers indexes.

Welding 85 347
Publisher: Welding Institute
Address: Abington Hall, Abington, Cambridge CB1 6AL, UK
Year: 1984
Availability: OP
Geographical area: United Kingdom.

Wer liefert was? 348
[Who supplies what?]
Publisher: Bezugsquellennachweis für den Einkauf
Address: Wer liefert was? GmbH, Normannenweg 18-20, 2000 Hamburg 26, German FR
Year: 1985
Pages: lxxx + 2000
Price: DM65.00
ISBN: 3-923878-02-8
Geographical area: Federal Republic of Germany.

Content: Alphabetical list of products in five languages, and manufacturers of all mentioned products arranged according to industry.
Next edition: March 1986.

Who owns whom 1985, United Kingdom and Republic of Ireland 349
Publisher: Dun and Bradstreet Limited
Address: 26-32 Clifton Street, London EC2B 2AQ, UK
Year: 1985
ISBN: £135.00
Geographical area: United Kingdom and Republic of Ireland.
Content: Volume 1 lists over 90 000 domestic and foreign subsidiaries and associates grouped under their 6000 parent companies with their full name and address. Volume 2 provides a complete alphabetical list of subsidiaries and associates and indicates their parent companies. There are two special sections: foreign investment - a worldwide listing of foreign companies with British subsidiaries and associates; consortia - full lists of members for North Sea oil groups, consortium banks, and many other types of company association.
Supplements are issued quarterly, at a small extra charge, to keep the user in touch with important changes in the ownership of UK companies.

Who's who in the water industry 350
Publisher: Turret-Wheatland Limited
Address: Penn House, Penn Place, Rickmansworth, Hertfordshire WD3 15N, UK
Year: 1985
Pages: 160
Price: £13.00
Geographical area: United Kingdom.
Content: Details of suppliers of plant, equipment, and services; also includes a full personnel directory of senior staff employed in the water industry. Published annually in December.
Next edition: December 1985; £14.00.

MEDICAL AND BIOLOGICAL SCIENCES

Annuaire dentaire* 351
[Dental directory]
Editor: B. Laloup
Publisher: Editions de Chabassol
Address: 30 rue de Gramont, 75009 Paris, France
Year: 1985
Pages: 1450

Price: F360.00
ISBN: 2-904132-01-5
Geographical area: France.
Content: The directory covers dental surgeons, dental technicians, manufacturers and suppliers of dental equipment, materials and preparations.
Next edition: May 1986; F400.00.

Annuaire des stations hydro-minérales, climatiques balnéaires et établissements médicaux Français 352

[Yearbook of French hydro-mineral, climatic and balneological resorts and medical establishments]
Publisher: Expansion Scientifique Française
Address: 15 rue Saint-Benoît, 75278 Paris Cedex 06, France
Year: 1979
Pages: 264
Availability: OP
ISBN: 2-7046-1017-7
Content: List of resorts classified by their therapeutic reputations; medicinal list; hydro-mineral resorts; climatic and balneological resorts; medical establishments.

Ärzte-adressbuch, Nordbaden, verzeichnis der praktizierenden ärzte, krankenanstalten 353

[Medical directory, North Baden, index of medical practitioners, hospitals]
Publisher: Thebal-Verlag
Editor: Wolfgang Ballenberger
Address: Alexanderstrasse 32, D-7000 Stuttgart 1, German FR
Year: 1980-81
Pages: 328
Price: DM26.00
ISSN: 0400-664X

Ärzte-adressbuch Nordwürttemberg, verzeichnis der praktizierenden ärzte, krankenanstalten 354

[Directory of North Wurttemberg, index of medical practitioners, hospitals]
Publisher: Thebal-Verlag
Editor: Wolfgang Ballenberger
Address: Alexanderstrasse 32, D-7000 Stuttgart 1, German FR
Year: 1985
Price: DM42.00

Ärzte-adressbuch Südwürttemberg: verzeichnis der praktizierenden ärzte, krankenanstalten 355

[Medical directory of South Wurttemberg: index of medical practitioners, hospitals]
Publisher: Thebal-Verlag
Editor: Wolfgang Ballenberger
Address: Alexanderstrasse 32, D-7000 Stuttgart 1, German FR
Year: 1984
Pages: 256
Price: DM34.00

ASLIB directory of information sources in the United Kingdom: volume 2 - social sciences, medicine and the humanities. 356

Editor: Ellen M. Codlin
Publisher: Aslib
Address: Information House, 26-27 Boswell Street, London WC1N 3JZ, UK
Year: 1984
Pages: 1000
Price: £70.00
ISBN: 0-85142-184-9
Content: There are 3583 entries, in alphabetical order of names, with references from changed or alternative names. Limited geographical indication of sources is possible because the place names of universities, public libraries, etc, are given in the alphabetical sequence.
The organizations include: very large, very small; professional, amateur, institutional, voluntary, academic, scientific, research, service, governmental, standardizing, qualifying; producers of data, statistics, abstracts.
The directory is indexed by subject and there is an abbreviation index.
Not available on-line.
Next edition: 1986-87.

British dental journal educational directory 1984 357

Publisher: British Dental Journal
Address: British Dental Association, 64 Wimpole Street, London W1M 8AL, UK
Pages: 36
Price: £1.00
Geographical area: United Kingdom and Republic of Ireland.
Content: Directory of dental schools; directory of postgraduate education; information on fellowships, scholarships, and bursaries.
Not available on-line.
Next edition: September 1986; £2.00.

Bundes apotheken register: verzeichnis der apotheken in der bundes republik deutschland und in Berlin (west-sektoren) 358
[Federal register of pharmacies in Germany FR and West Berlin]
Publisher: Deutscher Apotheker Verlag
Address: POB 40, Birkenwaldstrasse 44, D-7000 Stuttgart 1, German FR
Year: 1982
Pages: 767
Price: DM378.00
ISBN: 3-7692-0587-1
Next edition: 1985.

Calendar of the Pharmaceutical Society of Ireland 359
Publisher: Pharmaceutical Society of Ireland
Address: 37 Northumberland Road, Dublin 4, Ireland
Year: 1985
Pages: xxvi + 139
Geographical area: Republic of Ireland.

Chirugenverzeichnis 360
[Surgeons' register]
Editor: H. Junghanns
Publisher: Springer-Verlag
Address: Tiergartenstrasse 17, 6900 Heidelberg, German FR
Year: 1980
Pages: viii + 838
Price: DM143.00
ISBN: 3-540-09924-7
Geographical area: Germany.
Content: 2000 entries listing surgeons in Germany.

Dentists register, comprising the names and addresses of dental practitioners registered at 31 January 1985, together with the local list of names so registered and the list of bodies corporate carrying on the business of dentistry 361
Publisher: General Dental Council
Address: 37 Wimpole Street, London W1M 8DQ, UK
Year: 1985
Pages: lxii + 465
Price: £15.00
Geographical area: United Kingdom.
Content: Entries for 22 500 practitioners with addresses, etc.
Next edition: June 1986; £15.00.

Directory of agencies for the blind in the British Isles and overseas* 362
Publisher: Royal National Institute for the Blind
Address: 224 Great Portland Street, London EC1V 7JE, UK
Year: 1984
Pages: iv + 170
Price: £5.00
ISBN: 0-901797-09-X
Content: The emphasis of the directory is on social rather than medical welfare, with approximately 1000 entries. Appendices list national and local voluntary and statuatory agencies, grant-making agencies, eye hospitals, schools and colleges, rehabilitation day centres, accommodation, workshops and industrial units, services to deaf-blind people, special interest groups, Braille services, tape services, magazines, overseas agencies, and bibliography.
Not available on-line.

Directory of agencies offering therapy counselling and support for psychosexual problems 363
Editor: Francis Taylor
Publisher: British Association for Counselling
Address: 37a Sheep Street, Rugby, Warwickshire CV21 3DX, UK
Year: 1983
Pages: ii + 85
Price: £3.50
ISBN: 0-946181-05-5
Geographical area: United Kingdom.
Content: There are 141 entries, arranged geographically, showing names, addresses, opening hours, referrals, fees, type of psycho-sexual problems dealt with, and the therapy orientation, etc.
There is also a 34 page introductory essay by Dr Elphis Christopher with a comprehensive bibliography. (This is available separately from the publishers at £1.25.). A 1985 supplement is also available.
Not available on-line.

Directory of biological consulting practices 364
Publisher: Institute of Biology
Address: 20 Queensberry Place, London SW7 2DZ, UK
Year: 1976
Pages: 9
Price: £2.50
Content: Particulars of consulting practices and the disciplines they cover: part one - independent consulting practices; part two - subjects covered by consultancy services offered by universities or employed biologists accepting commissions.
Next edition: 1985-86.

Directory of British biotechnology 365
1984

Publisher: Cartermill Publishing Limited
Address: PO Box 33, St Andrews, Fife KY16 9EA, UK
Pages: 258
ISBN: 0-947573-00-3
ISSN: 0265-8275
Geographical area: United Kingdom.
Content: Lists profiles of biotechnological activities of 2185 products producers, 63 contract research establishments, 112 academic departments or units, and 45 venture capital firms. Sponsored by the UK Department of Trade and Industry's Biotechnology Unit this publication is the main source of who is active in biotechnology in the UK, and what they are doing.
Next edition: September 1986; £60.00; Longman Group Limited.

Directory of child guidance and 366
school psychological services

Publisher: Association for Child Psychology and Psychiatry
Address: 4 Southampton Row, London WC1B 4AB, UK
Year: 1982
Pages: 106
Price: £3.50
Geographical area: United Kingdom.

Directory of health services 1985 367

Publisher: Longman Group Limited
Address: Sixth Floor, Westgate House, The High, Harlow, Essex CM20 1NE, UK
Pages: 500
Price: £49.00
ISBN: 0-582-90360-2
Geographical area: United Kingdom.
Content: Provides full details of over 1300 health centres. Information given includes full address, telephone number, names and qualifications of medical practitioners, and services and facilities available. Dentists and community nursing are also covered. Fully indexed.

Directory of health, education and 368
research journals*

Editor: Lee Pratt
Publisher: Associated University Presses
Year: 1985
Price: £21.50
ISBN: 0-8386-3213-0
Geographical area: United Kingdom.
Content: Over 400 medical journals giving circulation, editor, publisher, scope, etc.

Directory of medical and health care 369
libraries in the United Kingdom and
Republic of Ireland 1982

Publisher: Library Association Publishing
Address: 7 Ridgmount Street, London WC1E 7AE, UK
Pages: 256
Price: £8.75
ISBN: 0-85365-5367
Content: The directory provides details of 600 medical and health care libraries. Details of each library include: opening hours; postal address; telephone and telex numbers; senior librarian; number of staff; borrowing qualifications; stock and special collections; and a new feature - information on data base retrieval facilities.

Directory of private hospitals and 370
health services 1985

Publisher: Longman Group Limited
Address: Sixth Floor, Westgate House, The High, Harlow, Essex CM20 1NE, UK
Pages: 600
Price: £25.00
ISBN: 0-582-90361-0
Geographical area: United Kingdom.
Content: A comprehensive guide to independent hospitals, clinics, nursing and rest homes in the UK. There are over 7000 entries giving full address, telephone number, names of senior staff, bed complement and facilities available. In addition, there is a geographical breakdown of private bed accommodation in NHS hospitals. Fully indexed.

Directory of research in biotechnology 371

Publisher: Science and Engineering Research Council
Address: Biotechnology Directorate, Polaris House, North Star Avenue, Swindon, Wiltshire SN2 1ET, UK
Year: 1984
Pages: xvii + 109
Price: Free
Geographical area: United Kingdom.
Content: The directory includes indexes of research programmes, membership of the Biotechnology Management Committee, and a list of SERC staff in the Biotechnology Directorate.
Next edition: October 1985; free.

Directory of schools of medicine and 372
nursing: British qualifications and
training in medicine, dentistry, nursing
and related professions

Editor: Laidon Alexander
Publisher: Kogan Page
Address: 120 Pentonville Road, London N1 9JN, UK
Year: 1984

Pages: 752
Price: £15.95
ISBN: 0-85038-788-4
Geographical area: United Kingdom.
Content: Guide to British qualifications and training in the health-care professions. It covers medicine and surgery, dentistry, nursing, the professions supplementary and related to medicine, alternative medicine, and health-care technology. It includes details of primary and advanced academic and professional qualifications in each area and the specializations within it: the training courses and programmes leading to the qualifications, helpful information on fees and the structure of training in each profession, indexes, maps, and a bibliography.

Fire directory 373

Editor: Val Hargreaves
Publisher: Unisaf Publications Limited
Address: Queensway House, Redhill, Surrey RH1 1QS, UK
Year: 1985
Pages: 308
Price: £21.75
ISSN: 0264-4827
Geographical area: United Kingdom.
Content: The directory covers public and industrial fire brigades, giving biographical details of officers in local authorities and industrial fire services.

Good medicine guide* 374

Editor: Vernon Coleman
Publisher: Thames and Hudson
Address: 30-34 Bloomsbury Street, London WC1B 3PQ, UK
Year: 1982
Price: £6.50
ISBN: 0-500-01270-9
Geographical area: United Kingdom.
Content: Information on medical services in Britain including drugs, procedures, facilities, insurance, associations, fringe medicine, medical abbreviations, vitamins, etc.

Guida di veterinaria e zootecnica 375

[Veterinary drugs: food additives and manufacturers guide, fourth edition]
Editor: Lucio Marini
Publisher: Organizzazione Editoriale Medico Farmaceutica SRL
Address: Via Edolo 42, PO Box 10434, 20125 Milano, Italy
Year: 1983
Pages: vi + 472

Availability: OP
ISBN: 88-7076-025-1
Geographical area: Italy.
Next edition: June 1985; L60.000.

Guida Monaci annuario sanitario 376

[Monaci health directory]
Publisher: Guida Monaci SpA
Address: Via Francesco Crispi 10, 00187 Roma, Italy
Year: 1985
Pages: 1500 (in 2 vols)
Price: L80 000
Geographical area: Italy.
Content: Detailed information on state health and medical organizations, hospitals, physicians, medical industries, and dealers. Alphabetical index.

Guide to postgraduate degrees, 377
diplomas and courses in medicine

Publisher: Council for Postgraduate Medical Education in England and Wales
Address: 7 Marylebone Road, London NW1 5HH, UK
Year: 1985
Pages: iv + 58
Price: £5.35
ISSN: 0265-2730
Geographical area: United Kingdom.
Content: Sections are devoted to each specialty. Details are given on higher degrees and diplomas, including the awarding body and, where possible, examination dates, qualifications needed, fees and other requirements. Published annually.
Next edition: January 1986; £5.00-6.00.

Hammaslaakarit 378

[Dentists]
Publisher: National Board of Health
Address: PB 224, Siltasaarenkatu 18A, Helsinki, Finland
Year: 1984
Pages: 147
Price: FMk29.00
ISSN: 0780-1807
Geographical area: Finland.
Content: Alphabetical listing of dentists in Finland.
Next edition: 1985; FMk29.00

Handbook of industrial fire protection 379
and security

Publisher: Trade and Technical Press Limited
Address: Crown House, London Road, Morden, Surrey SM4 5EW, UK
Year: 1976
Pages: 600

Price: £48.00 (UK); £54.00 (elsewhere)
ISBN: 0-85461-059-6
Geographical area: United Kingdom.
Content: Prevention, detection, materials and hardware, environmental, emergency, references for local and national bodies, together with a listing of trade names indexes and of equipment and materials.

Handbook of industrial safety and health 380

Publisher: Trade and Technical Press Limited
Address: Crown House, London Road, Morden, Surrey SM4 5EW, UK
Year: 1980
Pages: 650
Price: £52.00 (UK); £60.00 (elsewhere)
ISBN: 0-85461-075-8
Geographical area: United Kingdom.
Content: Management of safety and health, responsibilities and duties, legal aspects, occupational hazards and dangers, prevention, together with a listing of products and trade names, manufacturers. Buyers' guide section, editorial and advertisers indexes.

Handbuch für die sanitätsberufe Österreichs. Adress- und nachschlagewerk über die sanitätsbehörden, sanitätsberufe und sanitäts-einrichtigung in Österreich 381

[Handbook of Austrian health organizations: address and reference book of public health departments, health organizations, and health establishments in Austria]
Publisher: Verlag Dieter Göschl
Address: Andergasse 10, A-1170 Wien, Austria
Year: 1984-85
Pages: 780
Price: Sch410.00
ISBN: 3-85097-044-2
Next edition: December 1985; Sch430.00.

Health care directory* 382

Editor: Geoffrey King
Publisher: Savory, E.B., Milln and Company
Address: 3 London Wall Buildings, London EC2M 5PG, UK
Year: 1985
Geographical area: United Kingdom.
Content: Major healthcare companies with technical information of products etc.

Hospitals and health services year book 1985 383

Editor: N.W. Chaplin
Publisher: Institute of Health Services Management
Address: Department YB 85, 75 Portland Place, London W1N 4AN, UK
Pages: 1137
Price: £43.80 (UK); £47.80 (elsewhere)
Geographical area: United Kingdom.
Content: Entries list: government department officers; members of advisory committees; local authorities and their chief executives; directors of social services and education; Whitley Councils; specialist teaching hospitals; health service hospitals; health authorities; independent hospitals; suppliers of goods and services to hospitals.
Next edition: 1986.

L'informatore farmaceutico 384

[Italian directory of drugs and manufacturers, 45th edition]
Editor: Lucio Marini
Publisher: Organizzazione Editoriale Medico Farmaceutica SRL
Address: Via Edolo 42, PO Box 10434, 20125 Milano, Italy
Year: 1985
Pages: 1700 (in 3 vols)
Price: L170.00
ISBN: 88-7076-082-0
ISSN: 0392-3010
Geographical area: Italy.
Content: Volume one - 13 000 drugs with trade name, manufacturer, composition, dosage, etc and 6000 pharmaceutical substances currently used in Europe and overseas; volume two - 12 000 OTC drugs and cosmetics with trade name, manufacturer, composition, dosage, etc; volume three - 2000 addresses of manufacturers, 180 addresses of suppliers, and 110 addresses of producers and traders of pharmaceutical raw materials.
Bi-monthly cumulative updating supplements are included in the price.
Next edition: March 1986.

Lääkärit 385

[Doctors]
Publisher: National Board of Health
Address: PB 224, Siltasaarenkatu 18A, Helsinki, Finland
Year: 1984
Pages: 415
Price: FMk49.00
ISSN: 0780-1785

EUROPE INCLUDING USSR

Geographical area: Finland.
Content: Doctors in Finland listed alphabetically and by speciality.
Next edition: 1985; FMk49.00

List of hospitals and house officer posts in the United Kingdom which are approved or recognised for pre-registration service   386

Publisher: General Medical Council
Address: 44 Hallam Street, London W1N 6AE, UK
Year: 1985
Pages: ix + 63
Price: £10.00
Content: Approved hospitals are listed by (National Health Service) Regional Health Authority. Indexes are: alphabetical index of hospitals; alphabetical index of places. Quarterly amendment sheets (price £6.00 per copy) and an annual consolidated amendment sheet (price £6.00 per copy) are published.
Next edition: 1987.

Medical directory 1985, 141st edition   387

Publisher: Longman Group Limited
Address: Sixth Floor, Westgate House, The High, Harlow, Essex CM20 1NE, UK
Pages: 3500 (in 2 vols)
Price: £60.00
ISBN: 0-582-90357-2
Series: United Kingdom.
Content: An alphabetical listing of medical practitioners registered to practise in the UK giving details of name, address, qualifications, past and present appointments, and medical literary output. The directory also provides a complete list of NHS hospitals, and details of universities, medical schools, royal colleges, research institutes, medical societies, the nursing councils, coroners, and blood transfusion services.

Medical register   388

Publisher: General Medical Council
Address: 44 Hallam Street, London W1N 6AE, UK
Year: 1985
Pages: xvii + 4000 (in 2 vols)
Price: £54.00
Geographical area: United Kingdom.
Content: The register includes the names, addresses, and qualifications of all doctors in the Principal List of the register, arranged alphabetically. The names of provisionally registered doctors are distinguished by an asterisk. There is also general information, particularly with reference to the registration of medical practitioners who qualified overseas. The council publishes fortnightly lists of changes to the register.
Next edition: April 1986.

Medical research directory   389

Editor: J.E. Levy
Publisher: Wiley, John, and Sons Limited
Address: Baffins Lane, Chichester, Sussex PO19 1UD, UK
Year: 1983
Pages: liv + 730
Price: £89.25, $157.50
ISBN: 0-4711-0335-7
Geographical area: United Kingdom.
Content: Approximately 15 000 entries catalogue details of medical research in progress in the UK. Each entry provides an abstract of the research project, names of principal workers, funding sources, location, and expected date of completion. Entries are grouped according to subject area and institution. The directory also presents data on grant awarding bodies, geographical distribution of research, and research trends.
It is available on-line from Datastar.

Medizinische bibliotheken und informationsstellen in der Deutschen Demokratischen Republik   390

[Medical libraries and information services in the German Democratic Republic]
Publisher: Institut für Wissenschaftsinformation in der Medizin
Address: Rhinstrasse 107, 1136 Berlin, German Dr
Year: 1984
Pages: iii + 195
Availability: OP
Content: There are 444 entries, arranged according to the administrative structure (districts), with a subject index, a keyword index, and a register of searching facilities in medicine.
Not available on-line.
Next edition: 1989; DM9.10

Medizinische forschungsinstitute in der Sowjetunion   391

[Medical research institutes in the Soviet Union]
Editor: Professor Dr Heinz Müller-Dietz
Publisher: Freie Universität Berlin
Address: Osteuropa-Institut, Garystrasse 55, 1000 Berlin 33, German FR
Year: 1983
Pages: 219
ISBN: 3-92-1374-27-8
Series: Berichte des Osteuropa-Instituts an der Freien Universität Berlin

MIRDAB: microbiological resource databank catalogue　392

Editor: P. Bachman
Publisher: Elsevier Science Publishers, Biomedical Division
Address: PO Box 1527, 1000 BM Amsterdam, Netherlands
Year: 1984
Pages: vii + 612
Price: Dfl225.00
ISBN: 0-444-90387-9
Content: This first edition of MIRDAB contains a total of 601 records of animal cells (general, genetic mutants and hybridomas), plant cells and animal viruses. Entries within each catetgory of cell or virus line have been ordered and indexed to simplify the task of locating cell lines or viruses that meet the user's needs. There are two indexes in the catalogue: a consolidated cell/virus name index, and an applications index.
Next edition: 1986; Dfl250.00.

Opticians register, comprising the opticians act 1958, rules made by the council: register of, ophthalmic opticians and dispensing opticians, and lists of bodies corporate carrying on the business as ophthalmic or dispensing opticians*　393

Publisher: General Optical Council
Address: 41 Harley Street, London W1N 2DJ, UK
Year: 1984
Pages: lxiv + 615
Price: £20.00
Geographical area: United Kingdom.
Content: Register of ophthalmic opticians who both test sight and fit and supply optical appliances; register of ophthalmic opticians who test sight but do not fit and supply optical appliances; register of dispensing opticians; list of bodies corporate carrying on business as ophthalmic opticians; list of bodies corporate carrying on business as dispensing opticians; local list.
Next edition: June 1985; £22.00.

Organizations in the UK employing biologists　394

Publisher: Institute of Biology
Address: 20 Queensberry Place, London SW7 2DZ, UK
Year: 1981
Pages: 15
Availability: OP
Content: List of organizations in the UK which employ biologists. (Universities, polytechnics, colleges, schools and hospitals are not included because of the large number of these institutions and the fact that their addresses are readily available elsewhere.)
Next edition: 1985.

Public health laboratory service directory　395

Publisher: Public Health Laboratory Service
Address: 61 Colindale Avenue, London NW9 5DF, UK
Year: 1985
Pages: vi + 74
Price: £2.00
Geographical area: England and Wales.
Content: Names and addresses of all senior staff of the Public Health Laboratory Service, listed according to laboratory, together with associated information on reference experts and additional information. Reference facilities index; general index; index of personal names.

Register of pharmaceutical chemists 1985　396

Publisher: Pharmaceutical Society of Great Britain
Address: 1 Lambeth High Street, London SE1 7JN, UK
Year: 1985
Pages: vi + 990
Price: £44.00
Geographical area: Great Britain.
Content: Approximately 36 000 pharmacists registered by the Pharmaceutical Society of Great Britain; register of pharmaceutical chemists 1985; list of bodies corporate and their superintendents 1985; register of premises 1985.
Not available on-line.
Next edition: April 1986; £50.00.

Schweizerisches medizinisches jahrbuch　397

[Swiss medical yearbook]
Publisher: Schwabe and Company Limited
Address: Steinentorstrasse 13 Postfach, CH-4010 Basel, Switzerland
Year: 1985
Pages: 850
Price: SFr98.00
ISBN: 3-7965-0833-2
Geographical area: Switzerland and Liechtenstein.
Content: Lists of physicians, specialists, dentists, and pharmacies, both alphabetically and by canton and community.
Details of: authorities and central organizations in the Swiss Confederation; public health administrations and hospitals in the cantons; research and formation; public health leagues and disability insurance; professional associations. Subject index.

Seibt medizinische technik: bezugsquellennachweis für artz- und krankenhaus und zahnärztliche und zahntechnische arbeitsmittel und werkstoffe 398

[Seibt medical appliances: sources of supply reference for practice and hospital]
Publisher: Seibt Verlag GmbH
Address: Pilgersheimer Strasse 38, D-8000 München 90, German FR
Year: 1985
Price: DM88.00
Geographical area: West Germany.
Content: This is a directory of medical appliances giving sources of supply reference for medical practitioners and hospitals, and of medical and technical supplies and materials for the dental profession. Manufacturers and suppliers are listed with their address, telephone and telex numbers, and further details such as membership of trade associations are also given. The subject index is in German, English, French and Spanish.

Seibt pharma-technik: bezugsquellennachweis für die pharmazeutische technik und für labor- und apothekenausstatung 399

[Seibt pharmaceutical engineering: buyer's guide for pharmaceutical engineering and for laboratory and pharmaceutical equipment]
Publisher: Seibt Verlag GmbH
Address: Pilgersheimer Strasse 38, D-8000 München 90, German FR
Year: 1983-84
Price: DM74.00
ISBN: 3-922948-23-5
Geographical area: West Germany.
Content: This is a buyer's guide for pharmaceutical engineering and for laboratory and pharmaceutical equipment. In the guide's main section suppliers are listed, with their address, telephone and telex numbers, underneath a product classification which is given in German, English and French. There are two further sections to facilitate use of the guide: an alphabetical index of products in the above three languages, and a list of companies giving additional information on agents and sales outlets.

Terveydenhuollon organisaatio ja laitokset 400

[Organization and institutions of the Finnish health care]
Publisher: National Board of Health
Address: PB 224, Siltasaarenkatu 18A, Helsinki, Finland
Year: 1984
Pages: 238
Price: FMk36.00
ISSN: 0357-8607
Geographical area: Finland.
Content: Organization of health care in Finland; administrative structure of health care; list of hospitals, health centres, and other institutions.
Next edition: 1985.

Verzeichnis der ärzte für anaesthesiologie in der Bundesrepublik Deutschland, Österreich und der Schweiz 401

[Index of anaesthesiologists in Germany, Austria, and Switzerland]
Editor: Helmut Kronschwitz
Publisher: Springer-Verlag
Address: Postfach 10 52 80, Tiergartenstrasse 17, D-6900 Heidelberg 1, German FR
Year: 1985
Pages: 400
Price: DM78.00
ISBN: 3-540-13927-3
Geographical area: Germany, Austria, Switzerland.

Who's who in medicine, fifth edition 402

Publisher: Who's who - the international red series - Verlag GmbH
Address: PO Box 1150, D-8031 Wörthsee, München, German FR
Year: 1981
Pages: 735 (vol 1), 770 (vol 2)
Price: $120.00 (for the set)
ISBN: 3-921220-40-8 (vol 1), 3-921220-41-6 (vol 2)
Geographical area: German Federal Republic, Austria, Switzerland.
Content: Listing of 12 000 leading personalities in medicine and pharmaceuticals, giving name, address, date of birth, speciality, and career history. Index of related associations, societies, research institutions, and consulates.

PHYSICS, MATHEMATICS AND NUCLEAR SCIENCES

Annuaire CNRS mathématiques, sciences physiques 1981-82* 403

[CNRS yearbook of mathematics and physical sciences]
Publisher: Centre National de la Recherche Scientifique

Address: 15 quai Anatole France, 75700 Paris, France
Price: F150.00
ISBN: 2-222-03045-5
Geographical area: France.
Content: Research activities of CNRS and associated laboratories.

Introductory guide to information sources in physics 404

Editor: L.R.A. Melton
Publisher: Institute of Physics
Address: Techno House, Redcliffe Way, Bristol BS1 6NX, UK
Year: 1978
Pages: 44
Price: £2.50
ISBN: 0-85498-033-4

Nuclear industry almanac: volume 1, Western Europe* 405

Editor: Greenhalgh and Jeffs
Publisher: Nuclear Energy Intelligence
Address: Teal House, Moat Lane, Prestwood, Great Missenden, Buckinghamshire HP16 9DA, UK
Year: 1982
Price: £35.00
ISBN: 0-907643-00-0
Content: Guide to suppliers and industries of the nuclear power market.

Research fields in physics, at United Kingdom universities and polytechnics, seventh edition 406

Publisher: Institute of Physics
Address: Techno House, Redcliffe Way, Bristol BS1 6NX, UK
Year: 1984
Pages: 349
Price: £27.50
ISBN: 0-85498-042-3
Geographical area: United Kingdom.
Content: Details of research and development work in physics undertaken at universities and polytechnics in Britain. The entry for each institution/department outlines the various research programmes and lists the senior staff involved in each field, this information being cross-referenced by subject and name indexes.
Coverage of physics departments is supplemented by details of physics research undertaken in a large number of astronomy, mathematics, chemistry, engineering, materials science, and medical physics departments.
Next edition: April 1987; £30.00.

University postgraduate degrees by course and research in applied mathematics and pure mathematics in the United Kingdom 1985-86 407

Editor: John Heading
Publisher: Conference of Professors of Applied Mathematics
Address: Department of Applied Mathematics, University College of Wales, Penglais, Aberystwyth, Dyfed SY23 3BZ, Wales, UK
Year: 1984
Pages: 66
Price: Free (if airmail required please send 3 coupon-réponse international)
ISSN: 0307-9260
Content: There are 57 entries in the current edition, and this usually increases each year. UK universities are listed in alphabetical order, followed by some UK polytechnics. Each entry contains details of all pure and applied mathematics MSc degrees by course, and MSc and PhD degrees by research.
Next edition: September 1985 (published annually).

6 LATIN AMERICA AND THE CARIBBEAN

GENERAL SCIENCES

Directoría de establecimientos de   1
educación oficiales y privados:
educación media y superior
[Directory of establishments of private and official education: secondary and higher education]
Publisher: Departamento Administrativo Nacional de Estadística - DANE
Address: Centro Administrativo Nacional, Avenida el Dorado, Bogotá, DE, Colombia
Availability: OP
Geographical area: Colombia.
Content: The directory is produced under the Ministry of Education. The ministry also produces a directory of establishments for specialized education, and a directory of official educational establishments.

Directorio de bases de datos   2
disponibles en el país (versión
preliminar)
[Directory of databases available in the country (preliminary version)]
Publisher: Comisión Nacional de Investigación Científica y Tecnológica, Dirección de Información y Documentación
Address: Canada 308, Casilla 297 - V, Santiago, Chile
Year: 1985
Pages: ii + 61
Availability: $10.00
Series: Serie directorios, 20
Geographical area: Chile
Content: There are 61 entries. The databases are arranged in alphabetical order by institution, and within institution by database title. Entries include database name, acronym, year it was started, address, source of information, users, subjects, services, and other relevant details.
Indexes: subject, storage and information retrieval programs, type of database.

Directorio de la educación superior en   3
Colombia
[Directory of higher education in Colombia]
Editor: Eduardo Espinal-Arenas
Publisher: Instituto Colombiano para el Fomento de la Educación Superior
Address: División de Recursos Bibliográficos, Calle 17 3-40, Bogotá, DE, Colombia
Year: 1985
Pages: xxi + 169
Price: $10.00
Geographical area: Colombia.
Content: There are approximately 220 entries arranged geographically by city. Entry details include name of the institution, address, telephone, PO box, dean's name, type of institution (technology institute, university, etc), faculties, duration of the academic programme, schedules, legal status, and degrees awarded. Not available on-line.
Next edition: January 1986; approx. $12.00.

Directorio de servicios de información   4
y documentación en el Uruguay
[Directory of information and documentation services in Uruguay, second edition]
Editor: Centro Nacional de Documentación Científica, Técnica y Económica
Publisher: Biblioteca Nacional
Address: 18 de Julio 1790, Montevideo, Uruguay
Year: 1983
Pages: 220
ISBN: 84-8290-007-2

Content: There are 293 entries giving details of address, telephone, name of contact, types of service and people who use them, publications and computerization. Indexes are by acronym, abbreviation, name of organization, subject and location.

Directorio regional de unidades de información para el desarrollo*   5
[Regional directory of development information units]
Publisher: United Nations Economic Commission for Latin America
Address: Avenida Dag Hammarskjold, Casilla 179D, Santiago, Chile
Year: 1979
Availability: OP
Pages: (3 vols)

Directory of development research and training institutes in Latin America*   6
Publisher: Organization for Economic Cooperation and Development
Address: Publications Office, 2 rue André-Pascal, 75775 Paris Cedex 16, France
Year: 1984
Price: £14.00
ISBN: 92-64-02613-4

Directory of libraries and special collections on Latin America and the West Indies   7
Editor: Bernard Naylor, Laurence Hallewell, Colin Steele
Publisher: Athlone Press
Address: 44 Bedford Row, London WC1R 4LY, UK
Year: 1975
Pages: viii + 161
Price: $16.50
ISBN: 0-485-17705-6
Series: University of London, Institute of Latin American Studies Monographs
Geographical area: Latin American and the West Indies.
Content: Details of 146 libraries holding collections on Latin America and the West Indies are listed. An appendix lists other sources of information: international organizations, embassies, journals.

Environmental education: Caribbean   8
Publisher: Food and Agriculture Organization
Address: Via delle Terme di Caracalla, 00100 Roma, Italy
Year: 1985
Pages: vii + 89
Series: UNEP regional seas directories and bibliographies
Content: Compiled jointly by the Caribbean Conservation Association and the United Nations Environmental Programme, the directory covers environmental education in 26 countries.

Fundaciones científicas extranjeras: oportunidades de financiamiento para la comunidad científica   9
[Foreign scientific foundations : financing opportunities for the scientific community]
Publisher: Comisión Nacional de Investigación Científica y Tecnológica, Centro de Información y Orientación de Estudios en el Extranjero
Address: Canada 308, Casilla 297 -V, Santiago, Chile
Year: 1985
Pages: 110
Series: Serie directorios, 21
Price: $10.00
Content: The directory contains basic data on 26 foundations with information on their main activities. Entries contain : name of foundation, research area and aims, qualifications for grants, duration of grant, date and address for application. Index of foundation names.

Guía de bibliotecas especializadas y centros de documentación de Chile   10
[Guide to special libraries and documentation centres in Chile]
Publisher: Comisión Nacional de Investigación Científica y Tecnológica, Centro Nacional de Información y Documentación
Address: Canada 308, Casilla 297 - V, Santiago, Chile
Year: 1984
Pages: iii + 173
Price: $10.00
Series: Serie directorios, 1
Content: There are 372 entries arranged by region, and within region by province. Each entry contains title of organization and type, address, telephone, telex, director's name, hours and service, collections held, publications, personnel, and subject areas.
Indexes: subject, organization, acronym, geographic area.

Guía de escuelas y cursos de bibliotecología y documentación en America Latina   11
[Guide to library and documentation schools and courses in Latin America]
Publisher: Instituto Bibliotecnológico, Universidad de Buenos Aires

Address: Azcuenaga 280 - CC901, 1000 Capital Federal, Argentina
Year: 1979
Pages: x + 145

Guía informativa: sociedades científicas de Chile 12
[Scientific societies of Chile: information guide]
Publisher: Comisión Nacional de Investigación Científica y Tecnológica, Departamento de Fomento
Address: Canada 308, Casilla 297-V, Santiago, Chile
Year: 1983
Pages: iv + 57
Price: $5.00
Series: Serie directorios, 10
Content: The guide contains 69 entries arranged alphabetically by society name. Entries include the society name, aims, scientific activities, publications, board of directors. There is an alphabetical index of organization.
Next edition: Late 1985; $5.00.

Guía nacional de reuniones científicas 13
1982
[National guide to scientific events 1982]
Publisher: Comisión Nacional de Investigación Científica y Tecnológica, Dirección de Información y Documentación
Address: Canada 308, Casilla 297-V, Santiago, Chile
Year: 1983
Pages: v + 394
Series: Serie dirctorios, 16
Geographical area: Chile.
Content: There are 265 entries. Each entry contains general information on the meetings and bibliographic data of the reports or abstracts presented to the meeting. Indexes: author, organization, subject.
Next edition: January 1986.

Guía preliminar de las bibliotecas de 14
la universidad de Buenos Aires
[Directory of the libraries of the University of Buenos Aires]
Publisher: Biblioteca de la Universidad de Buenos Aires
Address: Azcuénaga 280, Buenos Aires, Argentina
Year: 1985
Pages: 35
Content: In Spanish. Details given include name, address, name of head and assistant librarian specialization, holdings and classification.
Next edition: 1986; $10.00.

Institutos, programas, proyectos: 15
temas de investigación en desarrollo
[Institutes, programmes, projects: developing research]
Publisher: Consejo Nacional de Investigaciones Científicas y Técnicas (PRODAT)
Address: Rivadavia 1906, Capital Federal, Argentina
Year: 1982
Pages: (7 vols)
Geographical area: Argentina.
Content: Published in six subject volumes, and a seventh with possible application area. Topics are as follows: Vol 1: physics and mathematics; Vol 2: natural sciences; Vol 3: chemical sciences; Vol 4: technology; Vol 5: (not given); Vol 6: medicine.

Inventaire des instituts de recherche 16
et de formation en matière de
développement en Amérique Latine
[Directory of development research and training institutes in Latin America]
Editor: OECD Development Centre
Publisher: Organization for Economic Cooperation and Development
Address: 2 rue André-Pascal, 75775 Paris Cedex 16, France
Year: 1984
Pages: xxxii + 337
Price: £14.00
ISBN: 92-64-02613-4
Content: The directory gives information on 436 development research and training institutes in 25 Latin American countries. The database includes general information relating to the institutes (name, address, etc), a short description of their postgraduate training and research programmes, of other activities (publications, documentation, etc) and of the facilities available (library, computer, conference rooms).
Selective information searches can be made on specific areas of interest.

Marine environmental centres: 17
Caribbean
Publisher: Food and Agriculture Organization
Address: Via delle Terme di Caracalla, 00100 Roma, Italy
Year: 1985
Pages: x + 214
Series: UNEP regional seas directories and bibliographies
Content: Compiled jointly by the Food and Agriculture Organization and the United Nations Environmental Programme, the directory lists 89 research centres in 26 countries.

Programas de post-grado ofrecidos por las universidades Chilenas   18
[Graduate programmes offered by Chilean universities]
Publisher: Comisión Nacional de Investigación Científica y Tecnológica, Centro de Información y Documentación de Estudios en el Extranjero
Address: Canada 308, Casilla 297-V, Santiago, Chile
Year: 1985
Pages: iii + 47
Price: $5.00
Series: Serie directorios, 19
Content: There are 102 entries arranged according to university, and by area. Indexes of universities and special subjects.

Science and technology in Latin America   19
Editor: Latin American Newsletters Limited
Publisher: Longman Group Limited
Address: Sixth Floor, Westgate House, The High, Harlow, Essex CM20 1NE, UK
Year: 1983
Pages: 363
Price: £49.00
ISBN: 0-582-90057-3
Series: Longman guide to world science and technology
Content: Covering the countries of Central and South America and the Latin countries of the Caribbean, it deals in detail with the current state of science policy and planning, finance and manpower in the area and describes research and development activities in the fields of agriculture and marine science, military science, medicine, nuclear science and industrial science and technology. There are two indexes: establishments mentioned and subject index, plus a directory of major institutions.

Scientific institutions of Latin America   20
Editor: Ronald Hilton
Publisher: California Institute of International Studies
Address: 766 Santa Ynez, Stanford, CA 94305, USA
Year: 1970
Pages: xx + 784
Price: $12.00
ISBN: 0-912098-08-2
Geographical area: Latin America and the Caribbean.
Content: All the scientific institutions (universities, etc) are covered on a country by country basis. There are also general chapters. In all, some 400 institutions are described on the basis of on-site visits.
Next edition: Not planned.

AGRICULTURE AND FOOD SCIENCE

Cassava directory: directory of persons interested in tropical root crops (preliminary edition)   21
Editor: James H. Cock, Miguel Angel Chaux, Jorge López S.
Publisher: Centro Internacional de Agricultura Tropical
Address: Apartado Aéreo 6713, Cali, Colombia
Year: 1984
Pages: viii + 200
Price: Free
Geographical area: World.
Content: 1611 names, addresses, and areas of interest of cassava workers. The names are listed in alphabetic order within each country. Alphabetical index. The publication is not available on-line, and there are no supplements.
Next edition: December 1985.

Index of Argentinian herbaria*   22
Editor: E.M. Zardini
Publisher: Journal: Taxon. (29/5-6), 731-41
Address: International Bureau for Plant Taxonomy, Tweede Transitorium, Kamer 1902, Uithof 3584 CS Utrecht, Netherlands
Year: 1980

Vadecum forestal para America Latina   23
[Forestry vadecum for Latin America]
Publisher: Food and Agricultural Organization of the United Nations
Address: Oficina Regional para America Latina y el Caribe, Providencia 871, Casilla 10095, Santiago, Chile
Year: 1966
Pages: 105
Availability: OP
Content: Lists international forestry studies committees, Latin American national park services, forestry services, schools of forestry, industrial and professional associations, and statistical institutions. Arrangement is by country and category. There is also a list of forestry periodicals originating in Latin America.
Next edition: Not planned.

LATIN AMERICA AND THE CARIBBEAN

ENERGY SCIENCES

Directorio del sector eléctrico 24
[Directory of electrical utilities, South America]
Publisher: Comisión de Integración Eléctrica Regional
Address: Bulevar Artigas 1040, Montevideo, Uruguay
Year: 1976
Geographical area: South America.

Latin American petroleum directory 25
Editor: William R. Leek Jr
Publisher: PennWell Publishing Company
Address: PO Box 1260, Tulsa, OK 74104, USA
Year: 1983
Pages: 200
Price: $45.00 (USA and Canada); $56.50 (export)
ISSN: 0193-8738
Geographical area: Central and South America, Mexico and the Caribbean.
Content: A country by country listing for Mexico, Central and South America. Includes government agencies and 4700 personnel in 1000 firms. Indexed by country and personnel.

INDUSTRY AND MANUFACTURING

Major companies of Argentina, Brazil, Mexico and Venezuela 1982* 26
Editor: S.J. Andrade
Publisher: Graham and Trotman
Year: 1981
Price: £60.00
Geographical area: Argentina, Brazil, Mexico and Venezuela.
Content: Entries include name, address, telephone, telex, names of directors and senior executives, financial details, subsidiaries, and brand names, of 3500 companies.

South America, Central America and the Caribbean, 1986* 27
Publisher: Europa Publications
Address: 18 Bedford Square, London WC1B 3JN UK
Year: 1985
Price: £45.00
ISBN: 0-946653-11-9
Content: Details of major commodities, regional organizations, statistics, background information, and directory of names and addresses.

7 NORTH AMERICA

GENERAL SCIENCES

Accredited institutions of post secondary education 1984-85 — 1

Editor: Sherry S. Harris
Publisher: Collier Macmillan Limited
Address: Macmillan Distribution Limited, Brunel Road, Houndmills, Basingstoke, Hampshire RG21 2XS, UK
Year: 1985
Pages: 400
Price: £20.95
ISBN: 0-02-901400-X
Geographical area: United States of America
Content: Published on an annual basis under the aegis of the American Council on Education, the guide provides detailed information on post-secondary institutions of all kinds in the United States - universities, professional, trade, technical, and business schools and two-and-four year colleges. Organized alphabetically by institutions within each state, the entries tell at a glance the date of first accreditation to or admission to candidate category, the date of latest renewal or affirmation of this status, the name of the accrediting body and which certificate and degree programmes have been accredited by the appropriate body. Each entry also includes the name of the chief executive officer, address and telephone number, type of control (public, private, and religious affiliations, type of institution (university, vocational college, etc), branch campuses and affiliated institutions, type of academic calendar and the autumn 1980 enrolment.

ALA yearbook of library and information services* — 2

Editor: Robert Wedgeworth
Publisher: American Library Association
Address: c/o Eurospan, 3 Henrietta Street, London WC2E 8LU, UK
Year: 1985
Price: £62.50
ISBN: 0-8389-0413-0
Geographical area: USA.
Content: Details of libraries listed alphabetically, with reports on regional and specialized library and information associations.

American library directory: a classified list of libraries in the United States and Canada with personnel and statistical data — 3

Editor: Jaques Cattell Press
Publisher: Bowker Publishing Company
Address: 58/62 High Street, Epping, Essex CM16 4BU, UK
Year: 1984
Pages: 1900 (in 2 vols)
Price: $130.00
ISBN: 0-8352-1891-0
Content: This geographically arranged directory lists comprehensively public, academic, special, government, and armed forces libraries in the USA, its possessions, and Canada. The following information is provided for each library: key personnel, including on-line services personnel; current library revenues, book budgets and salaries; projected revenues and expenditures for the coming fiscal year; the number of books, films, records, and other holdings including special collections; library systems and networks as well as library branches for each.

A comprehensive list of library schools is also included arranged by geographical location, with key personnel, number of yearly visiting faculty, type of school and training offered, entrance requirements, degrees and courses offered, and scholarship and tuition information for each.

Rounding out the volume is a state-by-state alphabetical index to library systems in the USA and an index to US armed forces and overseas libraries.
Next edition: October 1985; $135.00.

American men and women of science: physical and biological sciences, fifteenth edition   4

Editor: Jaques Cattell Press
Publisher: Bowker Publishing Company
Address: 58/62 High Street, Epping, Essex CM16 4BU, UK
Year: 1982
Pages: 8400
Price: £585.00
ISBN: 0-8352-1413-3
Geographical area: USA and Canada.
Content: Biographical and contact information on active American and Canadian scientists in hundreds of specilizations. More than 130 000 men and women in fields from pollution control to particle physics are listed in alphabetical sequence by surname.
Each biography aims to include: full name and address; birth place and date, year of marriage, number of children; principal specilization; undergraduate and graduate institutions attended; degrees earned, honorary degrees, and awards; professional experience and concurrent positions; memberships in scientific societies; research and development activities.
The publication is available on-line via Dialog and BRS.

American universities and colleges, twelfth edition   5

Editor: American Council on Education
Publisher: Walter de Gruyter
Address: Genthiner Strasse 13, 1 Berlin 30, German FR
Year: 1983
Pages: 2156
Price: DM285.00
ISBN: 3-11-008433-3
Content: Entries for nearly 1900 educational institutions and accredited professional schools, their status, history, organization, administration, including information about enrollment, fees, student aid, admission requirements, number and rank of faculty in each department, and graduate programmes. Aspects of student life and college community are described, including cultural opportunities, intercollegiate athletic programmes, percentage of students joining fraternities and sororities, and extent and nature of affiliation with religious organizations.
In addition, the twelfth edition presents up-to-date facts on graduation requirements, library resources, foreign study opportunities, notable research projects, cooperative programmes, number and origins of foreign students enrolled, and physical facilities for each of the institutions. Articles on individual professions report on admission requirements of accredited professional schools, the required programme of professional education, and state certification or licensing if applicable.

Biotechnology engineers: biographical directory   6

Editor: Dr Oskar R. Zaborsky, Donna K. Zubris
Publisher: OMEC Publishing Company
Address: 1128 16th Street NW, Washington, DC 20036, USA
Year: 1985
Pages: x + 360
Price: $29.95
ISBN: 0-931283-05-1
Geographical area: USA and Canada.
Content: 236 entries arranged in alphabetical order. Each entry describes the educational and professional background, research interests, publications and patents, and professional activities of a key US or Canadian biotechnology engineer. The directory is indexed by individual, organization, and research area, and includes two appendices, journal and professional society abbreviations.

Canadian conservation directory   7

Editor: Betty Pratt
Publisher: Canadian Nature Federation
Address: 75 Albert Street, Suite 203, Ottawa, Ontario K1P 6G1, Canada
Year: 1978
Pages: 96
Price: Can$2.00
Content: Details of name, address, membership, and publications. Both national and state organizations are included.

Conservation directory 1985   8

Editor: Rue Gordon
Publisher: National Wildlife Federation
Address: 1412 16th Street NW, Washington, DC 20036, USA
Year: 1985
Pages: 302
Price: $15.00
ISBN: 0-912086-56-9
Geographical area: USA.
Content: The directory lists 2492 organizations and 13 126 officials concerned with natural resource use and supervision, including: US federal departments, agencies, and offices; national, international, and interstate organizations and commissions; state agencies and citizens' groups; Canadian government agencies and citizens groups; colleges and universities that teach con-

servation programmes; fish and game commissioners and directors; national wildlife refuges, national forests, and national parks.
Next edition: January 1986; $15.00.

Consultants and consulting organizations directory, sixth edition* 9

Editor: Janice McLean
Publisher: Gale Research Company
Address: Book Tower, Detroit, MI 48226, USA
Year: 1984
Pages: 1293
Price: $290.00
ISBN: 0-8103-0356-6
Geographical area: North America.
Content: Details on about 8500 firms, individuals, and organizations active in 116 special fields. Entries include names, addresses, telephone numbers and details on services performed. Information contained in the directory may be approached by geographical location, by industries served, by subject area, by names of individual consultants, and by firm name.
The directory is updated by a periodical supplement of newly formed consulting bodies.

Consulting and laboratory services* 10

Publisher: Lundy, J.W., Enterprises
Address: Box 543, State College, PA 16801, USA
Year: 1984
Pages: 200
Price: $45.00
Geographical area: USA.
Content: The publication covers over 685 firms offering engineering and scientific consulting, testing, analytical, and research and development services in 20 midwestern and eastern states. Entries include: firm name, address, name of chief executive, telephone; number of professional employees; percentage of external work performed; ownership; affiliation with other companies or laboratories, description of services. Entries are arranged by service, then geographically.

Directory of audio-visual sources: history of science, medicine, and technology* 11

Editor: Bruce S. Eastwood
Publisher: Neale Watson Academic Publications Incorporated
Address: 156 Fifth Avenue, New York, NY 10010, USA
Year: 1979
Price: $20.00
Geographical area: USA.
Content: The publication includes lists of academic film rental libraries and of distributors of the films and other audiovisual materials described. Entries include library or firm name, address, with libraries arranged geographically; distributors alphabetically by acronym.
Next edition: 1985.

Directory of Canadian human services 1982-83 12

Editor: Canadian Council on Social Development
Publisher: Droit, Le
Address: 375 rue Rideau, Ottawa, Ontario, Canada
Year: 1983
Pages: ix + 770
Price: $17.50
ISSN: 0714-4687

Directory of Canadian universities 1984-85 13

Editor: Julia Hill-Downham, Kim Allen
Publisher: Association of Universities and Colleges of Canada
Address: 151 Slater Street, Ottawa, Ontario K1P 5N1, Canada
Year: 1984
Pages: li + 361
Price: $16.50
ISBN: 0-88876-083-3
ISSN: 0706-2338
Content: A guide to 71 universities in Canada, with information ranging from admission requirements to research facilities. A cross-Canada listing of degree, diploma, and certificate programmes is included. The publication is in English and French.
Next edition: December 1986.

Directory of Canadian university resources for international development 14

Publisher: Association of Universities and Colleges of Canada
Address: 151 Slater Street, Ottawa, Ontario K1P 5N1, Canada
Year: 1983
Price: $10.00 (Canada); $12.00 (USA); $15.00 (other countries)
ISBN: 0-88876-070-1
ISSN: 0822-1502
Content: Information on the areas of study specialization, and research as well as past experience in international cooperation of Canadian universities. It also includes a brief description of each institution, its policy on international development cooperation, the administrative channels for coordinating and organizing international activity, and special library resources of importance to international development and the name of a contact person on each campus. The directory is in English and French.

NORTH AMERICA

Directory of computer software 15

Editor: National Technical Information Service
Publisher: US Department of Commerce
Address: 5285 Port Royal Road, Springfield, VA 22161, USA
Year: 1985
Price: $40.00; $70.00 (overseas)
Geographical area: USA.
Content: The directory lists over 1300 computer programs, including 700 new programs from the National Energy Software Center, and is keyed by subject categories and by indexes by agency. The directory is available in both paper copy and microfiche.

Directory of computerized data files 1985 16

Editor: National Technical Information Service
Publisher: US Department of Commerce
Address: Database Services Division, Springfield, VA 22161, USA
Price: $40.00; $70.00 (overseas)
Geographical area: USA.
Content: The directory lists over 1000 numeric and source textual data files from some 50 federal agencies. Section I gives brief summaries of each file organized into 27 subject categories in economics, social sciences, and science and technology. Section II contains the agency index, together with the number and subject matter indexes.

Directory of federal technology resources, 1984 17

Editor: Edward J. Lehmann
Publisher: National Technical Information Service
Address: US Department of Commerce, 5285 Port Royal Road, Springfield, VA 22161, USA
Year: 1984
Pages: 200
Price: $25.00
Geographical area: USA.
Content: The directory covers all major federal laboratories.
Next edition: 1985.

Directory of federally supported information analysis centers 1979, fourth edition 18

Publisher: National Referral Center
Address: Washington, DC, USA
Year: 1979
Pages: 95
Price: $4.00
Geographical area: USA.

Directory of federally supported research in universities, 1983-84 19

Publisher: Canada Institute for Scientific and Technical Information
Address: National Research Council of Canada, Ottawa, Ontario K1A 0S2, Canada
Year: 1984
Price: $65.00 (in 2 vols)
Geographical area: Canada.
Content: The directory lists 14 341 scientific research projects reported by federal government agencies. Volume one lists research projects by granting agency, lists investigators, gives fiscal breakdowns by agency and by province. Volume two contains a keyword-out-of-context subject index arranged alphabetically by English and French terms appearing in the titles.
Much of the information is accessible on the IEC data base CISTI's CAN/OLE on-line retrieval system.

Directory of graduate research 20

Publisher: American Chemical Society
Address: 1155 Sixteenth Street NW, Washington DC 20036, USA
Year: 1983
Pages: xxiii + 1207
Price: $43.00 (US/Canada); $52.00 (other countries)
ISBN: 0-8412-0797-6
Geographical area: North America.
Content: The directory lists faculties, publications, and doctoral and master's theses in departments or divisions of chemistry, chemical engineering, biochemistry, pharmaceutical/medicinal chemistry, clinical chemistry and polymer science at universities in the United States and Canada. Entries cover approximately 12 000 faculty members and 700 departments, with an index of instructional staff.
The publications is available on-line.
Next edition: October 1985; $46.00.

Directory of industrial research and development facilities in Canada 1985 21

Publisher: Statistics Canada, Department of Supply and Services Canada
Address: Place du Portage Phase 111, 11 Laurier Street, Hull, Quebec, K1A 0S2, Canada
Price: Can$30.00 (Canada); Can$31.00 (export)
Content: This publication is intended to be the first issue of an annual series. The directory provides information on approximately 650 r&d units, and contains descriptive information on each unit such as areas in which r&d is being performed, specialized equipment, the number of scientists and engineers, as well as identi-

fying information such as institutional name, address, and name of contact person. Available in both English and French.
Similar directories are being prepared for federal government establishments and university affiliated centres.

Directory of international and regional organizations conducting standards-related activities. *   22

Publisher: National Bureau of Standards
Address: Commerce Department, Washington, DC 20234, USA
Year: 1983
Pages: 365
Price: $9.50
Geographical area: USA.
Content: About 275 government bodies, associations, and other groups which conduct standardization, certification, laboratory accreditation, or other standards-related activities. Entries include: name, acronym, address, telephone, telex, cable address, scope of interests, membership.

Directory of research grants 1985   23

Publisher: Oryx Press
Address: 2214 North Central at Encanto, Phoenix, AZ 85004, USA
Year: 1985
Pages: vii + 668
Price: $74.50
ISBN: 0-89774-147-1
Series: Directory of research grants
Geographical area: USA.
Content: This directory covers more than 3000 funding programmes supported by government agencies, private foundations, corporations, and professional organizations. It contains the specific detailed information about policies and programmes that applicants need to know in order to succeed in the competition for financial support.
This publication is also available on-line.
Next edition: February 1986; $74.50.

Directory of science resources for Maryland 1980-81   24

Publisher: Maryland Department of Economic and Community Development
Address: 45 Calvert Street, Annapolis, 21401, USA
Year: 1980
Pages: xvi + 239
Price: $7.00
Content: Listings of 546 research and development firms and science-orientated firms, as well as universities, colleges, and professional organizations, giving: company name; address; telephone number; total employment; chief executive officer; principal activities; facilities and equipment; special capabilities; and noteworthy accomplishments. Entries are indexed alphabetically, geographically, and by activities. There is also an index to information sources.

Directory of scientific resources in Georgia 1983-84   25

Editor: Harvey Diamond
Publisher: Georgia Tech Research Institute
Address: Georgia Institute of Technology, Atlanta, GA 30332, USA
Year: 1984
Pages: 182
Price: $10.00
Content: 664 entries, arranged under four sections: industrial research laboratories; consulting engineers; government laboratories; colleges and universities. Each entry includes: organization name; address; director's name; size of staff; fields of research or service; activity; some entries have special equipment notations. There is a subject index and a geographical index.
Not available on-line.

Directory of special libraries and information centres   26

Editor: Brigitte T. Darnay
Publisher: Gale Research Company
Address: Book Tower, Detroit, MI 48226, USA
Year: 1985
Pages: Volume 1 - 1700 (in 2 parts); volume 2 - 900
Price: Volume 1, $320.00; volume 2, $265.00; volume 3, $275.00
ISBN: Volume 1, 0-8103-1881-1; volume 2, 0-8103-1889-X; volume 3, 0-8103-0281-0
Geographical area: USA and Canada.
Content: Volume 1 contains over 17 000 entries offering full descriptions of special libraries, information centres, and similar units. 1000 entries are brand new, and all entries have been fully revised and updated and reset in new, modern typography. Includes information about computerized services and a listing of over 700 networks and consortia, plus a subject index.
Volume 2 contains geographical and personnel indexes in two sections: the geographic index lists by state or province all the institutions described in Volume 1; the personnel index is an alphabetical roster of all library personnel mentioned.
Volume 3 contains periodical supplements to Volume 1 to keep subscribers informed of new information facilities, also cumulative indexes.
Next edition: June 1987.

Directory of special libraries in Pittsburgh and vicinity 27

Publisher: Special Libraries Association, Pittsburgh Chapter
Address: SLA Directory Committee, Westinghouse Research Center, 1310 Beulah Road, Pittsburgh, PA 15235, USA
Year: 1984
Pages: x + 151
Price: $50.00 ($25.00 to non-profitmaking organizations)
Content: There are 131 entries which include name, address, telephone, collection size, subject strengths, hours, inter-library loan policy, database systems, staff name, title, telephone and professional/network memberships. The directory includes the following indexes: alphabetical by library name (with cross references) (followed by entries in alphabetical order by name); subject index by keyword to library name; personnel index by name, title to library name, and telephone number.
Next edition: 1986-87.

Directory of special libraries in the Toronto area, tenth edition 28

Publisher: Special Libraries Association, Toronto Chapter
Address: 801 Bay Street, 3rd Floor, Toronto, Ontario M7A 2B2, Canada
Year: 1985
Pages: 300
Price: $35.00 (to members); $40.00 (to non-members)
ISSN: 0710-9342
Geographical area: Toronto, Ontario.

Education and information science libraries 29

Editor: Brigitte T. Darnay
Publisher: Gale Research Company
Address: Book Tower, Detroit, MI 48226, USA
Year: 1983
Price: $145.00
ISBN: 0-8103-0434-1
Series: Subject directory of special libraries (volume 2)
Geographical area: USA and Canada.
Content: The publication is one of five volumes containing the same information as is found in volume 1 of *Directory of Special Libraries and Information Centres* (see separate entry) rearranged under subject sections.
A typical entry provides nearly two dozen useful facts about the library, including: official name; name of sponsoring organization or institution; address and zip code; name and title of person in charge; names and titles of other professional personnel; collection statistics; description of the subjects with which the library or collection is concerned; policies regarding use of the collection; services provided; and telephone number with area code. In addition to a subject index, each volume contains an alternative name index, which provides cross references from variant names for libraries.
Next edition: September 1987.

Encyclopedia of associations: regional, state, and local organizations - northeastern US 30

Editor: Katherine Gruber
Publisher: Gale Publishing Company
Address: Book Tower, Detroit, MI 48226, USA
Year: 1985 (first vol)
Price: $80.00
Geographical area: Northeastern USA.
Content: This is the first volume of seven under preparation to cover 50 000 regional, state, and local non-profit making membership organizations. Typical entries cover: name and address; acronym; telephone; executive officer; affiliations; year founded; membership; objectives; publications; convention/meeting data. The seven projected volumes will cover the following areas: north eastern; middle Atlantic; south eastern; Great Lakes; south west and south central; western; and north western and Great Plains states. The price will be $80.00 per volume, or $400.00 for the set.

Encyclopedia of associations: volume 1, national organizations of the US 31

Editor: Denise Akey
Publisher: Gale Research Company
Address: Book Tower, Detroit, MI 48226, USA
Year: 1985
Pages: 2023 (in 2 parts)
Price: $195.00
ISBN: 0-8103-1690-0
Geographical area: USA.
Content: Volume 1 contains more than 18 140 descriptions of organizations arranged in 17 subject categories, which are listed in alphabetical and keyword indexes.
Volume 2 of the publication contains geographical and executive indexes (price $175.00)
Volume 3. *New Associations and Projects* provides a periodical supplement to Volume 1 (subscription price $190.00).
Volume 4 covers International organizations.
In addition, the encyclopedia has also started introducing seven new volumes covering regional, state, and local organizations within the USA (see above).

Encyclopedia of governmental advisory organizations 32

Editor: Denise Allard Adzigian
Publisher: Gale Research Company
Address: Book Tower, Detroit, MI 48226, USA
Year: 1983
Pages: 964
Price: $390.00
ISBN: 0-8103-0254-3
Geographical area: USA.
Content: Each entry provides up to thirteen points of information about the organization covered, including official name, address, telephone number, executive secretary or director, history and authority, programme, findings and/or recommendations, membership, staff, subsidiary units, publications and reports, meetings, and remarks. A single alphabetical and keyword index provides access to entries of specific advisory bodies. Committees are indexed under their official name, popular name, all permutations of significant keywords in their name, and names of their reports, where appropriate.
Periodical updating supplements are published for an inter-edition subscription of $290.00.
Next edition: September 1985.

Encyclopedia of information systems and services, volume 2 - United States 33

Editor: John Schmittroth Jr
Publisher: Gale Research Company
Address: Book Tower, Detroit, MI 48226, USA
Year: 1985
Pages: 1400
Price: $190.00
ISBN: 0-8103-1541-6
Content: The encyclopedia covers about 2200 information organizations, systems, and services of international, national, or regional scope, that are located in the USA. Entries are divided into the following categories: information providers; information access services; information sources on the information industry; and support services. Entries furnish: name, address, and telephone number; year founded; head; staff; related organizations; description of systems or service; scope and/or subject matter; input sources; holdings and storage media; publications; microforms products and services; computer-based products and services; other services; clientele/availability; projected publications and services; remarks/addenda; contact person.
The encyclopedia contains the following indexes: master index; data bases; publications; software; function/services classifications; personal names; geographical; subjects.

See separate entry for: Encyclopedia of Information Systems and Services 1985-86 Sixth Edition, Volume 1 - International Volume.
The two volumes are also available as a set, price $325.00, ISBN 0-8103-1537-8.

Engineering, science, and computer jobs 1986 34

Editor: Christopher Billy, John Wells
Publisher: Peterson's Guides Incorporated
Address: PO Box 2123, 166 Bunn Drive, Princeton, NJ 08540, USA
Year: 1986
Pages: 700
Price: $15.95
ISBN: 0-87866-348-7
Series: Peterson's annual guides/career
Geographical area: USA
Content: The guide gives up-to-date information on over 1000 manufacturing, research, consulting, and government organizations currently hiring technical graduates.

Environmental directory 35

Publisher: Office of Environmental Quality Control
Address: 550 Halekauwila Street, Room 301, Honolulu, HI 96813, USA
Price: Free
Geographical area: State of Hawaii.
Content: Listing of government agencies and organizations involved with environmental matters and issues, giving expertise and interests.

Environmental impact statement directory: the national network of EIS - related agencies and organizations 36

Editor: Marc Landy
Publisher: Plenum Publishing Corporation
Address: 233 Spring Street, New York, NY 10013, USA
Year: 1981
Pages: ix + 367
Price: $102.00
ISBN: 0-306-65195-5
Geographical area: USA.
Content: The directory contains approximately 4000 entries, listing general directories, physical directories, and cultural directories.
Not available on-line.

Federal biotechnology funding sources 37

Editor: Dr Oskar R. Zaborsky, Brenda K. Young
Publisher: OMEC Publishing Company
Address: 1128 16th Street NW, Washington, DC 20036, USA

Year: 1984
Pages: x + 262
Price: $45.00
ISBN: 931283-23-X
Geographical area: USA.
Content: The publication provides the most up-to-date information available on US government programmes that fund extramural r&d relevant to biotechnology. Entries consist of 329 programmes and activities in 33 federal agencies and departments, arranged alphabetically. Each entry describes the programme and its research trusts. 282 contact individuals are named. The directory is indexed by organization/programme and contact individual. Small business programmes and general reference sections are appended.

Federal technology resources directory*   38

Publisher: National Technical Information Service
Address: 5285 Port Royal Road, Springfield, VA 22161, USA
Year: 1984
Price: $25.00
Geographical area: USA.
Content: The directory covers about 800 federal laboratories and engineering centres offering advice, specialized equipment, and facilities to assist non-governmental technicians, researchers, and inventors. Entries include: facility name; address; telephone; contact person; and capabilities.

Government programs and projects directory   39

Editor: Anthony T. Kruzas, Kay Gill
Publisher: Gale Research Company
Address: Book Tower, Detroit, MI 48226, USA
Year: 1983-84
Price: $135.00 (for the set)
ISBN: 0-8103-0422-8
Geographical area: USA.
Content: Consolidating information about federal programmes, this directory provides facts and figures on hundreds of programmes implemented, managed and supported by executive departments and independent agencies of the federal government. Entries furnish information gleaned from scattered reports and government publications as well as questionnaire responses and correspondence, including: name of programme or project; sponsoring department or agency; intermediate offices with programme responsibility; address of sponsoring agency; legislative authorization; funding; programme description; and sources of information; index. The publication consists of 1574 entries in three issues.

Government research directory   40

Editor: Kay Gill
Publisher: Gale Research Company
Address: Book Tower, Detroit, MI 48226, USA
Year: 1985
Pages: 500
Price: $295.00
ISBN: 0-8103-0463-5
Geographical area: USA.
Content: The information provided in GRCD was obtained from each unit through questionnaires, publications, and/or its spokesperson. Among the information elements found in the entries are: name, address, and telephone number; date established; director or chief administrator; place within governmental department; related contractor, state agency, university, or other cooperating organization; size and composition of staff; type of research activity; principal fields of research; where results are published; publications; recurring seminars, meetings, and special programmes; subsidiary branches or affiliated units. There are three indexes. The first provides access to the main entries by name, acronym, and subject keyword. The second is a classified agency index arranged in the sequence used in the *United States Government Manual*. New to the second edition of GRCD is the geographical index, which is organized by state and by city within state. An updating supplement is also published.
Next edition: September 1986; $290.00.

Graduate programs in engineering and applied sciences 1986   41

Editor: Amy J. Goldstein, Charles Granade
Publisher: Peterson's Guides Incorporated
Address: PO Box 2123, 166 Bunn Drive, Princeton, NJ 08540, USA
Year: 1985
Pages: 900
Price: $24.95
ISBN: 0-87866-346-0
Series: Peterson's annual guides to graduate study
Geographical area: USA and Canada.
Content: This directory, updated annually, is a guide to graduate and professional degrees offered in 45 academic areas including biomedical engineering, chemical engineering, computer and information sciences, environmental engineering, materials and polymer sciences, and operations research.
Next edition: 1986.

Graduate programs in the biological, agricultural, and health sciences 1986   42

Editor: Amy J. Goldstein, Barbara Ready
Publisher: Peterson's Guides Incorporated
Address: PO Box 2123, 166 Bunn Drive, Princeton, NJ 08540, USA

Year: 1985
Pages: 2050
Price: $28.95
ISBN: 0-87866-344-4
Series: Peterson's annual guides to graduate study
Geographical area: USA and Canada.
Content: This directory, updated annually, is a guide to graduate and professional degree offerings in 80 academic areas, including medicine, nursing, pharmacology, veterinary medicine, biology and biomedical sciences, animal sciences, food science and horticulture.
Next edition: 1986.

Graduate programs in the physical sciences and mathematics 1986   43

Editor: Amy Goldstein, Charles Granade
Publisher: Peterson's Guides Incorporated
Address: PO Box 2123, 166 Bunn Drive, Princeton, NJ 08540, USA
Year: 1985
Pages: 650
Price: $22.95
ISBN: 0-87866-345-2
Series: Peterson's annual guides to graduate study
Geographical area: USA and Canada.
Content: This directory, updated annually, is a guide to graduate and professional degree offerings in 20 academic areas, including astronomy, biometrics, chemistry, statistics, atmospheric sciences, geochemistry, geology, marine sciences/oceanography, mathematics, and physics.
Next edition: 1986.

Guide to American directories, eleventh edition*   44

Publisher: Klein, B. Publications
Address: PO Box 8503, Coral Springs, FL 33065, USA
Year: 1982
Price: £45.50
ISBN: 6533-5248
Geographical area: USA.
Content: Information on 7000 directories, arranged by subject.

Guide to scientific instruments*   45

Editor: Richard G. Sommer
Publisher: American Association for the Advancement of Science
Address: 1515 Massachusetts Avenue NW, Washington, DC 20005, USA
Year: 1984
Price: $14.00
Geographical area: USA.
Content: The publication includes 2000 manufacturers of instruments, apparatus, chemicals, furniture, software, etc, for use in all scientific disciplines in industrial, academic, governmental, and other laboratory facilities. Entries are arranged alphabetically, and include company name, address, and telephone, with product indexes. The principal content is 2400 categories of laboratory equipment, instruments, and apparatus, cross-referenced by their alternate names.
Next edition: 1985.

Guide to US government directories volume 2, 1980-84*   46

Editor: Donna Rae Larson
Publisher: Oryx Press
Address: 2214 North Central at Encanto, Phoenix, Arizona, USA
Year: 1985
Price: $55.00
ISBN: 0-89774-162-5
Content: Information on and annotated listing of US government directories published between 1980 and 1984.

Industrial research laboratories of the United States, nineteenth edition   47

Editor: Jacques Cattell Press
Publisher: Bowker Publishing Company
Address: 58/62 High Street, Epping, Essex CM16 4BU, UK
Year: 1985
Pages: x + 742
Price: £176.25
ISBN: 0-8352-2070-2
Geographical area: USA.
Content: The main body is comprised of nearly 6000 detailed entries arranged alphabetically, with laboratories appearing under the name of the parent company. Information includes the name, address, telephone number, telex and TWX, cable address, and where applicable, divisions or subsidiaries. The location of each laboratory is also provided with the names and titles of chief administrative and research personnel.

Also included are three indexes for easy access: the geographical index arranges the parent organizations and their facilities alphabetically by state and then by city; the personnel index alphabetically lists all administrative executives and researchers included in the main entries; and the classification index arranges by subject all research and development activities conducted in the facilities contained in the volume.
Next edition: 1987.

NORTH AMERICA

Leading consultants in technology, second edition 48

Publisher: Dick, J., Publishing (Research Publications)
Address: 12 Lunar Drive, Woodbridge, CT 06525, USA
Year: 1985
Pages: xv + 1996 (2 vols)
Price: $215.00
ISBN: 0-89235-089-X
Geographical area: North America.
Content: Contains 17 000 listings of the professional data of leading scientists and engineers who are available as technical consultants in North America. Arranged by state within technical discipline; all profiles are keyword indexed by 1500 key technical words cross-referenced by city and state; a master alphabetic index is also included. Not available on-line.
Next edition: November 1986; $260.00

National faculty directory 49

Publisher: Gale Research Company
Address: Book Tower, Detroit, MI 48226, USA
Year: 1984
Pages: 4114 (in 3 vols)
Price: $435.00 (for the set)
ISBN: 0-8103-0496-1
Geographical area: USA and Canada.
Content: Now identifying and locating about 600 000 individuals with faculty status, the directory is an up-to-date and reliable guide to who's where in the academic world. It provides comprehensive coverage of about 3030 junior colleges, colleges, and universities in the United States and 120 selected Canadian institutions. Arranged in one alphabetic sequence, individual entries furnish enough information to locate specific individuals: individual's name; department; institution name; street address if necessary, and city, state and zip code. Complementing the main listings is a geographical listing of the institutions covered.
Next edition: July 1985.

New biotechnology marketplace: USA and Canada* 50

Publisher: EIC Intelligence
Address: 48 West 38th Street, New York, NY 10018, USA
Year: 1984
Price: $295.00
Content: Profiles of biotechnology speciality companies.

Official directory of Canadian museums and related institutions* 51

Editor: Denis Roussel
Publisher: Canadian Museums Association

Address: 280 Metcalfe Street, Suite 202, Ottawa, Ontario K2P 1R7, Canada
Year: 1984
Pages: xx + 235
Price: $35.00
ISBN: 0-919106-16-1
Content: 1702 entries arranged alphabetically by city and province. The directory contains three indexes: index of institutions; index of categories of institutions; index of key personnel.
Next edition: January 1986; $35.00.

Ohio research and development facilities directory 52

Publisher: Department of Economic and Community Development
Address: Box 1001, Columbus, OH 43216, USA
Year: 1978
Pages: 178
Availability: OP
Geographical area: State of Ohio.
Content: Indexes under the following titles: industrial research and development facilities; educational research and development facilities; hospital research facilities; state government research facilities; federal government research facilities. Alphabetical index; geographical index; capability index.

Publishers directory 53

Editor: Linda S. Hubbard
Publisher: Gale Research Company
Address: Book Tower, Detroit, MI 48226, USA
Year: 1984
Pages: 1729 (in 2 vols)
Price: $210.00 (for the set)
ISBN: 0-8103-0412-0
Geographical area: USA and Canada.
Content: The new edition includes full entries describing nearly 9500 publishers based in the USA and Canada that are not listed in traditional sources. Included in the expanded edition are producers of books, classroom materials, reports, databases, software, and other print as well as non-print publications. A separate section covers distributors. In addition, the comprehensive publishers' index includes references to all publishers listed in *Literary Market Place*.There are also subject and geographical indexes. A softbound supplement is published between editions.
Next edition: September 1985; $240.00.

Research centers directory, ninth edition 54

Editor: Mary Michelle Watkins, James A. Ruffner
Publisher: Gale Research Company

Address: Book Tower, Detroit, MI 48226, USA
Year: 1984
Pages: 1308
Price: $310.00
ISBN: 0-8103-0459-7
Geographical area: USA and Canada.
Content: Material in RCD is arranged in sixteen subject sections, including agriculture, astonomy, business, education, engineering and technology, government and public affairs, and labour and industrial relations. Each entry furnishes some fifteen points of information about the research unit, including formal name, name of parent institution, address, telephone number, director's name, year founded, sources of support, annual budget, principal fields of activity, publications and special library facilities. The publication contains five enlarged indexes. New to the ninth edition are a special capabilities index and an acronyms of research centres index. The subject index has been expanded to include some 4000 subject headings, under which are listed some 30 000 references. The alphabetical index of research centres includes name changes as well as inactive and defunct research centres. The institutional index spotlights thousands of universities and other institutions that sponsor research in the USA and Canada.
Next edition: September 1985.

Research services directory, second edition 55

Publisher: Gale Research Company
Address: Book Tower, Detroit, MI 48226, USA
Year: 1982
Pages: 379
Price: $260.00
ISBN: 0-8103-0248-9
Geographical area: USA.
Content: The directory covers about 2000 for-profit organizations providing research services on a contract or free-for-service basis to a wide range of clients. The directory emphasizes small r&d firms and individuals specializing in particular areas and industries, including contract laboratories and consulting organizations. Detailed entries furnish information about the firm, its staff and clients, principal fields of research, activities performed, facilities, and more.
A supplement is available between editions to update the publication.
Next edition: June 1986; $325.00.

Schwendeman's directory of college geography of the United States 56

Editor: Dale Monsebroten
Publisher: Geographical Studies and Research Center
Address: Eastern Kentucky University, Richmond, KY 40475, USA
Year: 1985
Pages: iii + 127
Price: $4.00
ISSN: 0734-8185
Geographical area: USA.
Content: The directory includes information on enrolment data in regional, historical, philosophical, and technical courses.
Next edition: April 1986; $4.00.

Science and technology libraries: USA and Canada 57

Editor: Brigitte T. Darnay
Publisher: Gale Research Company
Address: Book Tower, Detroit, MI 48226, USA
Year: 1985
Price: $145.00
ISBN: 0-8103-1890-3
Series: Subject directory of special libraries and information centres (volume 5)
Content: The publication is one of five volumes containing the same information as is found in volume 1 of *Directory of Special Libraries and Information Centres* (see separate entry) rearranged under subject sections.
A typical entry provides nearly two dozen facts about the library, including: official name; name of sponsoring organization or institution; address with zip code; name and title of person in charge; names and titles of other professional personnel; collection statistics; description of the subjects with which the library or collection is concerned; policies regarding use of the collection; services provided; and telephone number with area code.
In addition to a subject index, each volume contains an alternative name index, which provides cross-references from variant names for libraries.
Next edition: September 1987.

Scientific and technical organizations and agencies directory 58

Editor: Margaret Labash Young
Publisher: Gale Research Company
Address: Book Tower, Detroit, MI 48226, USA
Year: 1985
Pages: 1000
Price: $125.00
ISBN: 0-8103-2100-9
Geographical area: USA.
Content: This directory gives sources of information in science and technology and covers a broad range of organizations, including government and private research centres, federal and state government agencies, engineering consultants, domestic assistance programmes, databases and computerized information sources, sci-tech book publishers, technical libraries and information centres, and more. In addition to full name, address, telephone number, and contact person, entries give founding date, number of members and staff size,

Scientific and technical societies of Canada 59

Publisher: Canada Institute for Scientific and Technical Information
Address: National Research Council of Canada, Ottawa, Ontario K1A 0S2, Canada
Year: 1984
Pages: 188
Price: Can$12.00
Geographical area: Canada.
Content: The publication lists nearly 500 national, provincial, and regional societies devoted to scientific or technical disciplines. Entries include officers, activities, publications. There is a keyword-out-of-context index using title of society keywords, and an index to publications of the societies.
Not available on-line.

Selected federal computer-based information systems 60

Editor: Saul Herner, Matthew J. Vellucci
Publisher: Information Resources Press
Address: 1700 N Moore Street, Suite 700, Arlington, VA 22209, USA
Year: 1972
Pages: ix + 215
Price: $15.00
ISBN: 0-87815-007-2
Geographical area: USA.
Content: Verbal and graphic descriptions of 35 major operating computer information systems in the US Federal Government. Uniform descriptions and flowcharts are used throughout, thus conveying a clear understanding of system structures and operations and facilitating intersystem comparisons.

Subject collections: a guide to special book collections and subject emphases as reported by university, college, public and special libraries and museums in the United States and Canada 61

Editor: Lee Ash
Publisher: Bowker Publishing Company
Address: 58/62 High Street, Epping, Essex CM16 4BU, UK
Year: 1985
Pages: 2144 (in 2 vols)
Price: £195.00
ISBN: 0-8352-1917-8

Geographical area: USA, Canada, Puerto Rico.
Content: This new edition covers collections housed in over 20 000 academic, public, and special libraries in the United States and Canada, as well as those located in 1000 museums in the United States, Canada, and Puerto Rico.
Collections are indexed under Library of Congress subject headings. Under each subject heading, the holding libraries are arranged alphabetically by state with the following data in each library entry: name and address of library; number of volumes within collection; holdings other than books (maps, pictures, slides, films, etc.); and photocopying and loan restrictions.

Subject directory of special libraries and information centers 62

Editor: Brigitte T. Darnay
Publisher: Gale Research Company
Address: Book Tower, Detroit, MI 48226, USA
Year: 1985
Pages: In 5 vols
Price: $625.00 (for the set)
ISBN: 0-8103-1890-3
Geographical area: USA and Canada.
Content: The five-volume publication contains the same information as is found in Volume 1 of *Directory of Special Libraries and Information Centres* (see separate entry) rearranged under subject sections. They are as follows: volume 1 - Business and Law Libraries; volume 2 - Education and Information Science Libraries; volume 3 - Health Science Libraries; volume 4 - Social Sciences and Humanities Libraries; volume 5 - Science and Technology Libraries. Information on volumes 2, 3, and 5 can be found in the appropriate sections of this chapter.
Next edition: September 1987.

United States and the global environment: a guide to American organizations concerned with the international environmental issues 63

Publisher: California Institute of Public Affairs
Address: PO Box 10, Claremont, CA 91711, USA
Year: 1983
Price: $25.00
ISBN: 0-912102-45-4
Series: Who's doing what
Content: In-depth profiles of over 100 US scientific, citizens', and university groups, as well as key government agencies.

US code name directory* 64

Publisher: DMS Incorporated
Address: DMS Building, 100 Northfield Street, Greenwich, CT 06830, USA

Price: $150.00
Geographical area: USA.
Content: Entries for over 17 000 code names and acronyms used by the Department for Defense and Industry.

US directory of environmental sources* 65

Publisher: Environmental Protection Agency
Address: 401 M Street NW, Room 2903B, Washington, DC 20460, USA
Year: 1981
Pages: 1000
Price: $67.00 (paper); $4.50 (microfiche)
Geographical area: USA.
Content: The directory covers 1495 United States environmental organizations ('sources', for purposes of the directory) registered with USIERC, the US National Focal Point of the United Nations Environmental Program's International Referral System (UNEP/INFOTERRA). Organizations include federal, state, and local government agencies; universities; industry; societies; laboratories; etc. Entries include organization name, address, telephone, subject interests, and description of activities, group served, etc, and are arranged at random, with sequential numerical identifiers which are used in subject, alphabetical, and geographical indexes.

Washington information directory 1985-86 66

Publisher: Congressional Quarterly Incorporated
Address: 1414 22nd Street NW, Washington, DC 200037, USA
Year: 1985
Pages: 965
Price: $39.95
ISBN: 0-87187-340-0
Series: Washington information directory
Content: The directory gives information on the following: communications and the media; economics and business; education and culture; employment and labour; energy; equal rights - minorities, women; government personnel and services; health and consumer affairs; housing and urban affairs; individual assistance programmes; international affairs; law and justice; national security; natural resources, environment, and agriculture; science, space, and transport; congress and politics.
Next edition: April 1986; $43.00.

Who's who in consulting: a reference guide to professional personnel engaged in consultation for business, industry, and government 67

Editor: Paul Wasserman, Janice McLean
Publisher: Gale Research Company
Address: Book Tower, Detroit, MI 48226, USA
Year: 1982-83
Price: $165.00
ISBN: 0-8103-0361-2
Geographical area: USA and Canada.
Content: Falling within the scope of *Who's Who in Consulting* are consultants that are currently active in either full-time or part-time consulting efforts for business, industry, or government. Those listed are either staff members of private consulting organizations or specialists in the academic community who accept occasional consulting assignments. While it is impossible to cover every individual engaged in consulting, the three issues of *Who's Who in Consulting: A Periodic Supplement* provide useful information on a significant portion of the consulting community in the United States and Canada. Foreign consultants are included, if their work is related to activities in the United States or Canada. Biographical entries are arranged alphabetically by the individual's name and typically include such information as: name of individual; business address, including telephone number; date and place of birth; education, colleges attended, degrees received; career; consulting specialities; honours or awards; fields of activity.

Who's who in frontier science and technology 68

Publisher: Marquis Who's Who Incorporated
Address: 200 East Ohio Street, Chicago, IL 60611, USA
Year: 1984
Pages: xvi + 846
Price: $84.50
ISBN: 0-8379-5701-X
Geographical area: North America.
Content: More than 16 000 entries, alphabetically arranged, containing biographical information. Title(s) of indexes: complete index to 350 subspecialties, including embryo transplants, evolutionary biology, foetal surgery, fibre optics, genetics, imaging technology, laser medicine, radiology, robotics, and satellite studies.
Next edition: November 1985; $112.00.

Who's who in technology today, fourth edition 69

Publisher: Dick, J., Publishing (Research Publications)
Address: 12 Lunar Drive, Woodbridge, CT 06525, USA
Year: 1984
Pages: xv + 4086 (5 vols)
Price: $468.00

ISBN: 0-943692-15-6
Geographical area: North America.
Content: Contains 32 000 biographical profiles of leading scientists, engineers, and technology managers in North America. Profiles are arranged alphabetically under discipline. All profiles are keyword indexed, based upon the principal expertise of the biographee. This results in a 130 000 line index using 1500 key technical words. A master alphabetic index is included in volume 5. Not available on-line.
Next edition: August 1986; $495.00

World meetings: United States and Canada 70

Publisher: Macmillan Publishing Company Incorporated
Address: 200D Brown Street, Riverside, NJ 08370, USA
Year: Quarterly
Price: $160.00 (per annum)
ISBN: 0-02-695270-X
Content: Details on all important scientific, technical, and medical meetings and conferences to be held in the USA and Canada during the next two years. Five separate indexes help to locate the main entry as well as names and addresses of sponsors, information contacts, content of meetings, exhibits and products on show, etc.

AGRICULTURE AND FOOD SCIENCE

American fisheries directory and reference book* 71

Address: 21 Elm Street, Camden, Maine 04843, USA
Year: 1979
Pages: 560

American Veterinary Medical Association directory 72

Editor: Jan LaFrana
Publisher: American Veterinary Medical Association
Address: 930 North Meacham Road, Schaumburg, IL 60196, USA
Year: 1985
Pages: 675
Price: $30.00
ISSN: 0066-1147
Geographical area: USA and Canada.
Content: Lists 39 945 members of the American Veterinary Medical Association and 3033 non-member veterinarians in alphabetical and geographical order with identification of practice activity and school and year of graduation. A 270-page reference section contains information on the AVMA, other veterinary and related associations, government agencies, specialty boards, veterinary schools, a digest of state practice acts, directory of information sources, and film, videotape and materials catalogues.
Next edition: January 1986; $35.00.

Annotated acronyms and abbreviations of marine sciences and related activities* 73

Editor: C.M. Ashby, A.R. Flesh
Publisher: National Oceanographic Data Centre
Address: National Science Foundation, 1800 G. Street NW, Washington, DC 20550, USA
Year: 1981
Pages: 349

Bird collections in the United States and Canada* 74

Editor: R.C. Banks, M.H. Clench, J.C. Barlow
Publisher: Journal: Auk 90, p 136-170
Address: Allen Press Incorporated, Box 368, Lawrence, KS 66044, USA
Year: 1973
Pages: 34

Canada's who's who of the poultry industry 75

Publisher: Farm Papers Limited
Address: 605 Royal Avenue, New Westminster, British Columbia V3M IJ4, Canada
Year: 1984
Price: Can$10.00
ISSN: 0068-8134
Geographical area: Canada.
Content: Listings, by province, including addresses, of departments of agriculture, poultry specialists, provincial university poultry departments, registered hatcheries, egg stations, poultry processors, feed manufactures, drug companies, poultry industry boards and associations, agricultural representatives for banks.
Next edition: June 1985; Can$10.00.

Canadian forest industries 76

Editor: Steve Pawlett
Publisher: Southam Communications Limited
Address: 1450 Don Mills Road, Don Mills, Ontario M3B 2X7, Canada
Year: 1985
Price: $3.75

Directory of animal disease diagnostic laboratories 77

Publisher: US Government Printing Office, Superintendent of Documents
Address: Washington, DC 20402, USA
Year: 1982
Pages: vii + 233
Price: $6.00
Geographical area: USA.

Directory of consultants in environmental science, first edition 78

Publisher: Dick, J., Publishing (Research Publications)
Address: 12 Lunar Drive, Woodbridge, CT 06525, USA
Year: 1985
Pages: x + 163
Price: $83.00
ISBN: 0-89235-094-6
Geographical area: North America.
Content: Contains professional listings of 1211 scientists and engineers in the fields of air, land, and water projects; environmental analysis; and agricultural analysis who are available for consulting assignments. Keyword indexed by 1500 technical terms and cross-referenced by city/state.
Next edition: January 1987; $95.00

Directory of food and nutrition information services and resources* 79

Editor: Robin C. Frank
Publisher: Oryx Press
Address: 2214 North Central at Encanto, Phoenix, Arizona, USA
Year: 1985
Price: £88.30
ISBN: 0-89774-078-5
Geographical area: North America.
Content: Food and nutrition information, services, and resources.

Directory of forest tree nurseries in the United States 80

Publisher: US Department of Agriculture, Forest Service
Address: 12th and Independence SW, PO Box 2417, Washington, DC 20013, USA
Year: 1976
Pages: 33
Price: Free
Content: Nurseries are listed alphabetically under states; bareroot and container operations are distinguished.

Directory of marine and freshwater scientists in Canada* 81

Publisher: Journal: Canadian Special Publication of Fisheries and Aquatic Sciences, 73
Address: Suite 603, 77 Metcalf Street, Ottawa, Ontario K1P 5LP, Canada
Year: 1984
Pages: 530

Directory of natural science centers 82

Editor: Mildred DeScherer
Publisher: Natural Science for Youth Foundation
Address: 763 Silvermine Road, New Canaan, CT 06840, USA
Year: 1984
Pages: v + 222
Price: $15.00
Geographical area: USA, Canada, and the Virgin Islands.
Content: Approximately 1200 entries, arranged as follows: alphabetically by state and then alphabetically within state according to location; name and address and telephone number (where possible), hours open, admission fee or not, additional information; seven-page xeroxed supplement.

Directory of the forest products industry 83

Publisher: Miller Freeman Publications Incorporated
Address: 500 Howard Street, San Francisco, CA 94105, USA
Year: 1982
Pages: 600
Price: $107.00
Geographical area: USA and Canada.
Content: Contents include: wood harvesting; primary forest products; producers; secondary forest products manufacturers; wholesalers; index by type of operation; index by lumber specialties; industry statistics; forest area maps; typical grade stamps; agencies and organizations; classified buyers' guide; general index.
Next edition: October 1985; $107.00.

Field offices of the forest service 84

Publisher: US Department of Agriculture, Forest Service
Address: PO Box 2417, Washington, DC 20013, USA
Year: 1978
Pages: 4
Geographical area: USA.
Content: Includes names and addresses of regional offices, research headquarters and state and private forestry areas.

Hawaii agricultural export products directory 85

Publisher: Department of Agriculture, State of Hawaii
Address: 1428 South King Street, Honolulu, HI 96814, USA
Price: Free
Content: Lists exporters and distributors of Hawaii-grown agricultural products, giving name, sales contact, address, telephone, types of products handled.

ICAR Inventory of Canadian agricultural research* 86

Publisher: Canadian Agricultural Research Council
Address: Central Experiment Farm, KW Neatby Building, Room 1137, Ottawa, Ontario K1A 0C6, Canada
Geographical area: Canada.
Content: The directory covers about 4250 agricultural research and development projects conducted by Canadian federal and provincial governments, universities, industry, and other organizations. Entries include: project title; location; name, address, and telephone number of sponsoring organization; names and titles of researchers; description of project.
Available on-line from CISTI under title 'ICAR'; updated annually.

Inventory of agricultural research 87

Editor: Department of Agriculture-Cooperative State Research Service
Publisher: Congressional Information Service Incorporated
Address: 4520 East-West Highway, Suite 800, Bethesda, MD 20814, USA
Year: 1985
Pages: xxv + 249
Price: $65.76
Series: American statistics index
Geographical area: USA.
Content: Copies are available for each year back to 1972.
The American Statistics Index, of which this publication is a part, is now available on-line via DIALOG Information Services Incorporated and System Development Corporation's ORBIT IV.
Next edition: 1986; $65.00.

List of California herbaria and working collections* 88

Publisher: Department of Food and Agriculture
Address: Division of Plant Industry, State of California, Sacramento, California, USA
Year: 1977
Pages: 48

Organization and functions of the Eastern Regional Research Center federal research 89

Publisher: United States Department of Agriculture, Eastern Regional Research Center
Address: 600 E Mermaid Lane, Philadelphia, PA 19118, USA
Year: 1982
Pages: 24
Geographical area: USA.
Content: Twenty-eight entries, arranged by laboratory. Not available on-line.
Next edition: 1985.

Technology 1984 90

Publisher: American Society of Agricultural Engineers
Address: 2950 Niles Road, St Joseph, MI 49085, USA
Year: 1985
Series: Agricultural engineering magazine
Geographical area: USA.
Content: New product and manufacturers' directory listing over 2000 manufacturers of agricultural equipments and products they supply.

CHEMICAL AND MATERIALS SCIENCE

ACS directory of graduate research (chemistry)* 91

Publisher: American Chemical Society
Address: 1155 Sixteenth Street NW, Washington, DC 20036, USA
Year: 1983
Pages: 1230
Price: $43.00
Geographical area: USA and Canada.
Content: The directory covers about 715 institutions offering master's and/or doctoral degrees in chemistry, chemical engineering, biochemistry, pharmaceutical/medicinal chemistry, clinical chemistry, and polymer science. Entries include: name of institution, address, telephone, department name, name of chairperson, degrees offered, fields of specialization; names and birth dates of faculty members, their educational backgrounds, special research interests, personal telephone number, and recent publications; names and thesis titles of recent graduates. They are classified by discipline, then alphabetically by keyword in institution name, and there is a personal name index. The directory also includes statistics by department on staff size, degrees granted, graduate students, postdoctoral appointments, etc.
Next edition: October 1985.

American chemists and chemical engineers   92

Editor: Dr Wyndham D. Miles
Publisher: American Chemical Society
Address: 1155 Sixteenth Street NW, Washington, DC 20036, USA
Year: 1976
Pages: x + 544
Price: $29.95 (US/Canada); $35.95 (export)
ISBN: 0-8412-0278-8
Geographical area: North America.
Content: A collection of biographies on 500 American chemists in various positions, including biographical references and index.

American Council of Independent Laboratories: directory   93

Publisher: Repro Incorporated
Address: 1725 K Street NW, Washington, DC 20006, USA
Year: 1984
Pages: 263
Price: Postage only
Geographical area: USA.
Content: Listing of member laboratories of American Council of Independent Laboratories, published biennially. Entries by laboratory - indexed geographically, alphabetically, and by discipline. Not yet available on-line. A comprehensive Index of Services is sent with each directory.
Next edition: April 1986.

Chemcyclopedia 85   94

Publisher: American Chemical Society
Address: 1155 Sixteenth Street NW, Washington, DC 20036, USA
Year: 1984
Pages: 320
Price: $40.00 (USA and Canada); $48.00 (export)
Geographical area: North America.
Content: Listing of chemicals giving trade names, available forms, special requirements, and potential applications. Chemicals are listed in categories consistent with accepted patterns of use. Indexes to chemical names, and by company/supplier (including address, telephone, and telex).

Chemical buyers guide   95

Editor: Arthur Kendrick
Publisher: Southam Communications Limited
Address: 1450 Don Mills Road, Don Mills, Ontario M3B 2X7, Canada
Year: 1985
Pages: 124
Price: Can$45.00
Series: Process industries Canada
Geographical area: Canada.
Content: Lists Canadian suppliers of commercial chemicals and new materials.
Next edition: 1986.

Chemfacts: Canada, first edition   96

Publisher: Chemical Data Services
Address: Quadrant House, The Quadrant, Sutton, Surrey SM2 5AS, UK
Year: 1982
Pages: 118
Price: $65.00
ISBN: 0-617-00373-4
Content: Survey of the present and planned plants for 72 chemical products made in Canada, including not only basic petrochemicals and plastics but also other downstream products, inorganics and fertilizers, and profiles 106 of the country's major chemical producers.

Chemical engineering faculties*   97

Editor: Michael E. Leesley
Publisher: American Institute of Chemical Engineers
Address: 345 East 47th Street, New York, NY 10017, USA
Year: 1984
Pages: $40.00
Geographical area: USA and Canada.
Content: The publication covers nearly 4000 faculty members in chemical engineering departments of about 150 United States and 220 foreign schools. Entries include: school name, address, telephone, department name and telephone; accreditation, student chapter name and name of advisor (if any); time required to complete undergraduate work; special study plans available; highest degree granted; number of degrees granted by level of degree; faculty member names with rank and other university titles; name and title of administrative officer to whom the department reports; and name and address of placement service supervisor. Entries are arranged geographically, with indexes to faculty member name (covers United States and Canada only).

College chemistry faculties   98

Editor: Marjorie A. Grant
Publisher: American Chemical Society
Address: 1155 16th Street NW, Washington, DC 20036, USA
Year: 1983
Pages: vii + 205
Price: $34.00
Geographical area: USA and Canada.
Content: Directory of two-year college, four-year college, and university teachers of chemistry, biochemistry,

medicinal/pharmaceutical chemistry and chemical engineering in the US and Canada, including indexes of institutions and individuals.
Next edition: October 1986; $50.00.

Directory of chemical engineering consultants* 99

Editor: Christine E. Burke
Publisher: American Institute of Chemical Engineers
Address: 345 East 47th Street, New York, NY 10017, USA
Year: 1984
Pages: 30
Price: $5.00
Geographical area: North America.
Content: The directory covers: chemical engineering consultants who are members of the institute; specialties include management services, environmental engineering services, energy technology, biochemical engineering, licensing, pollution control and other similar fields. Entries include firm name, address, telephone, names of principals, description of staff, type of organization, states in which licensed, specialties, and are arranged alphabetically within separate sections for full-time and part-time consultants.
Next edition: November 1985.

Directory of chemistry co-op 100

Publisher: American Chemical Society
Address: 1155 16th Street NW, Washington, DC 20036, USA
Year: 1984
Pages: viii + 280
Price: $20.00
Geographical area: USA and Canada.
Content: Directory of chemistry cooperative education programmes in the US and Canada, listing colleges and universities reporting such programmes, and giving summary statistics about the schools, providing cooperative contacts at the schools and describing each chemistry or chemistry-related cooperative programme at the institution. The directory has five indexes: geographical listing of schools; baccalaureate chemistry programmes; graduate chemistry programmes; associate programmes; other programmes.
Next edition: 1986; $25.00.

Directory of consultants in plastics and chemicals, first edition 101

Publisher: Dick, J., Publishing (Research Publications)
Address: 12 Lunar Drive, Woodbridge, CT 06525, USA
Year: 1985
Pages: x + 318
Price: $94.00
ISBN: 0-89235-095-4

Geographical area: North America.
Content: Contains professional listings of 2765 scientists and engineers in the fields of plastics development, chemistry research, and chemical engineering who are available for consulting assignments. Keyword indexed by 1500 technical terms and cross-referenced by city/state.
Next edition: January 1987; $95.00

Directory of graduate programs in chemical engineering in Canadian universities 102

Publisher: Canadian Society for Chemical Engineering
Address: 151 Slater Street, Suite 906, Ottawa, Ontario K1P 5H3, Canada
Year: 1983
Pages: i + 70
Price: Can$6.00
Content: The entries consist of summaries of the graduate programmes in 15 Canadian universities.
Not available on-line.
Next edition: October 1985; Can$12.00.

Metalworking directory - a Dun's industrial guide* 103

Publisher: Dun and Bradstreet
Address: 27 Paul Street, London EC2, UK
Year: 1985
Price: £380.00
Geographical area: USA.
Content: Details of products, processes, and purchases of over 68 000 manufacturing plants and distributors.

Research in chemistry at private undergraduate colleges* 104

Editor: Brian Andreen
Publisher: Council on Undergraduate Research
Address: 6840 East Broadway Boulevard, Tucson, AZ 85710, USA
Year: 1985
Pages: 360
Price: $20.00
Geographical area: USA.
Content: The publication covers about 125 private college chemistry departments with about 800 faculty members. Entries are arranged alphabetically, and include: college name, address; department chairman; enrolment, number of faculty, number of staff, facilities, size of library, equipment available, funding; for faculty: members name, research interests, current publications, and grants received.

Twenty-first directory of chemical engineering research in Canadian universities 105

Publisher: Canadian Society for Chemical Engineering
Address: 151 Slater Street, Suite 906, Ottawa, Ontario K1P 5H3, Canada
Year: 1983
Pages: viii + 144
Price: Can$12.00
Geographical area: Canada.
Content: The introduction consists of a foreword, a code for research areas, a table giving distribution of research topics, and table of contents covering six main topics as follows: teaching staffs in chemical engineering in Canadian universities; research areas and corresponding research directors; current research programmes; list of graduate students; list of post doctorals; list of research associates.
Not available on-line.
Next edition: February 1985; Can$25.00.

Directory of chemical producers 106

Editor: Janet R. Hardy
Publisher: SRI International
Address: 333 Ravens Wood Avenue, Menlo Park, CA 94025, USA
Year: 1985
Pages: 1088
Price: $775
Geographical area: USA.
Content: Detailed descriptions of 1500 companies showing divisional structure, plant locations, and products manufactured at each site. More than 10 000 commercial chemical products are listed, showing producers and locations; 250 capacity tables are included. Two supplements are published during the year to update the directory.
Next edition: April 1986; $800.

EARTH AND SPACE SCIENCES

Astronomy directory (western states)* 107

Editor: Trudi Hauptman
Publisher: Astronomy Society of the Pacific
Address: 1290 24th Avenue, San Francisco, CA 94122, USA
Year: 1983
Pages: 20
Price: $3.00
Geographical area: Western USA.
Content: The directory covers planetariums, observatories, amateur astronomy groups, astronomy courses, stores specializing in astronomy equipment, and other astronomy services. About 100-150 entries in each of six separate editions cover all states in the western half of the United States. Entries include: institution or company name, address, telephone, contact person; institutional listings give time, date, cost and similar data for courses and events; commercial listings indicate merchandise carried, discounts available to society members. Entries are arranged by type of institution or service.

California water resources directory: a guide to organizations and information resources 108

Publisher: California Institute of Public Affairs
Address: PO Box 10, Claremont, CA 91711, USA
Year: 1983
Price: $20.00
ISBN: 0-912102-60-8
Content: Guide to organizations concerned with the field of water resources in California.

Current geological research in Texas: a directory of faculty and student research* 109

Editor: J.G. Price
Publisher: Bureau of Economic Geology
Address: Austin, Texas USA
Year: 1983
Pages: 29

Directory of geoscience departments - United States and Canada* 110

Publisher: American Geological Institute
Address: 4220 King Street, Alexandria, VA 22302, USA
Year: 1984
Pages: 200
Price: $16.50
Content: The directory covers about 800 colleges and universities with departments of geology, geochemistry, geophysics, and other geosciences, their staffs, and courses, including field courses. Entries are geographically arranged, and include: institution name and location, department names with addresses, telephone, faculty, and personnel. There are indexes to faculty, speciality, and institution name.
Next edition: November 1985.

NORTH AMERICA

Directory of information resources in the United States: geosciences and oceanography   111

Publisher: Library of Congress
Address: Washington, DC 20540, USA
Year: 1981
Pages: 375
Price: $8.50

Directory of North American geoscientists engaged in mathematics, statistics and computer applications   112

Editor: Dr Don Myers
Publisher: University of Arizona, Mathematics Department
Address: Tucson, AZ 85721, USA
Year: 1983
Pages: xx + 126
Price: $10.00
Geographical area: North America.
Content: 720 entries, giving name, address, phone, university degrees, specialties, professional affiliations, and photograph. Indexes are arranged by state, province, and specialty.
Not available on-line.

Graduate programs in physics, astronomy and related fields 1984-85   113

Publisher: American Institute of Physics
Address: 335 East 45th Street, New York, NY 10017, USA
Year: 1984
Pages: xvi + 912
Price: $17.50
ISBN: 0-88318-459-1
Geographical area: North America.
Content: There are 298 departments from 235 institutions included in this book. Entries are organized geographically with separate parts for institutions in the United States, Canada, and Mexico. Within these parts, entries are organized alphabetically by state or province and within each state or province alphabetically by the name of the institution. Each entry includes the following information: general; number of faculty in department; admission, financial aid, and housing; graduate degree requirements; personnel engaged in separately budgeted research; separately budgeted research expenditures by source of support; separately funded and managed laboratories; extension centres and summer programmes; faculty; research specialties and staff.
There are appendices covering the following: geographical listing of departments; alphabetical listing of departments; alphabetical listing of graduate programmes by highest degree granted; research specialties of doctoral programmes in physics, astronomy, and related fields; areas of concentration of master's programmes in physics, astronomy, and related fields; summary of data from the previous graduate programmes book.
Not available on-line.
Next edition: October 1985; $17.50.

Satellite directory*   114

Publisher: Phillips Publications Incorporated
Address: 7315 Wisconsin Avenue, Suite 1200 N, Bethesda, MD 20814, USA
Year: 1984
Price: £200.00
ISBN: 0-934960-14-3
Geographical area: Mainly USA and Canada.
Content: The directory is mainly USA in scope, but includes Canadian and some European organizations, and gives information on suppliers and regulatory bodies, orbital assignments and spacecraft technology.

Sea technology buyers guide/directory 1984-85   115

Publisher: Compass Publications Incorporated
Address: 1117 North 19th Street, Arlington, VA 22209, USA
Year: 1984
Pages: 155
Price: $19.50
Geographical area: USA.
Content: Updated technical articles and industry reports, listing of manufacturers, cross index of products and services available, ocean research vessels, geophysical survey vessels, and educational institutions. This volume is a standard reference for the ocean community and covers ships, structures, propulsion systems, power plants, electrical equipment, ordnance, detection systems, navigation and positioning, communications, instrumentation, test equipment, computers, data processing, basic materials, marine hardware, deck gear, life support, and safety equipment.
Next edition: September 1985; $19.50.

USA oilfield service, supply and manufacturers' directory, 1985*   116

Publisher: PennWell Publishing Company
Address: PO Box 1260, Tulsa, OK 74101, USA
Year: 1985
Price: £106.50
Geographical area: USA.
Content: Entries are arranged in three sections: supply companies; service companies; manufacturing companies.

US directory of marine scientists*   117

Editor: Nancy Maynard
Publisher: National Academy of Sciences
Address: 2101 Constitution Avenue, Washington, DC 20418, USA
Year: 1982
Pages: 330
Price: $7.50
Geographical area: USA.
Content: There are 3020 individuals listed, with entries arranged geographically, and giving name, office address, and areas of specialization. There are indexes to personal name and special research interest.
Next edition: Not planned.

ELECTRONICS

ARPANET directory*   118

Publisher: ARPANET Network Information Center
Address: SRI International, 333 Ravenswood, Menlo Park, CA 94025, USA
Year: 1984
Geographical area: USA.
Content: The directory covers users and host organizations on ARPANET, a military network managed by the Defense Communications Agency. Entries include: for users - user name, on-line and off-line mail addresses, telephone, and host affiliation of users; for hosts - name, address.
Next edition: 1986.

Computer and telecommunications buyers' guide   119

Publisher: Treasure Island Publishing Incorporated
Address: 1111 Fort Street Mall, Honolulu, HI 96813, USA
Price: One year's subscription - $20.00 (USA); $25.00 (elsewhere)
Geographical area: State of Hawaii.
Content: Alphabetical listing by category, brand names, and vendors in the State of Hawaii. The guide is printed as the January issue of *The Prinout* magazine and updated monthly.

Data base directory, 1984-85*   120

Editor: American Society for Information Science
Publisher: Knowledge Industry Publications
Address: 701 Westchester Avenue, White Plains, NY 10604, USA
Price: £135.00
ISBN: 0-86729-081-1
Geographical area: North America.
Content: Entries for all public on-line data bases available, giving file size, frequency of update, cover restraints on access or usage. Arranged alphabetically, with subject producer, and vendor indexes.

Data processing professionals directory   121

Publisher: Data Processing Professionals Directory Incorporated
Address: 1136 Union Mall, Suite 103, Honolulu, HI 96813, USA
Pages: $12.00
Geographical area: State of Hawaii.

Directory of consultants in computer systems, third edition   122

Publisher: Dick, J., Publishing (Research Publications)
Address: 12 Lunar Drive, Woodbridge, CT 06525, USA
Year: 1985
Pages: x + 346
Price: $83.00
ISBN: 0-89235-08703
Geographical area: North America.
Content: Contains professional data on 2500 consultants and consulting firms involved with computer systems and software. Arranged alphabetically by state; data includes experience, hardware, software applications, languages, operating systems, years of experience, etc. Keyword index by hardware brand, applications, languages, etc, which is further cross-referenced by city/state.
Next edition: February 1986; $95.00

Directory of consultants in electronics, first edition   123

Publisher: Dick, J., Publishing (Research Publications)
Address: 12 Lunar Drive, Woodbridge, CT 06525, USA
Year: 1985
Pages: x + 328
Price: $94.00
ISBN: 0-89235-097-0
Geographical area: North America.
Content: Contains professional listings of 2726 scientists and engineers in the fields of electronics, control systems, computer science, magnetics, and electrical engineering, who are available for consulting assignments. Keyword indexed by 1500 technical terms and cross-referenced by city/state.
Next edition: January 1987; $95.00

NORTH AMERICA

ISA directory of instrumentation, volume 6   124

Publisher: Instrument Society of America
Address: 67 Alexander Drive, PO Box 12277, Research Triangle Park, NC 27709, USA
Year: 1985
Pages: 1612
Price: £120.00
ISBN: 0-87664-796-4
Geographical area: USA.
Content: This is the industry's most comprehensive reference for control and/or instrumentation practitioner. Individual sections cover: products - alphabetically arranged by generic product categories followed by an alphabetical listing of the companies who manufacture each product; manufacturers - an alphabetical listing of manufacturers, with name, address, telephone and telex/TWX, and key offices internationally; trade names - arranged alphabetically and cross-referenced by manufacturer; sales offices, representatives - all manufacturers are arranged alphabetically with address, telephone, and telex/TWX, followed by sales offices and representatives in the USA and Canada as well as territory serviced; specifications - detailed information on the products; manufacturers' representatives - located in the USA and Canada with complete address, telephone, and telex/TWX and list of manufacturers they represent arranged geographically, then alphabetically or product lines they represent; standards and practices - including some of the ISA's most requested standards.
Next edition: March 1986.

Klein, S., directory of computer graphics suppliers: hardware, software, systems and services*   125

Editor: Malcolm Stiefel
Publisher: Technology and Business Communications Incorporated
Address: 730 Boston Post Road, Suite 25, Sudbury, MD 01776, USA
Year: 1985
Pages: 225
Price: $60.00
Geographical area: USA.
Content: The directory covers more than 500 manufacturers of hardware and systems and suppliers of software and services relevant to the production of charts, diagrams, maps, and other graphic materials from computer data bases. Entries include supplier name, address, telephone, name of contact, description of products and services. Listings for suppliers of software and services and hardware manufacturers also include year founded, names of officers, and number of employees.
Next edition: Spring 1986.

North American on-line directory*   126

Publisher: Bowker
Address: PO Box 5, Epping, Essex, CM16 4BU, UK
Price: £88.50
ISBN: 0-8352-1879-1
Geographical area: USA.
Content: Details of more than 1500 data base producers, on-line vendors, telecommunications networks, library networks, government agencies, etc, classified by field of activity, subject matter, and geographical location.

On-line database search services directory   127

Editor: John Schmittroth Jr, Doris Morris Maxfield
Publisher: Gale Research Company
Address: Book Tower, Detroit, MI 48226, USA
Year: 1984
Price: $120.00
ISBN: 0-8103-1698-6
Geographical area: North America.
Content: Entries in the directory furnish up to 17 points of information, including: full organization and service name, address, and telephone; key contact person; year search services started; number of staff conducting searches; on-line systems accessed; frequently searched databases; subject areas searched; associated services (such as document delivery); service availability; fee policy; search request procedure; names of search personnel. The indexes are: organization index; index to organizations by on-line systems used; index to organizations by data-bases searched; index to organizations by subject areas searched; search personnel index; geographical index. Cross-indexes of alternate names are provided before the on-line systems and databases indexed to assist users in locating the standardized version of the name used in the indexes.

Optical industry and systems purchasing directory*   128

Editor: Teddi C. Laurin
Publisher: Optical Publishing Company Incorporated
Address: Box 1146, Berkshire Common, Pittsfield, MD 01202, USA
Year: 1985
Pages: 1000 (in 2 vols)
Price: $60.00
ISSN: 0191-0647
Geographical area: North America.
Content: Over 2000 manufacturers and suppliers are listed in volume 1, the buyers' guide. Volume 2 is an optical industry encyclopedia and dictionary. Entries are arranged alphabetically, and include company name, address, telephone number, names of executives and technical personnel, and description of products, and services. There are geographical and product indexes.
Next edition: March 1986.

Robotics industry directory* 129

Editor: Philip Flora
Publisher: Technical Database Corporation
Address: 1300 South Frazier, Conroe, TX 77305, USA
Year: 1985
Pages: 355
Price: $35.00 (directory only); $50.00 (including updates)
ISSN: 0278-159X
Geographical area: USA.
Content: The directory covers: companies with manufacture industrial robots, controllers, hydraulic and pneumatic components; robotics consultants; and suppliers of automated guided-vehicle systems. Entries are arranged alphabetically, and include company name, address, telephone, names of executives, trade and brand names, products or services, marketing contact, price of product. There are product and services indexes. The publication is annual, with monthly updates.
Next edition: February 1986.

Telecommunications systems and services directory 130

Editor: Martin Connors
Publisher: Gale Research Company
Address: Book Tower, Detroit MI 48226, USA
Year: 1985
Pages: 1000
Price: $240.00
ISBN: 0-8103-1697-8
Geographical area: North America.
Content: The directory covers more than 70 interstate and international long-distance telephone services; 100 local, national, and international data communications services; 110 teleconferencing services and systems; 80 electronic mail services and systems; 25 videotex and teletext operations; and more than 100 facsimile, telegram, and related services. Each entry includes (as applicable) general description, specific applications, means of access, equipment required or provided, rate structure, geographic locations served, availability, and complete contact information, including toll-free telephone numbers. TSSD also includes a glossary of more than 300 terms, acronyms, standards, concepts, and governmental rulings related to the telecommunications field. Cumulative indexes provide access by organization and system/service names, products, and acronyms; geographic location; type of service or system; and personal names listed as contacts or heads of the particular system or service.
Next edition: August 1986; $250.

Video register 1985-86, eighth edition 131

Publisher: Knowledge Industry Publications Incorporated
Address: 701 Westchester Avenue, White Plains, NY 10604, USA
Year: 1985
Pages: 440
Price: $59.50
ISBN: 0-86729-181-8
Geographical area: USA.

Who's who in electronics 132

Editor: Kathi Graeser
Publisher: Harris Publishing Company Incorporated
Address: 2057-2 Aurora Road, Twinsburg, OH 44087, USA
Year: 1985
Pages: xvi + 1016
Price: $83.00
ISBN: 0-916512-72-X
Geographical area: USA.
Content: The publication contains data, arranged both alphabetically and geograhically, on 8500 manufacturers of electronic equipment, 4200 distributors, and 3800 independent representatives.
Next edition: January 1986; $110.00.

Who's who in microcomputing 133

Editor: DataPro Research Corporation
Publisher: McGraw-Hill Book Company (UK) Limited
Address: Shoppenhangers Road, Maidenhead, Berkshire SL6 2QL, UK
Year: 1983
Price: £31.95
ISBN: 0-07-015405-8
Geographical area: United States of America.
Content: The directory contains some 2000 entries.

ENERGY SCIENCES

AEE directory of energy professionals* 134

Publisher: Fairmont Press Incorporated
Address: 4025 Pleasantdale Road, Suite 320, Atlanta, GA 30340, USA
Year: 1982
Pages: 280
Price: $28.00
Geographical area: North America.
Content: The directory covers: engineers, architects, and consultants in energy engineering and conservation who are members of the Association of Energy Engineers, and manufacturer and supplier members; Department of Energy regional offices and directors; state

Bio-energy directory (biomass)* 135

Editor: Paul F. Bente
Publisher: Bio-Energy Council
Address: 1625 I Street NW, Suite 825A, Washington, DC 20006, USA
Year: 1984
Pages: 630
Price: $95.00
Geographical area: USA.
Content: The directory covers more than 600 biomass-to-energy programmes being pursued in government and industry; surveys sources of biomass materials and energy, microbial and thermal conversion projects, including alcohol technology and testing of biofuels. Entries include programme name, objective, name of operating group, address, names of personnel, sponsor name and its financial commitments, summary of operation, classified by type of operation. Indexes are by organization name and geographical.

California energy directory: a guide to organizations and information resources 136

Publisher: California Institute of Public Affairs
Address: PO Box 10, Claremont, CA 91711, USA
Year: 1980
Pages: 88
Price: $16.50
ISSN: 0-912102-51-9
Content: Entries for over 600 public and private organizations with maps, organization charts, indexes, and users' guide.

Directory of consultants in energy technologies, first edition 137

Publisher: Dick, J., Publishing (Research Publications)
Address: 12 Lunar Drive, Woodbridge, CT 06525, USA
Year: 1985
Pages: x + 140
Price: $83.00
ISBN: 0-89235-091-1
Geographical area: North America.
Content: Contains professional listings of 1100 scientists and engineers in the fields of photovoltaics, wind energy, fossil fuels, and mining who are available for consulting assignments. Keyword indexed by 1500 technical terms and cross-referenced by city/state.
Next edition: January 1987; $95.00.

Directory of federal energy data sources: computer products and recurring publications 138

Publisher: Federal Energy Administration
Address: Washington, DC, USA
Year: 1976
Pages: 84
Geographical area: USA.
Content: Entries are under the following section headings: electronics and electrical engineering; energy conversion (non-propulsive); computers, control, and information theory; energy policies, regulations, and studies; library and information sciences.

Energy: a guide to organizations and information resources in the United States 139

Publisher: California Institute of Public Affairs
Address: PO Box 10, Claremont, CA 91711, USA
Year: 1978
Pages: 222
Price: $20.00
ISBN: 0-912102-33-0
Series: Who's doing what
Content: Describes 1500 governmental and private organizations by subject.

Energy information directory 140

Publisher: Energy Information Agency
Address: National Energy Information Center, Department of Energy, 1000 Independence Avenue SW, Washington, DC 20585, USA
Year: 1984
Pages: 84
Price: Free
Geographical area: USA.
Content: The directory lists, by subject area, programme offices of the Department of Energy and other government agencies concerned with energy.
Next edition: 1986.

Energy research programs directory* 141

Publisher: Bowker
Address: PO Box 5, Epping, Essex CM16 4BU, UK
Year: 1981
Price: £89.00
ISBN: 0-8352-1242-4
Geographical area: USA, Canada, and Mexico.
Content: Entries provide address, telephone and telex, personnel, and research capabilities of organizations involved in energy-related research work. There is a geographical index and an index to all research activities by energy source classification.

---

energy offices. Entries include: for members - individual or company name, address, telephone number, services or products; for agencies - name, address, telephone number, name of contact or chief executive. Entries are arranged by activity.

Inventory of power plants in the US 142

Publisher: Energy Information Agency
Address: National Energy Information Center, Department of Energy, 1000 Independence Avenue SW, Washington, DC 20585, USA
Year: 1985
Pages: 96
Content: Current status of electricity generating plants - active, standby, and plants projected to go into commercial operation within the next ten years.
Next edition: 1986.

Keystone coal industry manual 1985 143

Publisher: Keystone
Address: 1221 Avenue of the Americas, New York, NY 10020, USA
Price: $120.00
Geographical area: USA and Canada.
Content: The directory gives details of coal companies and mines producing 95 per cent of the industry tonnage; directories of coal-burning utility, coke, cement and industrial plants; coal seams descriptions by state, coal reserves by company; directories of coal sales and export firms, rail and river transportation facilities, tidewater and lake docks, consultants coal mine financing organizations, Two large fold out maps are included.

Oil directories: producing and drilling in the USA 144

Publisher: Midwest Oil Register Incorporated
Address: Drawer 7248, Tulsa, OK 74105, USA
Year: 1985
Pages: 7 vols
Geographical area: USA.
Content: The prices of the seven volumes are as follows: Texas - $30.00; Oklahoma - $20.00; California - $10.00; Kansas - $10.00; Michigan, Illinois, Indiana, Kentucky - $10.00; Louisianna, Arkansas, Mississippi, Georgia, Florida - $15.00; Rocky Mountain region, Four Corners and New Mexico - $15.00.
The directories list key operating and home office personnel, from presidents to foremen, companies, partnerships, independent producers, drilling contractors, and the type of drilling equipment owned. The 135 000 entries give names, titles, telephone numbers, and mail addresses.

Oil directory of Alaska 145

Publisher: Midwest Oil Register Incorporated
Address: PO Box 700597, Tulsa, OK 74170, USA
Year: 1985
Pages: 250
Price: $10.00
Content: The directory gives details of company name; address; telephone number; what kind of company; personnel showing titles.
Not available on-line.
Next edition: Not available after 1985 edition.

Oil directory of Canada 146

Publisher: Midwest Oil Register Incorporated
Address: PO Box 700597, Tulsa, OK 74170, USA
Year: 1985
Pages: 454
Price: $20.00
Content: The directory gives details of company name; address; telephone number; what kind of company; personnel with titles.
Not available on-line.

Oil directory of Houston: Texas 147

Publisher: Midwest Oil Register Incorporated
Address: PO Box 700597, Tulsa, OK 74170, USA
Year: 1985
Pages: 486
Price: $15.00
Content: The directory gives details of company name; address; telephone number; what kind of company; personnel showing titles.
Not available on-line.
Next edition: May 1986; $15.00.

USA oil industry directory 148

Editor: William R. Leek
Publisher: PennWell Publishing Company
Address: PO Box 1260, Tulsa, OK 74101, USA
Year: 1985
Pages: 628
Price: $95.00 (USA and Canada); $118.00 (elsewhere)
ISSN: 0082-8599
Content: The directory lists major integrated companies, independent producers, fund companies, retail and wholesale petroleum marketers, state and federal agencies and petroleum-related associations. Besides a company index, a geographical cross-reference and a subject index, this directory includes: a separate personnel index supplement referencing over 32 000 personnel; More than 5000 companies, their addresses, telephone and telex numbers; the updated *Oil & Gas Journal 400* report.
Next edition: January 1986; $95.00 and $118.00.

USA oilfield service, supply and manufacturers directory 1985, third edition* 149

Publisher: PennWell Publishing Company
Address: PO Box 1260, Tulsa, OK 74101, USA

Year: 1985
Pages: 450
Price: $85.00 (USA and Canada); $106.50 (elsewhere)
Content: The directory is divided into three sections: supply companies - all companies involved in wholesale and/or retail sale of products used in all phases of the petroleum industry; service companies - companies providing services such as site preparation, catering, haulage equipment, tool rental, mud service, etc; manufacturing companies - those engaged in the engineering, design, and construction of equipment used in the oil industry.
This volume can be purchased together with *The USA Oil Industry Directory* for $160.00 (USA and Canada) and $200.00 (elsewhere).

ENGINEERING AND TRANSPORTATION

Directory of consultants in robotics and mechanics, first edition — 150

Publisher: Dick, J. Publishing (Research Publications)
Address: 12 Lunar Drive, Woodbridge, CT 06525, USA
Year: 1985
Pages: x + 239
ISBN: 0-89235-096-2
Geographical area: North America.
Content: Contains professional listings of 1958 scientists and engineers in the fields of robotics, mechanics, aeronautics, and manufacturing who are available for consulting assignments. Keyword indexed by 1500 technical terms and cross-referenced by city/state.
Next edition: January 1987; $95.00.

Directory of transportation libraries in the United States and Canada — 151

Editor: Transportation Division, Special Libraries Association
Publisher: ATE Information Services
Address: 617 Vine Street, Cincinnati, OH 45202, USA
Year: 1984
Pages: 175
Price: $11.00
Content: Listings for 146 transport libraries of all types, each with address and telephone number, contact person, and a description of the library's collection. Listings are in alphabetical order by name of institution and are accompanied by subject and geographical indexes and a separate list of state department of transportation libraries.
Not available on-line.
Next edition: 1986; $15.00.

Urban mass transit: a guide to organizations and information sources — 152

Publisher: California Institute of Public Affairs
Address: PO Box 10, Claremont, CA 91711, USA
Year: 1979
Pages: 148
Price: $25.00
ISBN: 0-912102-38-1
Series: Who's doing what
Geographical area: USA.
Content: Descriptions of 837 government and private organizations concerned with urban transport in the US, with key international and foreign groups also covered.

Who's who in engineering 1985 — 153

Editor: Gordon Davis
Publisher: American Association of Engineering Societies
Address: 415 Second Avenue NE, Washington DC 20002, USA
Pages: xvi + 900
Price: $200.00
ISBN: 0-87615-014-8
Content: Over 14 000 biographical entries of engineers of distinction and achievement throughout the world. Book contains listing of recipients of major engineering awards, and is cross-indexed by state (US), country, and area of specialization.
Next edition: October 1985; $200.00.

INDUSTRY AND MANUFACTURING

Alabama directory of mining and manufacturing 1985-86 — 154

Editor: Richard W. McLaney
Publisher: Alabama Development Office
Address: State Capitol, Montgomery, AL 36130, USA
Year: 1985-86
Pages: 800
Price: $35.00
Content: Entries cover almost 5000 firms and give the following information: company name; address; contact name and telephone number; employment; product; resources used in production; parent company name and address. The directory is divided into six sections: alphabetical, geographical; product; resource; parent company; international.
Next edition: January 1987

Alaska petroleum and industrial directory 1984 155

Publisher: Manufacturers' News Incorporated
Address: 4 East Huron Street, Chicago, IL 60611, USA
Price: $60.00
Content: The directory contains 25 000 companies and 75 000 executives giving the company name, geographical location, zip code, telephone number, product, and personnel index.
Next edition: October 1985.

Arizona directory of manufacturers 156

Publisher: Manufacturers' News Incorporated
Address: 4 East Huron Street, Chicago, IL 60611, USA
Year: 1985
Pages: 196
Price: $55.00
Content: The directory lists 4000 manufacturers and 8000 executives, giving company name, address, telephone, officers, products manufactured, and the number of exployees. Entries are listed alphabetically, geographically, and by Standard Industrial Classification.
Next edition: March 1986.

Arkansas directory of manufacturers 157

Publisher: Manufacturers' News Incorporated
Address: 4 East Huron Street, Chicago, IL 60611, USA
Year: 1984
Pages: 324
Price: $45.00
Content: The directory lists 2700 firms, giving company name, telephone number, officers' names, number of employees, year established, and export and import activities. Entries are listed alphabetically, geographically, and by Standard Industrial Classification.
Next edition: May 1985.

California manufacturers register 158

Publisher: Manufacturers' News Incorporated
Address: 4 East Huron Street, Chicago, IL 60611, USA
Year: 1985
Pages: 950
Price: $135.00
Content: Entries for 16 000 manufacturers and 90 000 executives, listed alphabetically, by city and town, by product, geographically, numerically, by Standard Industrial Classification, number of employees, and zip codes.
Next edition: January 1986.

Canadian trade index 159

Editor: Nancy L. O'Hara
Publisher: Canadian Manufacturers' Association
Address: 1 Yonge Street, Suite 1400, Toronto, Ontario, M5E 1J9, Canada
Year: 1985
Pages: 1200
Price: Can$84.00
ISBN: 0-919102-05-0
Content: The index gives a profile of Canadian companies and their products. It is broken down into seven sections: alphabetical list which includes company head office, address, telephone number, senior operating executives, principal products manufactured, export interest, etc; a geographical section which lists plants and head offices; a classified section which lists over 10 000 product headings under which each company is listed (along with the town in which they are located); farm products listing of producers of agricultural products; French or Spanish glossary to assist readers; a services section which includes customs brokers, airlines, transport specialists, Canadian Government foreign trade service, banks, exporting trading houses etc; a trade mark section. The publication is available on-line.
Next edition: January 1986; Can$90.00.

Chicago geographic edition 1985 160

Publisher: Manufacturers' News Incorporated
Address: 4 East Huron Street, Chicago, IL 60611, USA
Year: 1985 $13 480
Price: $59.95
Content: Entries are broken down in zip code sequence, and then by square block within the zip. There are names of companies, addresses, officers, telephone numbers, numbers of employees, etc.
Next edition: March 1986; $65.00.

Connecticut manufacturing directory 161

Publisher: Connecticut Labor Department, Office of Research and Information
Address: 200 Folly Brook Boulevard, Wethersfield, CT 06109, USA
Year: 1984
Pages: xvii + 331
Price: $20.00
Content: The directory lists 6000 Connecticut-based manufacturers or corporate headquarters of manufacturers arranged alphabetically, alphabetically by town and by firm, and by Standard Industrial Classification. A product index appears at the back of the book. Supplements are published quarterly at a subscription price of $7.00 which includes sales tax, postage, and handling.

Delaware directory of commerce and industry 162

Publisher: Manufacturers' News Incorporated
Address: 4 East Huron Street, Chicago, IL 60611, USA

Year: 1985-86
Pages: 225
Price: $50.00
Content: Entries cover 6000 firms and 8500 executives, with indications of import and export activities. Both services and manufacturing businesses are covered, and they are listed alphabetically, geographically, and by Standard Industrial Classification.
Next edition: March 1987

Directory: affiliates and offices of Japanese firms in the USA*   163

Publisher: JETRO
Address: c/o North Oxford Academic, 242 Banbury Road, Oxford OX2 7DR, UK
Year: 1982
Price: £50.00
Content: State-by-state directory, giving name, address, telephone, and other information on firms.

Directory of central Atlantic states - manufacturers   164

Publisher: Manufacturers' News Incorporated
Address: 4 East Huron Street, Chicago, IL 60611, USA
Year: 1985
Pages: 1062 $15 $69.00
Geographical area: States of Maryland, Delaware, Virginia, West Virginia, North and South Carolina.
Content: The directory covers 20 000 companies, listed by state alphabetically within each city. Entries cover addresses, zip codes, telephone numbers, chief officers, number of employees, products, and Standard Industrial Classification codes.
Next edition: June 1986.

Directory of Colorado manufacturers 1985   165

Editor: Gerald Allen
Publisher: University of Colorado, Bureau of Business Research
Address: Campus Box 420, Boulder, CO 80309, USA
Year: 1985
Pages: xii + 397
Price: $50.00
ISBN: 0-89478-085-9
Content: 4504 entries, listed alphabetically, geographically, and by product (using Standard Industrial Classification codes).
Next edition: July 1986; $55.00.

Directory of electric light and power companies   166

Publisher: Midwest Oil Register Incorporated
Address: PO Box 700597, 1381 E. Skelly Drive, Tulsa, OK 74170, USA
Year: 1985
Pages: 740
Price: $25.00
Geographical area: USA.
Content: The directory gives details of company name, address, telephone number, whether producer or distributor, number of customers, amount of sales, plus personnel with titles.
Not available on-line.

Directory of Florida industries   167

Publisher: Manufacturers' News Incorporated
Address: 4 East Huron Street, Chicago, IL 60611, USA
Year: 1984-85
Pages: 658
Price: $55.00
Content: The directory covers 8000 manufacturers, 16 000 executives, giving addresses, telephone numbers, names of executives, numbers of employees, etc. The companies are listed alphabetically, geographically, and by Standard Industrial Classification and products.
Next edition: October 1985.

Directory of gas utility companies   168

Publisher: Midwest Oil Register Incorporated Drawer 7248, Tulsa, OK 74105, USA
Year: 1985
Pages: 528
Price: $25.00
Geographical area: USA.

Directory of industries   169

Publisher: Committee of 100
Address: PO Box 420, Tampa, FA 33601, USA
Year: 1985
Pages: cxxx + 128
Price: $12.00
Geographical area: Tampa Metropolitan Area.
Content: 1200 entries of manufacturing industries, giving: name of firm; location; principal officer; telephone number; type of product produced; import-export code; employment (male-female). Entries are listed alphabetically and by product.
Not available on-line.
Next edition: January 1986; $15.00.

Directory of Japanese firms and representatives in Hawaii   170

Publisher: Hawaii State Department of Planning and Economic Development
Address: PO Box 2359, Honolulu, HI 96804, USA
Price: Free
Content: Lists local contact, address, and main activity of major Japanese firms in Hawaii.

Directory of Kansas manufacturers and producers 171

Publisher: Manufacturers' News Incorporated
Address: 4 East Huron Street, Chicago, IL 60611, USA
Year: 1985
Pages: 453
Price: $44.00
Content: 4200 firms are catered, listed alphabetically, by counties, and cities. Entries give address, telephone numbers, and research and development sections.
Next edition: February 1986.

Directory of manufacturers, State of Hawaii 172

Publisher: State of Hawaii Department of Planning and Economic Development
Address: PO Box 2359, Honolulu, HI 96804, USA
Content: Entries classified by county, giving name of principal executive, number of employees, products, and type of sales.

Directory of Maryland manufacturers 1985-86 173

Editor: Marilyn J. Corbett
Publisher: Maryland Department of Economic and Community Development
Address: 45 Calvert Street, Annapolis, MD 21401, USA
Year: 1985
Pages: 656
Price: $30.00
Content: The directory lists 2674 manufacturing plants in the state. Firms have been classified under major groups 20 through 39, in accordance with the 1977 supplement to the *Standard Industrial Classification Manual* published by the US Department of Commerce, Office of Federal Statistical Policy and Standards. Each firm's manufacturing processes and/or products have been assigned one or more Standard Industrial Classification (SIC) four-digit industry numbers. The directory is arranged in seven main sections: alphabetical firm listing; city-county index; manufacturers listed by county, city, and community; manufacturers listed by industry; exporters and importers by SIC; index to products; industrial parks.
Next edition: July 1987.

Directory of Nebraska manufacturers 1984-85 174

Publisher: Nebraska Department of Economic Development
Address: PO Box 94666, 301 Centennial Mall South, Lincoln, NB 68509, USA
Pages: 186
Price: $12.00
Content: The directory is a cross-referenced publication listing nearly 2000 manufacturing firms in Nebraska. It contains an alphabetical entry of all companies, a listing by community, and a product listing using Standard Industrial Classification numbers.
Next edition: June 1986.

Directory of New England manufacturers 175

Publisher: Manufacturers' News Incorporated
Address: 4 East Huron Street, Chicago, IL 60611, USA
Year: 1985-86
Pages: 1192
Price: $99.00
Geographical area: States of Connecticut, Maine, Massachusetts, New Hampshire, Rhode Island, Vermont.
Content: Information given includes addresses, zip codes, telephone numbers, officers, products, and numbers of employees of over 20 000 firms, listed alphabetically, geographically, by product, and Standard Industrial Classification.
Next edition: September 1986.

Directory of New Mexico manufacturing and mining 176

Publisher: Manufacturers' News Incorporated
Address: 4 East Huron Street, Chicago, IL 60611, USA
Year: 1985
Pages: 177
Price: $38.00
Content: The directory contains two sections; manufacturing and high technology. Entries include addresses, zip codes, counties, telephone numbers, dates established, numbers of employees, owners, and managers.
Next edition: April 1986.

Directory of North Carolina manufacturing firms 177

Publisher: Manufacturers' News Incorporated
Address: 4 East Huron Street, Chicago, IL 60611, USA
Year: 1985-86
Pages: 650
Price: $58.00
Content: The directory covers 6700 companies and 7000 executives, giving addresses, zip codes, telephone numbers, key offices, numbers of employees, and year established, Standard Industrial Classification codes, and import and export activities.
Next edition: February 1987.

NORTH AMERICA

Directory of scientific and technological capabilities in Canadian industry (1977) 178

Publisher: Statistics Canada, Department of Supply and Services Canada
Address: Place du Portage Phase 111, 11 Laurier Street, Hull, Quebec, K1A OS2, Canada
Year: 1978
Pages: viii + 342
Availability: OP
Content: There are 951 entries arranged by industry with name of company, name of president, address, company telephone, telex, products group is concerned with, facilities, personnel, activities performed by group; alphabetical index.
Next edition: July 1985; $30.000; Statistics Canada.

Directory of Texas manufacturers, 1985, 35th edition 179

Publisher: Bureau of Business Research, University of Texas at Austin
Address: PO Box 7459, Austin, TX 78713, USA
Pages: xxii + 1269 (in 2 vols)
Price: $110.00 (includes monthly supplement)
ISBN: 0-87755-288-6
Content: Provides information on nearly 15 000 Texas manufacturing plants. A monthly supplement, Texas Industrial Expansion, in newsletter form informs about proposed new and expanding manufacturing facilities.
Next edition: January 1986; $110.00.

Directory of the canning, freezing, preserving industries 180

Editor: James J. Judge
Publisher: Judge, James J., Incorporated
Address: PO Box 550, Westminster, MD 21157, USA
Year: 1984
Pages: 592
Price: $80.00
Geographical area: USA and Canada.
Content: Lists 1800 commercial processors (canned, glass, frozen, dehydrated, freeze-dried, etc) of fruits, vegetables, juices, meats, seafoods, mayonnaise, dressings, pickles, jams and jellies, soups, prepared foods and specialties in the US and Canada.
Next edition: April 1986; $90.00.

Directory of top computer executives 181

Publisher: Applied Computer Research Incorporated
Address: PO Box 9280, Phoenix, AZ 85068, USA
Year: 1985
Pages: vi + 434
Price: Single copy $175.00 (both editions), $95.00 (one edition); annual subscription $275.00 (both editions), $150.00 (one edition)
ISSN: 0193-9920
Geographical area: USA.
Content: The directory is published every six months in two editions, covering the eastern and western regions of the USA. It provides information on: more than 8000 medium to large scale data processing installations; a cross-reference index alphabetically by company name; list of the top computer executive, plus second level managers; manufacturer and model numbers of the major mainframes installed, custom listings. A quarterly newsletter, 'Sources', of marketing ideas for the information industries is supplied free to subscribers.
Next edition: October 1985.

Directory of US and Canadian marketing surveys and services 1985 182

Publisher: Rauch Associates Incorporated
Address: PO Box 6802, Bridgewater, NJ 08807, USA
Pages: 500
Price: $167.00
ISBN: 0-932157-01-9
Content: Over 3000 entries.

Directory of Utah manufacturers 183

Publisher: Manufacturers' News Incorporated
Address: 4 East Huron Street, Chicago, IL 60611, USA
Year: 1984
Pages: 166
Price: $29.00
Content: The directory lists 2200 firms, giving addresses, zip codes, and employee figures. Entries are listed geographically, and by Standard Industrial Classification.
Next edition: July 1985.

Georgia manufacturing directory 184

Publisher: Manufacturers' News Incorporated
Address: 4 East Huron Street, Chicago, IL 60611, USA
Year: 1984
Pages: 612
Price: $50.00
Content: The directory details 5500 plants and 5500 executives, listing manufacturing, mining, agricultural, and processing plants. It gives addresses, zip codes, telephone numbers, officers, number of employees.
Next edition: April 1986.

Harris Illinois industrial directory 185

Publisher: Harris Publishing Company Incorporated
Address: 2057-2 Aurora Road, Twinsburg, OH 44087, USA
Year: 1985
Pages: 1106
Price: $115.00

Content: Approximately 17 500 manufacturers are listed in four sections: alphabetical; geographical; product; and Standard Industrial Classification. Information given includes: company name; address and telephone number; overseas trade involvement; names of key executives; employment figures; plant size; annual sales; years in business.
Next edition: 1986.

Harris Indiana industrial directory   186

Publisher: Harris Publishing Company Incorporated
Address: 2057-2 Aurora Road, Twinsburg, OH 44087, USA
Year: 1985
Pages: 470
Price: $75.00
Content: Approximately 7500 manufacturers are featured in four sections: alphabetical; geographical; product; and Standard Industrial Classification. Information includes: company name; address and telephone number; overseas trade involvement; names of key executives; employment figures; plant size; annual sales; years in business.
Next edition: 1986.

Harris Michigan industrial directory   187

Publisher: Harris Publishing Company Incorporated
Address: 2057-2 Aurora Road, Twinsburg, OH 44087, USA
Year: 1985
Pages: 670
Price: $105.00
Content: The directory features over 12 000 manufacturers in four sections: alphabetical; geographical; product; and Standard Industrial Classification. Information given includes: company name; address; telephone number; names of top officers; overseas trade involvement; employment figures; annual sales.
Next edition: 1986.

Harris Pennsylvania industrial directory   188

Publisher: Harris Publishing Company Incorporated
Address: 2057-2 Aurora Road, Twinsburg, OH 44087, USA
Year: 1985
Pages: 800
Price: $105.00
Content: The directory lists approximately 14 500 manufacturers by name, product, location, and Standard Industrial Classification code. Information given includes: name; address; telephone number and telex; employment; products manufactured; overseas trade involvement; key personnel; sales volumes.
Next edition: 1986.

Hawaii business directory   189

Publisher: Hawaii Business Directory Incorporated
Address: 1164 Bishop Street, Suite 1410, Honolulu, HI 96813, USA
Price: $150.00
Content: Listing of more than 37 000 businesses giving name and address, telephone, products and services, key personnel, number of employees, annual sales volume, subsidiaries, branches, affiliates, etc.

Hawaii directory of manufacturers   190

Publisher: Manufacturers' News Incorporated
Address: 4 East Huron Street, Chicago, IL 60611, USA
Year: 1983-84
Pages: 90
Price: $40.00
Content: The directory lists 3500 firms, alphabetically by brand and trade names. It gives addresses, telephone numbers, zip codes, name of chief executives, year established, sales codes, and parent company.
Next edition: December 1985.

Idaho manufacturers directory   191

Publisher: Manufacturers' News Incorporated
Address: 4 East Huron Street, Chicago, IL 60611, USA
Year: 1985
Pages: 168
Content: 1250 manufacturers, listed alphabetically, geographically and by product. Entries include company name, address, zip code, telephone, and chief officers.

Illinois manufacturers directory   192

Publisher: Manufacturers' News Incorporated
Address: 4 East Huron Street, Chicago, IL 60611, USA
Year: 1985
Pages: 1118
Price: $115.00
Content: Entries are broken down geograhically by city and town, and then by company within each city and town, and give name of company, address, zip code, telephone number, officers, number of employees, products, and Standard Industrial Classification.
Next edition: February 1986; $120.00.

Indiana industrial directory   193

Publisher: Harris Publishing Company
Address: 2057-2 Aurora Road, Twinsburg, OH 44087, USA
Year: 1985
Price: $49.50

Indiana manufacturers' directory — 194

Publisher: Manufacturers' News Incorporated
Address: 4 East Huron Street, Chicago, IL 60611, USA
Year: 1985
Pages: 560
Price: $85.00
Content: Entries are broken down geographically by city and town, and then alphabetically by company within each city and town. Company names, addresses, zip codes, telephone numbers, officers, number of employees, and products are given.
Next edition: May 1986; $90.00.

Industrial Alabama — 195

Publisher: Alabama Development Office
Address: c/o State Capital, Montgomery, AL 36130, USA
Content: There are four sections, geographical, alphabetical, products, and an advertising index.

Industrial directory of the commonwealth of Pennsylvania, 1984-85* — 196

Publisher: Harris Publishing Company
Address: 2057-2 Aurora Road, Twinsburg, OH 44087 USA
Pages: 800
Price: $83.00
ISBN: 0-916-512-47-9
Content: Geographical, alphabetical, purchasing, and Standard Industrial Classification sections are included.

Industrial research laboratories of the United States — 197

Content: See entry under general sciences.

Iron and steel works directory of the United States and Canada, 1984 — 198

Publisher: American Iron and Steel Institute
Address: 1000 16th Street NW, Washington, DC 20036, USA
Year: 1985
Pages: 359
Price: $25.00

Kentucky directory of manufacturers — 199

Editor: Mildred J. Keefer
Publisher: Kentucky Department of Economic Development
Address: 23rd Floor, Capital Plaza Tower, Frankfort, KY 40601, USA
Year: 1985
Pages: xiv + 262
Price: $20.00
Content: Entries are arranged by the following categories: alphabetic listing of companies; geographical listing of companies; companies listed under specific Standard Industrial Classification codes. There are three indexes: to products; to city/county; and to parent company.
Next edition: February 1986; $25.00.

Maine marketing directory — 200

Publisher: Manufacturers' News Incorporated
Address: 4 East Huron Street, Chicago, IL 60611, USA
Year: 1984
Pages: 100
Price: $24.00
Content: The directory covers 2000 firms and 3000 executives, listed by name, location, and Standard Industrial Classification. Telephone numbers, and total number of employees are given.
Next edition: July 1986.

Massachusetts directory of manufacturers — 201

Publisher: Manufacturers' News Incorporated
Address: 4 East Huron Street, Chicago, IL 60611, USA
Year: 1985-86
Pages: 390
Price: $62.50
Content: Entries for more than 8500 companies, giving company name, parent company, address, zip code, telephone number, year established, products and Standard Industrial Classification, bank relationship, accountants, etc.
Next edition: February 1986.

Michigan manufacturers directory — 202

Publisher: Manufacturers' News Incorporated
Address: 4 East Huron Street, Chicago, IL 60611, USA
Year: 1985
Pages: 970 (in 2 vols)
Content: The directory lists over 15 000 manufacturers in 717 cities, in alphabetically and geographically arranged volumes. It includes addresses, zip codes, telephone numbers, products, officers, numbers of employees, and year established.
Next edition: March 1986.

Michigan manufacturers directory — 203

Publisher: Pick Publications Incorporated
Address: 8543 Puritan, Detroit, MI 48238, USA
Year: 1985
Pages: 728
Price: $123.00
ISBN: 0-936526-X

Content: 15 333 manufacturers arranged geographically by city, alphabetically, and by products, products are SIC (Standard Industrial Classifications) the standard SIC is a four digit number. A supplement is included with seven digit classifications on 3224 manufacturers.
Entries include: company name, division, subsidiary of who; full street and mailing address; branch plant locations; all products made by SIC; officers from president to purchasing agent by name and title; telex, TWX, cable code numbers; employment, date established; telephone number; square footage of plant; annual sales volume; import/export interest designation.
Not available on-line.
Next edition: March 1986; $125.00.

Minnesota Manufacturers Register 1985   204

Publisher: Manufacturers' News Incorporated
Address: 4 East Huron Street, Chicago, IL 60611, USA
Pages: 336
Price: $67.00
Content: Entries are broken down geographically by city and town, and then by company within each city and town, and give name of company, address, zip code, telephone number, officers, number of employees, products, and Standard Industrial Classification.
Next edition: September 1985; $70.00.

Mississippi manufacturers directory   205

Publisher: Manufacturers' News Incorporated
Address: 4 East Huron Street, Chicago, IL 60611, USA
Year: 1985
Pages: 322
Price: $60.00
Content: Entries for 2580 manufacturers and 5000 executives, colour coded, and arranged alphabetically, geographically, and by product. Entries give addresses, zip codes, telephone numbers, and products.
Next edition: March 1986.

Missouri directory of manufacturing and mining industrial services   206

Publisher: Manufacturers' News Incorporated
Address: 4 East Huron Street, Chicago, IL 60611, USA
Year: 1984
Pages: 640
Price: $80.00
Content: Manufacturers are listed alphabetically with address, zip code, and telephone number. Geographical listing is also given, by city and county.
Next edition: September 1985.

Montana manufacturers and products directory   207

Publisher: Manufacturers' News Incorporated
Address: 4 East Huron Street, Chicago, IL 60611, USA
Year: 1985
Pages: 80
Price: $24.00
Content: The directory gives addresses, zip codes, telephone numbers, chief officer, and number of employees of 850 manufacturers. Entries are listed alphabetically, geographically, by Standard Industrial Classification, and by product.
Next edition: June 1987.

Nevada industrial directory   208

Publisher: Commission on Economic Development
Address: Capitol Complex, Carson City, NV 89710, USA
Year: 1985
Content: Alphabetical listing of industries; geographical listing; standard industrial code listing.

New Hampshire marketing directory   209

Publisher: Manufacturers' News Incorporated
Address: 4 East Huron Street, Chicago, IL 60611, USA
Year: 1984
Pages: 406
Price: $24.00
Content: Entries for 1500 firms, giving addresses, telephone numbers, and total numbers of employees, listed alphabetically, geographically, by product, and Standard Industrial Classification.
Next edition: July 1986.

New Jersey directory of manufacturers   210

Publisher: Manufacturers' News Incorporated
Address: 4 East Huron Street, Chicago, IL 60611, USA
Year: 1984-85
Pages: 524
Price: $82.50
Content: The directory lists 9000 firms and 45 000 executives alphabetically, geographically, and by Standard Industrial Classification. Entries include name of company, address, zip code, and telephone number.
Next edition: December 1985.

New York manufacturers directory   211

Publisher: Manufacturers' News Incorporated
Address: 4 East Huron Street, Chicago, IL 60611, USA
Year: 1984-85
Pages: 854

Content: Addresses, zip codes, telephone numbers, numbers of employees, etc, are given for 20 000 firms, listed alphabetically, geographically, and by product.
Next edition: July 1986.

Ohio industrial directory 212

Publisher: Harris Publishing Company Incorporated
Address: 2057-2 Aurora Road, Twinsburg, OH 44087, USA
Year: 1985
Price: $105.00
ISBN: 0-916-512-61-4
Content: The publication contains data, arranged alphabetically and geographically, on 15 000 manufacturers, as well as county and state statistics.
Next edition: November 1985; $105.00.

Oklahoma directory of manufacturers and products 213

Editor: Barbara Clements
Publisher: Economic Development Department, State of Oklahoma
Address: PO Box 53424, Oklahoma City, OK 73152, USA
Year: 1985
Price: $35.00
Content: Includes an alphabetical list of firms, geographical list by cities, a listing by Standard Industrial Classification code, a section on mineral producers and a product index.

Plastics directory and buyers' guide 1984* 214

Editor: Sandra Cruickshanks
Publisher: Southam Communications Limited
Address: 1450 Don Mills Road, Don Mills, Ontario M3B 2X7, USA
Year: 1984
Pages: 210
Price: $25.00
Geographical area: North America.
Next edition: 1985.

Post's pulp and paper directory 215

Publisher: Miller Freeman Publications Incorporated
Address: 500 Howard Street, San Francisco, CA 94105, USA
Year: 1985
Pages: 718
Price: $80.00
Geographical area: USA and Canada.
Content: Mill listings, USA and Canada; industry executives; converting plants, USA and Canada; producers of over 300 paper, pulp, and converted paper grades; market pulp directory by grades, producers and sales organizations; capacity and production statistics; associations, schools, and information sources; buyer's guide to equipment, supplies, and services; mill maps for each state and province.
Next edition: October 1985; $90.00.

Rhode Island directory of manufacturers 216

Publisher: Manufacturers' News Incorporated
Address: 4 East Huron Street, Chicago, IL 60611, USA
Year: 1983-84
Pages: 267
Price: $18.00
Content: Name, location, telephone number, number of employees, executive officer, etc. of 2500 firms, listed alphabetically, geographically, and by Standard Industrial Classification.
Next edition: January 1986.

Texas trade and professional associations and other selected organizations 1985 217

Editor: Rita J. Wright, Laurie Gamel, Mildred Anderson
Publisher: Bureau of Business Research, University of Texas at Austin
Address: PO Box 7459, Austin, TX 78713, USA
Pages: viii + 79
Price: $8.00
ISBN: 0-87755-292-4
Content: Provides names of officers, addresses, telephone numbers, number of members, and titles and frequency of association publications.
Next edition: January 1987

US industrial directory* 218

Publisher: Cahners Publishing Company
Address: 270 Saint Paul Street, Denver, CO 80206, USA
Year: 1985
Price: £137.50
Content: Alphabetical and classified lists of manufacturers, suppliers, and equipment.
Next edition: January 1986.

Vermont business phone book 1984-85 219

Publisher: Manufacturers' News Incorporated
Address: 4 East Huron Street, Chicago, IL 60611, USA
Year: 1984
Pages: 86
Price: $15.00

Content: Entries list names of companies, locations, telephone numbers, number of employees, chief executive officers, export and impot activities, and Standard Industrial Classifications.
Next edition: August 1985.

Virginia industrial directory 1984-85   220

Editor: John R. Broadway
Publisher: Virginia Chamber of Commerce
Address: 611 East Franklin Street, Richmond, VA 23219, USA
Year: 1984
Pages: 378
Price: $60.00
Content: Over 4000 entries covering manufacturing and mining firms, giving company name, mailing address, location, names of selected executives, and telephone number. Each entry, where possible, also provides information on the firm's size, its parent company, and its major offices. Entries are listed alphabetically, geographically, and by products manufactured or mined (using Standard Industrial Classification codes). There is also a product index.
Next edition: January 1986; $65.00.

Walker's manuals of western US corporations 1985*   221

Publisher: Walker's Manuals
Address: c/o Graham and Trotman Limited, Sterling House, 66 Wilton Road, London SW1V 1DE, UK
Year: 1985
Pages: 1850
Price: £275.00, $440.00 (basic work only); £125.00, $200.00 (monthly updating supplement service)
ISBN: 0-86010-721-3 (basic work); 0-86010-722-1 (monthly updating supplement service)
Geographical area: Western States of the USA.
Content: Complete in two volumes with a monthly updating supplement service, this directory covers business and financial information on nearly every publicly-owned corporation in the Western States of America. It provides a detailed database on these companies, including information on address, telephone number, nature of business, acquisitions, officers and directors, income and expenditure statements, balance sheets, shareholders, employees and auditors.
This title is not available from Graham and Trotman in the USA or Canada.

Washington manufacturers register   222

Publisher: Manufacturers' News Incorporated
Address: 4 East Huron Street, Chicago, IL 60611, USA
Year: 1984-85
Pages: 250
Price: $75.00

Content: Companies are listed alphabetically by city and town, giving addresses, key executives, products, Standard Industrial Classification, telephone numbers, number of employees, sales volume, import and export.
Next edition: February 1986.

Who's who in finance and industry   223

Publisher: Marquis Who's Who Incorporated
Address: 200 East Ohio Street, Chicago, IL 60611, USA
Year: 1983
Pages: xiv + 945
Price: $84.50
ISBN: 0-8379-0323-8
Geographical area: North America.
Content: More than 21 000 entries, which include a broad range of information, including educational background, career history, civic and political activities, professional and social memberships, writings and awards, names of family members, and home and office addresses.
Next edition: July 1985; $119.00.

Wisconsin manufacturers register   224

Publisher: Manufacturers' News Incorporated
Address: 4 East Huron Street, Chicago, IL 60611, USA
Year: 1985
Pages: 440
Price: $77.00
Content: Entries are broken down geographically by city and town, and then by company within each city and town, and give name of company, address, zip code, telephone number, offices, number of employees, products, and Standard Industrial Classification number.
Next edition: July 1985; $82.00.

Wyoming directory of manufacturing and mining   225

Publisher: Manufacturers' News Incorporated
Address: 4 East Huron Street, Chicago, IL 60611, USA
Year: 1985-86
Pages: 48
Price: $15.00
Content: Includes firm name, address, chief officer, employee figures, telephone number, product, and marketing area.
Next edition: January 1987.

MEDICAL AND BIOLOGICAL SCIENCES

American malacologists: a national register of living professional and amateur conchologists* 226

Editor: R. Tucker Abbott
Publisher: American Malacologists
Address: Box 2255, Melbourne, FL 32901, USA
Year: 1984
Pages: 610
Price: $18.00
Content: Over 1000 professional and amateur malacologists (mollusc and shellfish experts, palaeoconchologists, and advanced shell collectors). Entries are alphabetically arranged and include name, office and home addresses, personal and career data, writings, club memberships, travels, collection size, research activities, and honours. Indexes to place of residence, occupation, and area of research.

American medical directory* 227

Publisher: PSG Publishing Company Incorporated
Address: 545 Great Road, Littleton, MA 01460, USA
Year: 1982
Pages: 4 vols
Price: $310.00
Geographical area: USA.
Content: The directory covers about 445 000 physicians in the United States and United States physicians in foreign countries. Entries include name, address, year licensed, medical school, type of practice, primary and secondary specialties, and board certifications, arranged geographically, with federal service and United States physicians abroad in separate section. Volume 1 comprises an alphabetical index.

American Psychiatric Association biographical directory, seventh edition 228

Publisher: Bowker Publishing Company
Address: 58/62 High Street, Epping, Essex CM16 4BU, *Year:* 1978
Price: £62.75
ISBN: 0-8352-0977-6
Geographical area: USA.

American Psychological Association, membership register 229

Editor: John A. Lazo
Publisher: American Psychological Association
Address: 1700 17th Street NW, Washington, DC 20036, USA
Year: 1984
Pages: viii + 800

ISSN: 0737-1446
Geographical area: USA.
Content: The publication provides a current record of the association membership, including mailing addresses, telephone numbers, membership status, and divisional affiliations. It is published three years out of four, excepting the year in which the directory is published.
Not available on-line.
Next edition: April 1986; $25.00 (to non-members), $15.00 (to members).

American Speech and Hearing Association directory 230

Editor: Frederick T. Spahr
Publisher: American Speech-Language-Hearing Association
Address: 10801 Rockville Pike, Rockville, MD 26852, USA
Year: 1983-84
Pages: lxxiii + 576
Price: $20.00 (to members); $32.00 (to non-members)
ISBN: 0-910329-08-7
Geographical area: USA and Canada.
Content: This directory contains an alphabetical list and a geographical list of ASHA members, bylaws and code of ethics of the association, a list of recognized state associations, a list of the members and the public, as well as names of all members in the National Office records as of November 30 1982.
The 1985 ASHA Directory Supplement contains an alphabetical list and a geographical list of ASHA members (as of August 15, 1984) bylaws and code of ethics of the association and a list of recognized state associations.
Next edition: 1986.

Canadian hospital directory 231

Editor: Eleanor Sawyer
Publisher: Canadian Hospital Association
Address: 17 York Street, Suite 100, Ottawa, Ontario K1N 9J6, Canada
Year: 1984
Pages: 400
Price: $60.00
ISSN: 0068-8932
Content: Lists information on 1231 hospitals across Canada; over 300 educational programmes for health personnel; over 700 health care associations and health organizations and 1200 manufacturers/suppliers and their health care products; plus statistical information on bed distribution, number of patient days, etc in a statistical compendium section.
Next edition: October 1985; $60.00.

Canadian medical directory   232

Editor: Barbara P. Hutchison
Publisher: Southam Communications Limited
Address: 1450 Don Mills Road, Don Mills, Ontario M3B 2X7, Canada
Year: 1984
Pages: viii + 780
Price: $65.00
Content: The directory contains listings of approximately 46 000 physicians, divided into three distinct sections. The white section lists all the doctors across Canada biographically, according to surname, given names etc, it shows their office address, (where applicable) their year of graduation, medical degrees, specialist certification, hospital, teaching and industrial appointments if applicable, and their numbers and area codes. The blue section is a breakdown of the doctors alphabetically by province and town, showing names and specialist certifications. The buff section contains general information of interest to the medical profession, such as hospitals, medical faculties with departmental heads, lists of graduates of medicine, government licensing agencies, etc.
Next edition: June 1985; $56.00 to MDs, $65.00 to others.

Clinical and public health laboratory directory   233

Publisher: US Directory Service
Address: PO Box 011565, Miami, FL 33101, USA
Year: 1978
Pages: 316
Price: $125.00
ISBN: 0-916524-09-04
Geographical area: USA.
Content: Entries cover 15 830 clinical and public health laboratories, giving name and address with zip code.

Clinical pharmacology: a guide to training programs   234

Editor: Barbara C. Ready
Publisher: Peterson's Guides Incorporated
Address: PO Box 2123, 166 Bunn Drive, Princeton, NJ 08540, USA
Year: 1985
Pages: 150
Price: $9.95
ISBN: 0-87866-385-1
Series: USA and Canada.
Content: The guide describes the US and Canadian postdoctoral programs that are devoted to the training of clinical pharmacologists.

Directory: health systems agencies, state health planning and development agencies, statewide health coordinating councils*   235

Publisher: Health and Human Services Department
Address: Rockville, MD 20857, USA
Year: 1983
Pages: 25
Price: Free
Geographical area: USA.
Content: The directory covers about 245 state health coordinating councils, state health planning and development agencies, and health systems agencies established under the National Health Planning and Resources Development Act. Entries include; agency or council name, address, telephone, name of director. Health systems agency listings also include name and address of chairman of governing body, congressional districts covered, and type of organization (nonprofit, etc). Entries are arranged geographically within the type of agency.

Directory of consultants in biotechnology, first edition   236

Publisher: Dick, J., Publishing (Research Publications)
Address: 12 Lunar Drive, Woodbridge, CT 06525, USA
Year: 1985
Pages: x + 256
Price: $94.00
ISBN: 0-89235-090-3
Geographical area: North America.
Content: Contains professional listings of 2140 scientists and engineers in the fields of genetics, biomedical science/engineering, biology, and microbiology who are available for consulting assignments. Keyword indexed by 1500 technical terms and cross-referenced by city/state.
Next edition: January 1987; $95.00

Directory of medical libraries in New York state   237

Publisher: University of the State of New York, State Education Department
Address: New York State Library, Gifts and Exchange, Cultural Education Center, Empire State Plaza, Albany, NY 12230, USA
Year: 1982
Pages: i + 237
Price: $3.00
Content: 232 entries, alphabetically arranged by name of parent organization with address. Each entry includes the name of the head librarian (or person overseeing the library), the interlibrary loan librarian as well as telephone and teletype numbers. Statistical data on holidays, budget, and staff are based on 1981 reported

figures. Also included is information on subject specialization, clientele served, user services provided, and interlibrary loan policies. There are no indexes. Appendix 1 is a directory of head librarians arranged alphabetically.

Directory of medical specialists   238

Publisher: Marquis Who's Who Incorporated
Address: 200 East Ohio Street, Chicago, IL 60611, USA
Year: 1983
Pages: xxxi + 4414
Price: $215.00
ISBN: 0-8379-0521-4
Geographical area: USA.
Content: Entries cover more than 290 000 specialists certified by the 23 boards of the American Board of Medical Specialties, containing information on education, career history, memberships, date of certification, date of birth, teaching positions, hospital affiliations, type of practice, office address, and office telephone number.
Physicians can be located in the directory by geographical location and specialty. All professional sketches are grouped alphabetically within each specialty, city, state, and foreign country. In addition, the purpose, function, and requirements of each specialty board are outlined in the introduction to each of the 23 specialties.
Next edition: October 1985; $235.00.

Directory of names and addresses and research specialisations for current members of the Botanical Society   239

Publisher: Botanical Society of America
Address: c/o Secretary, Department of Biology, University of Indiana, Bloomington, IN 47405, USA
Year: 1982
Pages: 83
Availability: OP
Geographical area: USA.
Content: The directory includes names, addresses, and research specializations for current members of the Botanical Society of America. A list of officers and committees is included. The directory is arranged alphabetically by name but also includes a geographical index of members by state. There are approximately 3000 entries.
Next edition: 1986.

Directory of personnel responsible for radiological health programs: 1985   240

Publisher: Conference of Radiation Control Program Directors Incorporated
Address: 71 Fountain Place, Frankfort, KY 40601, USA
Pages: iii + 41
Price: $5.00
Geographical area: USA.
Content: Conference staff, executive board, including state and federal liaisons and Council Chairpersons to the Conference working groups (page 1); pages 2-23 is a state-by-state listing of personnel as requested by state radiation control programme director to be listed, listing their names, titles, addresses and appropriate telephone numbers; pages 24-35 are listings of federal agencies concerned with radiation control, also listing pertinent personnel, titles, addresses and telephone numbers; pages 36-38 is a listing of the conference membership at January 1985; and pages 39-41 is an alphabetical roster of those individuals contained in the book and on which page they may be found. In the front of the directory is a table of contents and an introduction.
Next edition: January 1986.

Directory of psychologists registered in the province of Ontario   241

Publisher: Ontario Board of Examiners in Psychology
Address: 101 Davenport Road, Toronto, Ontario M5R 3P1, Canada
Year: 1985
Pages: 124
Price: No charge for one copy
ISBN: 0316-0793
Next edition: March 1986; no charge.

Encyclopedia of medical organizations and agencies   242

Editor: Anthony T. Kruzas
Publisher: Gale Research Company
Address: Book Tower, Detroit, MI 48226, USA
Year: 1983
Pages: 768
Price: $170.00
ISBN: 0-8103-0347-7
Geographical area: USA.
Content: Current information on some 10 000 major public and private agencies in medicine and related fields that are concerned with information, funding, research, education, planning, advocacy, and service. Entries describing more than 3000 national and international associations, 1250 state associations, 1500 federal and state agencies, 2000 medical and allied health schools, 225 foundations, 1400 research centres, and 200 databases are arranged into 78 chapters covering specific areas of modern health care and medicine, including biomedical engineering, child health, dentistry, hypnosis, mental health, etc.

ESA membership directory 1984-85   243

Publisher: Entomological Society of America
Address: 4603 Calvert Road, College Park, MD 20740, USA

Pages: 74
Price: $3.50
Geographical area: North America.
Content: There are 7765 Entomological Society of America members listed, alphabetically, along with address, telephone number (for most) and ESA Section (A: systematics, morphology, evolution; B: physiology, biochemistry, toxicology; C: ecology, behaviour; and bionomics; D: medical and veterinary entomology; E: extension and regulatory entomology; and F: crop protection entomology). This directory is a supplement to *Bulletin of the ESA* and is published each even-numbered year.
Next edition: June 1986; $3.50.

Florida medical directory 244

Publisher: Florida Medical Association Incorporated
Address: 760 Riverside Avenue, PO Box 2411, Jacksonville, FL 32203, USA
Year: 1985
Pages: 212
Price: $16.50
Content: The directory is published annually and lists all physicians who are members of the Florida Medical Association with biological information, specialty and mailing address.

Genetic engineering; biotechnology sourcebook* 245

Editor: Robert G. Pergolizzi
Publisher: Macmillan Publishing Company
Address: Brunel Road, Houndsmill, Basingstoke, Hants RG21 2XS, UK
Year: 1982
Price: £57.50
ISBN: 0-333-34148-1
Geographical area: USA.
Content: Information on 1529 current and recent genetic engineering and biotechnology research projects in the USA.

Guide to graduate education in speech-language pathology and audiology, 1985 246

Publisher: American Speech-Language-Hearing Association
Address: 10801 Rockville Pike, Rockville, MD 20852, USA
Pages: ix + 268
Price: $33.50 (to non-members); $17.50 (to members)
ISBN: 0-910329-20-6
Geographical area: USA and Canada.
Content: This guide contains information about the location, size, faculty, and requirements of 240 colleges and universities offering master's degrees in speech-language pathology and audiology. The guide is divided into two sections: section I contains the profiles of graduate programmes that are accredited by the Education Standards Board of the American Speech-Language-Hearing Association; section II contains the profiles of graduate programmes that are not accredited by the ESB of ASHA. Following the programme descriptions is a matrix that summarizes the degree requirements for each programme to facilitate comparison.

Guide to graduate study in botany in the United States and Canada 247

Publisher: Botanical Society of America
Address: c/o Secretary, Department of Biology, Indiana University, Bloomington, IN 47405, USA
Year: 1983
Pages: 84
Price: $5.00
Content: This guide lists 82 plant science departments in the United States and 11 in Canada which offer the PhD degree in some area of the plant sciences. Each departmental listing includes the name and address of the institution, name of the department with number of faculty, current graduate enrolment, fields of specialization represented in the department, and name, academic background, area of specializaion, and titles of recent PhD theses directed for all botanical faculties in the department.

Guide to professional services in speech pathology and audiology 248

Publisher: American Speech-Language-Hearing Association
Address: 10801 Rockville Pike, Rockville, MD 20852, USA
Year: 1983-84
Pages: v + 201
ISBN: 0-910329-11-7
Geographical area: USA and Canada.
Content: This guide provides a description of each service programme accredited by the Professional Services Board of ASHA as well as other clinical programmes. The information provided by the director of each programme is presented alphabetically by state and city in the following order: name, address, and telephone number of programme; director, staff, and language(s); accreditation status; services; referral sources. This guide also lists: certified members in private practice; industrial hearing conservation specialists; resource personnel. This information is also arranged alphabetically by state and city.
Next edition: May 1985.

Health and welfare directory 13th edition　249

Publisher: Federation for Community Planning
Address: 1001 Huron Road, Cleveland, OH 44115, USA
Year: 1982
Pages: vi + 122
Price: $6.50
Geographical area: Cuyahoga County, Ohio.
Content: The directory is a descriptive reference to 933 health and social service agencies in the Cuyahoga County area. This 13th edition was prepared in May 1982 and includes a services locator index.
Next edition: 1986.

Health organizations of the United States, Canada, and the world, fifth edition　250

Editor: Paul Wasserman, Marek Kaszubski
Publisher: Gale Research Company
Address: Book Tower, Detroit, MI 48226, USA
Year: 1981
Pages: 411
Price: $85.00
ISBN: 0-8103-0466-X
Content: Contains complete details for more than 1600 voluntary associations, professional societies, and other groups concerned with health, medical, hospital, pharmaceutical, and related fields. Listed in the first part of the directory are the national organizations of the USA and Canada as well as international organizations. Part two is a subject-classified listing of the organizations and societies.

Health sciences information in Canada: associations　251

Publisher: Canada Institute for Scientific and Technical Information
Address: Ottawa, Ontario K1A 0S2, Canada
Price: $18.00
Content: The publication lists nearly 500 Canadian associations in all areas of health science, giving: name, address, telephone number; name of person in charge; year established; purpose of organization; affiliations with other organizations; membership, size, and criteria for membership; serial publications including title, frequency, ISSN, editor, year of first issue. Entries are grouped by province, indexes are provided by subject, name of organization, and titles of major serial publications.

Health sciences information in Canada: libraries　252

Publisher: Canada Institute for Scientific and Technical Information
Address: Ottawa, Ontario K1A 0S2, Canada
Availability: $15.00
ISSN: 0708-9465
Content: Individual entries include: name of organization; name, address, and telephone/telex number of library; name of person in charge; services offered to external users; language of service; collection size and subject specialty. There is also listing of new data bases available, classification/subject headings used, size of faculties, and number of employees. Entries are grouped by province; indexes are provided by subject, name of organization, and name of person in charge.

Health sciences libraries: USA and Canada　253

Editor: Brigitte T. Darnay
Publisher: Gale Research Company
Address: Book Tower, Detroit, MI 48226, USA
Year: 1985
Price: $145.00
Series: Subject directory of special libraries and information centres (volume 3)
Content: The publication is one of five volumes containing the same information as is found in volume 1 of *Directory of Special Libraries and Information Centres* (see separate entry) rearranged under subject sections. A typical entry provides nearly two dozen facts about the library, including: official name; name of sponsoring organization or institution; address with zip code; name and title of person in charge; names and titles of other professional personnel; collection statistics; description of the subjects with which the library or collection is concerned; policies regarding use of the collection; services provided; and telephone number with area code. In addition to a subject index, each volume contains an alternative name index, which provides cross-references from variant names for libraries.
Next edition: September 1987.

Marquis who's who in cancer: professionals and facilities　254

Publisher: Marquis Who's Who Incorporated
Address: 200 East Ohio Street, Chicago, IL 60611, USA
Year: 1985
Pages: xxviii + 802
Price: $125.00
ISBN: 0-8379-6501-2
Geographical area: North America.
Content: Entries covering approximately, 6000 individuals, 1100 centres, arranged geographically by job type (physicians, scientists, associated professionals, and cancer centers). Individuals' entries list name, degree(s), primary clinical emphasis, treatment modalities emphasized, cancer types emphasized, organ or tissue type interest areas, research emphasis, published works, background (born, education, internships and residen-

cies, licensures), certification, career history, current employment (academically ranked appointments, types of practice), awards and honours, professional memberships, address, telephone. Entries for centres give name of centre or unit, address, telephone, types of cancer treated, treatment modalities emphasized, research centres/special programmes, patient information, staff, admissions limitations, admissions contact. There are indexes to: primary clinical emphasis; treatment modalities; cancer type; research emphasis; alphabetical index of professionals. Alphabetical index of centres.

Marquis who's who in rehabilitation: professionals and facilities 255

Publisher: Marquis Who's Who Incorporated
Address: 200 East Ohio Street, Chicago, IL 60611, USA
Year: 1985
Pages: xxx + 429
Price: $125.00
ISBN: 0-8379-6601-9
Geographical area: North America.
Content: Entries cover approximately 4100 individuals and 700 centres, arranged geographically by job type. Individuals' entries include: name; professional certification; clinical emphasis; publications; background; career; current employment; awards and honours; professional memberships; address and telephone number. Centre's entries include: name; address; telephone number; treatment categories; special centres, services, and laboratories; patient information; staff; admissions contact.
There are indexes to clinical emphasis, research emphasis, occupations, professionals (alphabetically) and centres (also alphabetically).

Medical and health information directory, third edition volume 1 256

Editor: Anthony T. Kruzas, Kay Gill, Karen Backus
Publisher: Gale Research Company
Address: Book Tower, Detroit, MI 48226, USA
Year: 1984
Pages: 1925
Price: $160.00
ISBN: 0-8103-0269-1
Geographical area: USA and Canada.
Content: Completely revised, updated, and expanded, the third edition of the directory is published in three sequential volumes. Each volume is a self-contained reference focusing on particular segments of the medical and health care delivery and information system. Entries typically provide concise, up-to-date information, including names, addresses, key personnel, and descriptions of services, facilities, and functions. Where appropriate, individual sections have their own indexes to help retrieve specific information.

Volume 1: Organizations, Agencies and Institutions (published November 1984) includes national and state professional and voluntary associations, federal and state agencies, foundations and grant-awarding organizations, health insurance providers, pharmaceutical companies, medical and allied health schools, consultants, publishers, and research centers. Volume 2: Libraries, Publications, Audiovisuals, and Data Base Services (ready October 1985) includes journals, newsletters, annual reviews, abstracting and indexing services, audiovisuals, directories, libraries, and data base services. Volume 3: Health Services (ready September 1986) includes clinics, hospitals, treatment centres, rehabilitation facilities, care programmes, and counselling/diagnostic services.
Next edition: Volume 2, September 1985; Volume 3, September 1986.

New directory of medical schools 257

Editor: Alex Sandri White, E. Pokress
Publisher: Aurea Publications
Address: PO Box 176, Allenhurst, NJ 07711, USA
Year: 1983
Pages: 148
Price: $11.95
ISBN: 0-685-22749-9
Geographical area: USA and Canada.
Content: The directory lists the universities and colleges that feature training in the following areas: medicine; dentistry; chiropody and podiatry; pharmacy; optometry; public health; physical therapy and massage; nursing; occupational therapy; veterinary medicine. It also lists institutes of higher learning in osteopathy, chiropractice, and naturopathy.

Oncology directory* 258

Publisher: Marquis Who's Who
Address: c/o Henry Thompson, London Road, Sunningdale, Berkshire, UK
Year: 1983
Price: £40.00
Content: Details of 9000 members of the American Clinical Oncology Society.

Reference issue: American annals of the deaf 259

Editor: William U. Craig, Helen B. Craig
Publisher: Convention of American Instructors of the Deaf and Conference of Educational Administrators Serving the Deaf
Address: 814 Thayer Avenue, Silver Spring, Maryland, MD 20910, USA
Year: 1984
Pages: vii + 288
Price: $16.50

ISSN: 0002-726X
Geographical area: USA and Canada.
Content: The publication contains approximately 2000 entries, covering selected topics of interest; educational programmes and services; supportive and rehabilitation programmes and services - federal programmes for the deaf, state programmes for the deaf, independent programmes for the deaf, social/recreational programmes and agencies, supportive organizations and agencies; research programmes ad services.
Not available on-line.
Next edition: May 1985; $16.50.

Research programs in the medical sciences 260

Editor: Jaques Cattell Press
Publisher: Bowker Publishing Company
Address: 58/62 High Street, Epping, Essex CM16 4BU, UK
Year: 1981
Pages: 816
Price: £94.50
ISBN: 0-8352-1293-9
Geographical area: USA.
Content: This first edition is an alphabetical compilation of manufacturing and industrial service companies, academic and non-profit making organizations and independent institutes, all doing research within the United States in the medical sciences.
It contains three complete indexes: a geograhical index, listing each facility alphabetically by city within states; a personnel index listing all names in the main text in alphabetical sequence; plus an index to all research activities. For each research facility the directory provides name and title of director, librarian, a code indicating for whom the research is being done and a detailed statement of current fields of research.

Scientific directory and annual bibliography 261

Editor: Betty MacVicar
Publisher: National Institutes of Health
Address: 9000 Rockville Pike, Bethesda, Maryland 20205, USA
Year: 1985
Content: An annual document, this is a reference resource for biomedical researchers. It presents in broad outlines the NIH organizational structure, the professional staff, and their scientific and technical publications covering work done at NIH. It also includes the National Institute of Mental Health, a former component, now a part of the Alcohol, Drug Abuse, and Mental Health Administration. Material is arranged by components, with the directory and bibliography entries together at the laboratory or branch level within each component. It is not available on-line. Indices: journal and abbreviations, name, and subject. Contains over 5000 citations of work performed by NIH scientists and professional staff.

US medical directory, sixth edition 1983-84 262

Publisher: US Directory Service
Address: PO Box 011565, Miami, FL 33101, USA
Pages: 880
Price: $89.95
ISBN: 0-916524-20-5
ISSN: 0091-8393
Content: Detailed information on: medical doctors; hospitals; nursing facilities; laboratories; medical information sources; poison control centres; medical schools; and a buyers' guide.

PHYSICS, MATHEMATICS AND NUCLEAR SCIENCES

Directory of consultants in lasers and physics, first edition 263

Publisher: Dick, J., Publishing (Research Publications)
Address: 12 Lunar Drive, Woodbridge, CT 06525, USA
Year: 1985
Pages: x + 256
Price: $94.00
ISBN: 0-89235-093-8
Geographical area: North America.
Content: Contains professional listings of 2236 scientists and engineers in the fields of optics, lasers, physics, and acoustics who are available for consulting assignments. Keyword indexed by 1500 technical terms and cross-referenced by city/state.
Next edition: January 1987; $95.00.

Directory of physics and astronomy staff members 264

Publisher: American Institute of Physics
Address: 335 East 45th Street, New York, NY 10017, USA
Year: 1984
Pages: vi + 414
Price: $30.00 (institutional); $10.00 (personal)
ISBN: 0-88318-458-3
Geographical area: North America.
Content: Alphabetical list of approximately 30 000 names, addresses and telephone numbers; alphabetical list of academic institutions including approximately

2400 departments from 2100 institutions, giving addresses and telephone numbers; alphabetical list of 850 research and development organizations gives the name of the company or parent institution, the division or laboratory name, the mailing address and telephone number of US federally funded research development centres, government, industrial, and not-for-profit laboratories in the USA and Canada; geographical list of academic institutions - within each country, listings are alphabetical by state or province and, within the state or province, alphabetical by institution; alphabetical list of research and development organizations by type.
Not available on-line.
Next edition: 1985; $30.00.

Research, training, test and production 265 reactor directory*

Editor: R. Robert Burn
Publisher: American Nuclear Society
Address: 555 North Kensington Avenue, LaGrange Park, IL 60525, USA
Year: 1983
Pages: 820
Price: $350.00
Geographical area: USA.
Content: The directory covers about 125 operating reactors and 137 shutdown reactors (arranged in separate sections, then alphabetically). Entries include organization name, reactor name, location, name and telephone number of reactor administrator.

AUTHOR, EDITOR AND COMPILER INDEX

Abbott, R. Tucker   7.226 usa
Abtahi, M.   1.70 fra; 1.72 fra
Adzigian, Denise Allard   7.32 usa
Agrawal, Usha   3.4 ind
Ahlfeld, Helmut   1.282 gfr
Akey, Denise   7.31 usa
Aleksander, I.   1.238 uni; 1.255 uni
Alexander, Laidon   5.372 uni
Alger, M.S.M.   1.134 uni
Allen, Gerald   7.165 usa
Allen, Kim   7.13 can
Allen, N.   5.98 uni
American Council on Education   7.5 gfr
Anderson, Alun M.   3.42 uni
Anderson, I.G.   1.8 uni; 5.29 uni
Anderson, Mildred   7.217 usa
Andrade, S.J.   6.26 uni
Andreen, Brian   7.104 usa
Anthony, L.J.   1.229 uni
Anton, Gayle J.   1.131 usa
Arab Petroleum Research Center   3.66 uni
Archbold, T.   1.227 uni
Armstrong, Michael   5.324 uni
Arnett, Ross H.   1.342 usa
Ash, Lee   7.61 uni
Ashby, C.M.   7.73 usa
Austin, John   5.222 uni
Australian Water Resources Council   4.23 aus

Bachman, P.   5.392 net
Backus, Karen   7.256 usa
Bailey, J.   1.260 uni
Baker, F.W.G.   1.87 fra
Baldeschwieler, John D.   3.59 usa
Ballenberger, Wolfgang   5.353 gfr; 5.354 gfr; 5.355 gfr
Bank, H.   5.82 gfr
Banks, R.C.   7.74 usa
Bannerman, Carol   1.26 usa
Bannouri, Rabii   2.11 tun
Barber, M.J.   5.283 uni
Barker, Michael J.C.   5.23 uni
Barlow, J.C.   7.74 usa
Barnard, Carmel   1.311 uni
Barry, Margaret   5.35 ire

Barty, Euan   3.62 hng
Baven, Peter M.   5.109 net
Bays, Grace   5.338 uni
Bazaz, Mohoan C.   3.94 ind
Bearse, Stacy V.   1.200 usa
Bednowitz, A.L.   1.159 net
Behrens, Dicter   1.121 gfr
Bell, B.   1.347 uni
Bennett, Richard J.   1.6 uni
Berg-Madsen, Vivianne   1.167 usa
Bessenyei, Helvi M.   1.111 usa
Bevan, Alun   5.35 ire
Bhat, S.G.   3.8 ind
Biggs, Lynne   1.157 uni
Billy, Christopher   7.34 usa
Biskup, Peter   4.3 uni
Blanc, G.   1.2 fra
Boehm, K.   5.49 uni
Bradfield, Valerie J.   1.228 uni
Brauer, Wilfried   5.204 gfr
Bricault, G.C.   3.87 uni; 3.89 uni
British Calibration Service   5.9 uni
Broadway, John R.   7.220 usa
Brown, James W.   1.22 usa
Brown, Marjorie J.   5.79 uni
Brown, Pat   5.247 uni
Brown, R.P.   5.167 uni
Browne, Fred   5.316 uni
Bryant, P.   1.151 uni
Bundy, Carol   2.1 uni
Bunsell, A.R.   5.157 uni
Burke, Christine E.   7.99 usa
Burkett, Jack   5.70 uni; 5.71 uni
Burn, R. Robert   7.265 usa
Bushell, C.M.   1.246 uni
Buttress, F.A.   1.74 uni

Cairns, Tom   5.28 uni
Campbell, Eila M.J.   1.174 uni
Caplin, Steve   5.74 uni
Carp, E.   1.169 swi
Carpenter, Julian R.   5.136 uni
Carr, Jennifer   2.27 uni; 3.40 uni
Casley, Christine   5.260 uni

AUTHOR, EDITOR AND COMPILER INDEX

Centro Nacional de Documentación Científica, Técnica y Económica   6.4 uru
Cesbron, Fabien   1.181 fra
Chandra Kumar, S.   1.165 tha
Chanyarak, Karnjana   3.16 tha
Chaplin, N.W.   5.383 uni
Chaudier, LouAnn   1.67 usa
Chaux, Miguel Angel   6.21 col
Chehab, Riyadh A.   1.243 leb
Chehab, Saadeddine   1.114 leb
Chelliah, T.   3.55 sin
Christodoulou, A.   1.7 uni
Chunchwell, Jan W.   1.67 usa
Clark, P.K.   1.174 uni
Cleevely, R.J.   1.182 usa
Clements, Barbara   7.213 usa
Clench, M.H.   7.74 usa
Clutton, A. Elizabeth   1.174 uni
Coan, E.V.   1.69 usa
Cock, James H.   6.21 col
Cocke, Carolyn   5.25 uni
Codlin, Ellen M.   5.4 uni; 5.356 uni
Cole, H.A.   5.332 uni
Coleman, Vernon   5.374 uni
Collier, Yvonne   5.216 uni
Colligon, J.S.   5.266 uni
Committee of Directors of Polytechnics   5.85 uni
Connell, Norman   1.280 uni; 1.289 uni; 1.304 uni; 1.309 uni
Connors, Martin   7.130 usa
Cooke, Henry   1.281 uni
Coombs, J.   1.329 uni; 5.211 uni
Coope, Brian   1.267 uni
Cooper, Alan   5.93 uni
Copp, D.J.B.   5.89 uni
Coppard, Susan   5.139 uni
Corbett, Marilyn J.   7.173 usa
Cordero, Raymond   1.281 uni; 1.296 uni
Cousins, J.A.   5.133 uni
Cox, A.L.   5.295 uni
Crafts-Lighty, Anita   1.325 uni
Craig, Helen B.   7.259 usa
Craig, T.   1.7 uni
Craig, William U.   7.259 usa
Crane, Eva   1.90 uni
Cruickshanks, Sandra   7.214 usa
Curtis, Denis   5.32 uni

Darnay, Brigitte T.   7.26 usa; 7.29 usa; 7.57 usa; 7.62 usa; 7.253 usa
Darrington, Hugh   5.276 uni
Darwin, K.   1.210 usa
DataPro Research Corporation   7.133 uni
Davies, Helen   2.14 uni
Davis, Gordon   7.153 usa
Dawes, John   5.108 uni
Day, Alan J.   1.62 uni
Dean, D.S.   5.332 uni
Deighton, Suzan   1.188 uni
Denyer, Bryan A.   1.235 uni
Derz, Friedrich   1.122 gfr; 1.129 gfr; 1.130 gfr
DeScherer, Mildred   7.82 usa
Deanette, J.B.   5.111 uni
Diamond, Harvey   7.25 usa
Didelot, J.   1.233 swi

Diment, Judith A.   5.181 uni
Doescher, Rex A.   1.171 usa
Domay, Friedrich   5.59 gfr
Donougher, Danielle   1.262 uni
Dowman, I.J.   5.179 uni
Dunmore, Jane   5.78 uni
Dwinarto, Sudarisman   3.20 ino

Eastwood, Bruce S.   7.11 usa
Eberhard, R.   5.16 uni
Eddowes, Derek   5.326 uni
Edwards, Donald   5.229 uni
Eisberg, Neil   5.243 uni
Elster, Robert J.   1.302 usa
Ermers, A.   4.14 aus
Espinal-Arenas, Eduardo   6.3 col
Ethridge, James M.   1.11 usa; 1.12 usa
Evan, Frederica   1.187 usa
Evans, Peter A.   1.111 usa

Fang, Josephine R.   1.39 uni
Fenat-Haessig, M.   5.82 gfr
Festing, M.P.W.   1.335 uni
Firth, C.   1.217 uni; 1.254 uni
Firth, F.   5.239 uni
Fitch, Jennifer M.   1.172 uni
Fitzgerald, D.J.   5.133 uni
Flesh, A.R.   7.73 usa
Flora, Philip   7.129 usa
Fong, Khoo Guan   3.10 sin
Foong, Sng Yok   3.10 sin
Foster, Janet   5.8 uni
Found, Peter   3.70 uni
Fourie, J.A.   2.9 saf
Frank, Robin C.   7.79 usa
Fransen, Hans   2.12 saf
French, Victor H.   5.115 uni
Furnival, J.   5.210 uni

Gamel, Laurie   7.217 usa
Gebhardt, Walther   5.104 gfr
Gerbrands, Jan J.   5.198 net
Gilat, Geula   3.15 isr
Gilbertson, David   1.286 uni
Gilder, Lesley   5.93 uni
Gill, Kay   1.44 usa; 7.39 usa; 7.40 usa; 7.256 usa
Gillispie, Charles Coulston   1.10 usa
Gimson, Garry   3.67 uni
Godbolt, S.   1.327 uni
Goddard, Jennifer   4.11 aus
Goldstein, Amy J.   7.41 usa; 7.42 usa; 7.43 usa
Goodman, Doreen   4.3 uni
Gordon, Rue   7.8 usa
Gordon Cook, J.   1.133 uni
Götzsche, Christian   5.7 den
Gough, B.E.   1.73 usa
Graeser, Kathi   7.132 usa
Graham, David M.   1.249 usa
Granade, Charles   7.41 usa; 7.43 usa
Grant, Marjorie A.   7.98 usa
Greenhalgh and Jeffs,   5.405 uni
Gregory, P.H.   1.92 uni
Grimes, Brian W.   1.193 uni
Grimes, Denis J.   1.193 uni; 1.194 uni

AUTHOR, EDITOR AND COMPILER INDEX

Gruber, Katherine   1.23 usa; 7.30 usa
Gurnsey, John   1.188 uni

Haacke, Wolfhart   5.204 gfr
Hale, Dean   1.256 usa
Hall, James L.   1.57 uni; 5.79 uni; 5.80 uni
Hall, Jean F.   1.341 uni
Hallam, Saskia   5.47 uni
Hallewell, Laurence   6.7 uni
Hannam, Harry   2.17 uni
Hardy, J.   1.173 uni
Hardy, Janet R.   7.106 usa
Hargreaves, Val   5.373 uni
Haron, Edward   5.83 pol
Harris, P.M.   5.178 uni
Harris, Sherry S.   7.1 uni
Harrison, David   3.67 uni
Harvey, A.   1.173 uni
Harvey, Joan M.   1.63 uni; 1.64 uni; 2.18 uni; 5.105 uni
Harvey, Nigel   1.89 uni
Hauptman, Trudi   7.107 usa
Heading, John   5.407 uni
Hedges, J.   1.151 uni; 5.213 uni
Heikkilä, Marjatta   5.107 fin
Henderson, C.A.P.   5.81 uni
Henderson, Faye   1.268 net
Henderson, G.P.   5.24 uni; 5.50 uni
Henderson, S.P.A.   5.24 uni
Henton, M.P.   5.182 uni
Herner, Saul   7.60 usa
Hill-Downham, Julia   7.13 can
Hilton, Ronald   6.20 usa
Holopainen, Viljo   5.134 fin
Homer, Patricia I.   1.315 usa
Honssen, J.J.   5.88 net
Howard, C.E.   1.196 swi
Howard, R.A.   3.54 usa
Howlett, J.   5.66 uni
Hsieh, Enid   4.11 aus
Hubbard, Linda S.   7.53 usa
Hudson, Kenneth   1.17 uni
Husain, Shahrukh   3.1 uni
Hutchison, Barbara P.   7.232 can
Hutton, George   1.291 uni

International Cancer Research Data Bank   1.316 usa
International ISBN Agency, Staatsbibliothek Preussischer Kulturbesitz, Berlin   1.49 gfr
Irons, R.I.   4.27 nze

Jack, G.A.   5.209 uni
Jackson, William G.   1.131 usa
Jacobsen, T.D.   1.338 usa
Jacques Cattell Press   7.3 uni; 7.4 uni; 7.47 uni
Japan Chemical Week,   3.61 jap
Jequier, N.   1.2 fra
Johansson, C.M.A.   5.208 uni
Johnson, Robert W.   4.19 aus
Johnston, Donald E.   1.88 usa
Judge, James J.   7.180 usa
Junghanns, H.   5.360 gfr
Juniper, Richard   1.154 uni

Kaplan, Barbara   1.330 usa
Kareh, Rene P.   1.153 leb
Kaszubski, Marek   7.250 usa
Kawata, S.   3.82 jap
Kay, Ernest   1.239 uni
Keefer, Mildred J.   7.199 usa
Kelly, A.   5.157 uni
Kelly, Brian W.   1.194 uni
Kendrick, Arthur   7.95 can
Kesarwani, S.K.   3.8 ind
Kheng, Lau Siew   3.10 sin
Kiger, R.W.   1.338 usa
Kim, S.T.   3.34 kor
King, Geoffrey   5.382 uni
Kirby, C.   5.186 uni
Knight, Jenny   5.22 uni; 5.54 uni
Kono, Rokuro   3.38 jap
Kothari, H.   3.50 ind; 3.51 ind
Kronschwitz, Helmut   5.401 gfr
Kruzas, Anthony T.   1.44 usa; 7.39 usa; 7.242 usa; 7.256 usa
Kusbandarrumsamsi, Hendrarta   3.20 ino

Laemmel, C.   5.255 swi; 5.281 swi; 5.288 swi; 5.312 swi; 5.314 swi
LaFrana, Jan   7.72 usa
Laidlow, J.C.   1.227 uni
Laloup, B.   5.351 fra
Lamb, Malcolm   3.13 uni
Lambert, C.M.   5.102 uni
Landy, Marc   7.36 usa
Larson, Donna Rae   7.46 usa
Latin American Newsletters Limited   6.19 uni
Laverick, A.   5.272 uni
Lawn, M.   2.27 uni
Lazo, John A.   7.229 usa
Le Bas, M.J.   5.191 uni
Lecznar, E.W.   5.202 uni
Lee, Joachim   1.185 sin
Leek, William R.   1.213 usa; 1.214 usa; 6.25 usa; 7.148 usa
Leesley, Michael E.   7.97 usa
Lehmann, Edward J.   7.17 usa
Leighton, D.   1.297 uni
Lengenfelder, Helga   1.75 gfr; 1.77 gfr; 5.58 gfr; 5.60 gfr
Leroy, Jean   1.176 usa
Levick, G.R.T.   1.93 net
Levy, J.E.   5.389 uni
Lewanski, Richard C.   5.106 uni
Lichtenfels, J.R.   1.324 uni
Liinamaa, Matti   5.107 fin
Lilly, R.M.   1.338 usa
Lindqvist, Rune   5.282 swe
Lindsay, E.J.   5.189 uni
Lines, E.   1.152 uni
Lockwood, S.J.   5.126 uni
Logan, B.G.R.   1.148 uni
Long, Chris   5.321 uni
López S., Jorge   6.21 col
Łos, Leon   5.99 pol
Lynch, Henry T.   1.331 usa

Macafee, Graham   4.21 aus
Mackay, Ross   4.26 aus; 4.28 aus
MacKechnie, J.   1.227 uni
MacVicar, Betty   7.261 usa

AUTHOR, EDITOR AND COMPILER INDEX

Mager, Chris   5.139 uni
Marini, Lucio   5.375 ita; 5.384 ita
Martin, Harriet   5.341 uni
Maxim, Barry   5.123 uni
Maynard, Nancy   7.117 usa
M'Boungou-Mayengué, Daniel   2.6 gha
McGowan, V.F.   1.345 aus
McIver, Glenys   4.22 aus
McLaney, Richard W.   7.154 usa
McLean, Janice   7.9 usa; 7.67 usa
McMillan, Carl H.   1.259 uni
Meenan, A.   5.36 uni
Melton, L.R.A.   5.404 uni
Miles, Wyndham D.   7.92 usa
Milesi, Edwina   5.34 uni
Millard, Patricia   5.342 uni
Miller, Claire   1.294 uni
Miller, Ian   5.94 uni
Miller, R.J.   1.242 uni
Mitchell, Brenda   1.303 uni
Mitchell, E.   1.152 uni
Modica, C.R.   1.322 usa
Molegraaf, Rudi   5.75 net
Monsebroten, Dale   7.56 usa
Montgomerie, G.   1.252 uni
Mori, Masao   3.39 jap
Morris, B.   5.49 uni
Morris, R.   1.293 uni
Morris Maxfield, Doris   7.127 usa
Morton, Leslie T.   1.327; 1.341 uni
Mostow, G.D.   1.356 usa
Muir, C.S.   1.320 uni
Mukherjee, B.   3.53 ind
Müller-Dietz, Heinz   5.391 gfr
Münch, Siegfried   5.204 gfr
Murdoch, Alison   5.256 uni
Myers, Arnold   5.227 uni
Myers, Don   7.112 usa

National Information and Statistics on Science and Technology   2.10 egy
National Technical Information Service   7.15 usa
National Technical Information Service   7.16 usa
Naylor, Bernard   6.7 uni
Nelson, Don   1.288 uni
Newill, A.E.   5.197 uni
Ng, Elizabeth W.   3.2 hng; 3.7 hng
Nicholls, Ann   1.17 uni
Nisbet, John   1.85 uni
Nisbet, Stanley   1.85 uni
Noyes, Robert   1.323 usa
Nurse, Milton   1.260 uni; 1.294 uni

O' Reilly,   1.164 net
Odgers, Elizabeth   4.8 aus
OECD Development Centre   2.13 fra; 3.30 fra; 6.16 fra
O'Hara, Nancy L.   7.159 can
Ohashi, Hitoshi   3.65 jap
Olanvoravuth, Ninnat   3.27 tha
Oldham, Tony   5.176 uni
Olney, P.J.S.   1.339 uni
Ong, A.S.H.   3.40 sin
Opitz, Helmut   1.76 gfr; 1.78 gfr
Osborn, Peter   5.214 uni

Osborne, C.W.   1.48 uni
Ostroff, Jesse   1.109 usa

Parker, C.C.   1.34 uni
Pawlett, Steve   7.76 can
Pawlik, Kurt   1.333 net
Paylore, Patricia   1.162 usa
Peach, W.   1.261 uni
Pearce, Patricia   5.53 uni
Pergolizzi, Robert G.   7.245 uni
Pierce, Gerald   3.57 uni
Pirazzoli, P.   1.168 swe
Pokress, E.   7.257 usa
Pollock, Gisella Linder   1.128 usa
Poppe, Barbara B.   1.170 usa
Pratt, Betty   7.7 can
Price, J.G.   7.109 usa
Pritchard, M.H.   1.324 usa
Pudney, Matthew   5.101 uni

Ramsay, Graeme W.   4.31 nze
Rasmussen, Douglas J.   3.5 hng; 3.60 hng; 3.64 hng
Rasmussen, K.   5.127 uni; 5.128 uni
Rauchle, N.M.   4.24 aus
Read, Paula   1.260 uni
Ready, Barbara C.   7.42 usa; 7.234 usa
Reed, David   5.333 uni
Rhys Jones, Roderick   1.236 uni
Rimington, G. Philip   1.113 uni
Roberts, Stephen   5.93 uni
Roberts, Steven   1.197 uni
Robertson, James   5.195 uni; 5.269 uni; 5.327 uni
Robinson, B.   1.260 uni
Robinson, Kim   1.125 uni
Rockwell, Cynthia   1.41 ita
Roland, M.   5.82 gfr
Rosenau, Fred   1.268 net
Ross, J.F.   4.33 aus
Roussel, Denis   7.51 can
Rowland, K.T.   5.339 uni
Ruffner, James A.   7.54 usa
Ruscoe, Q.W.   4.10 nze

Sakiyama, Seikoh   3.33 usa
Salvi, L.A.   5.206 ita
Sandor, Ruth   1.58 usa
Sandri White, Alex   7.257 usa
Sansom, Bob   5.185 uni
Sardar, Ziauddin   3.43 uni
Sawyer, Eleanor   7.231 can
Schärer, Martin R.   5.95 swi
Schiavone, Giuseppe   1.43 uni
Schmittroth Jr, John   1.1 usa; 1.24 usa; 7.33 usa; 7.127 usa
Seingry, G. Francis   5.48 bel; 5.140 bel; 5.265 bel
Sellar, Lindsay   5.18 uni
Sendov, Blagovast Hristov   5.12 bul
Serjeantson, Richard   1.281 uni; 1.287 uni; 1.290 uni
Shanin-Cohen, Naomi   3.14 isr
Sharpe, R.S.   5.332 uni
Sharpe, T.   5.184 uni
Shaw, A.   1.237 usa
Shaw, D.F.   1.351 uni
Sheehy, Eugene P.   1.31 uni

AUTHOR, EDITOR AND COMPILER INDEX

Shennan, I.   1.168 swe
Shepherd, P.A.   5.202 uni
Sheppard, Julia   5.8 uni
Shields, J.   5.240 uni
Shuter, Janet   5.65 uni
Sigel, Efrem   1.187 usa
Singh, Pritam   4.31 nze
Sittig, Marshall   1.323 usa
Smart, J.E.   5.73 uni
Smith, Harry   1.313 uni
Smith, Wendy M.   1.219 uni
Sommer, Richard G.   7.45 usa
Songe, Alice H.   1.39 uni
Sors, Andrew I.   1.179 uni
Spahr, Frederick T.   7.230 usa
Steadman, D.   5.213 uni
Steadman, Dorothy   5.280 uni
Steed, Colin   5.264 uni
Steele, Colin R.   1.52 uni; 6.7 uni
Steenkamp, N.S.   2.21 saf
Stiefel, Malcolm   7.125 usa
Su Eng, L.   3.40 sin
Šulc, Branka   5.72 yug
Surgenor, Christopher   1.190 uni
Szentirmay, Paul   4.9 nze
Szentirmay, Thiam Ch'ng   4.9 nze

Takahashi, Hiroshi   3.81 jap
Tan, B.K.   1.165 tha
Tang, Tong B.   3.41 uni
Taylor, Francis   5.363 uni
Thomas, W.J.K.   5.133 uni
Thompson, Dick   5.101 uni
Thompson, Susan   1.310 uni
Thomson, H.M.   5.311 uni
Thomson, Keith W.   4.1 nze
Tiratsoo, J.N.H.   1.220 uni
Titoko, Judith   4.15 fij
Toase, Mary   5.87 uni
Tomlinson, Janet   1.188 uni
Torrens, H.S.   5.190 uni
Tratner, A.   1.210 usa
Trzyna, Thaddeus C.   1.69 usa; 1.119 usa
Turley, R.V.   1.34 uni
Turney, John   5.97 uni
Tyler, D.A.   5.332 uni

Vanderlin, Jane   1.58 usa
Varley, J.   1.354 uni
Vellucci, Matthew J.   7.60 usa
Verougstraete, Janine   5.67 bel

Vestdal, Jón E.   5.238 ice
Vieth, Thomas   5.205 gfr
Vokac, Libena   5.77 nor
von Hulst, H.   1.108 ita
Vozzo, Steven F.   1.149 usa

Wagner, G.   1.320 uni
Wain, B.J.   5.295 uni
Wainwright, David   5.94 uni
Walker, Anthony   1.226 uni
Wanklyn, Margaret   4.11 aus
Warner, Robin   3.69 uni
Warring, R.H.   5.287 uni; 5.293 uni; 5.325 uni; 5.329 uni; 5.331 uni; 5.336 uni; 5.346 uni.
Wasserman, Paul   7.67 usa; 7.250 usa
Watkins, Mary Michelle   7.54 usa
Watson, D.G.   5.193 uni
Way, Anna   1.253 uni
Webb, Penny   5.101 uni
Weber, Molly   5.188 uni
Wedgeworth, Robert   7.2 uni
Weida, William A.   1.131 usa
Wells, John   7.34 usa
West, J.   5.332 uni
Weston, Richard   5.246 uni
Whiteside, R.M.   5.217 uni; 5.318 uni; 5.320 uni
Williams, Moelwyn   5.39 uni
Willmore, A.P.   1.184 uni
Willmore, S.R.   1.184 uni
Wilson, C.W.J.   1.357 uni; 5.37 uni
Wilson, E.   5.155 uni
Wilson, Keith   1.225 uni
Wood, Allan   1.65 uni
Wood, D.N.   1.173 uni
Wood, Donna   1.301 usa
Wright, Rita J.   7.217 usa
Wyatt, H.V.   1.326 uni

Young, Brenda K.   7.37 usa
Young, Margaret Labash   7.58 usa

Zaborsky, Oskar R.   7.6 usa; 7.37 usa
Zanina, Putri   3.86 may
Zardini, E.M.   6.22 net
Zell, Hans M.   2.1 uni
Zubris, Donna K.   7.6 usa

DIRECTORY TITLES INDEX

abbreviations of organizations, seventh edition, World guide to   1.74 uni
ABC Belge pour le commerce et l'industrie, Belgisch ABC voor handel en industrie;   5.249 bel
Abstracting and indexing services directory, first edition   1.1 usa
Acarologists of the world   1.88 usa
Accredited institutions of post secondary education 1984-85   7.1 uni
Achema-jahrbuch 1983-85   1.121 gfr
acid deposition researchers, North American and European edition, International directory of   1.149 usa
ACS directory of graduate research (chemistry)   7.91 usa
Adhesives Euro-guide, second edition   5.241 uni
Adhesives handbook   5.240 uni
(ADIPA): a directory of members, Association of development research and training institutes of Asia and the Pacific   3.2 hng
Adresboek van de oostvlaamse industrie   5.242 bel
Adressbuch der Deutschen tierarzteschaft   5.119 gfr
Adressbuch des deutschsprachigen buchhandels   5.1 gfr
Adressbuch deutscher chemiker   5.141 gfr
AEE directory of energy professionals   7.134 usa
Aerosol Manufacturers' Association: 24th annual report 1984, British   5.251 uni
Aerosol review   5.243 uni
aerospace directory, Interavia   1.233 swi
Aerospace research index: a guide to world research in aeronautics, meteorology, astronomy, and space science   1.184 uni
Africa: a bibliography, Libraries in West   2.14 uni
Africa, Directory for scientific research organizations in South   2.4 saf
Africa, Directory of development research and training institutes in   2.7 fra
Africa, Directory of industrial and technological research institutes in   2.24 aut
Africa, Directory of scientific and technical societies in South   2.8 saf
Africa, Marine research centres: 2.15 ita
Africa south of the Sahara 1984-85   2.3 uni
Africa, 34th edition, Companies and suppliers in   2.28 fra
African book world and press: a directory, third edition   2.1 uni

African experts, Directory of   2.5 aut
African international organizations directory 1984-85   2.2 gfr
aging, International directory of organizations concerned with   1.332 aut
agribusiness buyers' guide, Asian   3.52 uni
agricultural and food research, Index of   5.130 uni
Agricultural education: full time and sandwich courses serving England and Wales   5.120 uni
agricultural education, Handbook of   3.56 ind
agricultural engineering institutions, International directory of   1.108 ita
agricultural export products directory, Hawaii   7.85 usa
agricultural, horticultural and fishery co-operatives in the United Kingdom, sixth edition, Directory of   5.124 uni
agricultural information sources in Asia and Oceania, Guide to   1.93 net
agricultural organizations, Directory of European   5.125 uni
Agricultural research centres: a world directory of organizations and programmes, seventh edition   1.89 uni
agricultural research, ICAR Inventory of Canadian   7.86 can
Agricultural research service and institutes and units of the agricultural research service   5.121 uni
agricultural residues for the production of panels, pulp and paper, Information sources on utilization of   1.138 aut
Agricultural science in the Netherlands (including the former guide, Wageningen, centre of agricultural science) 1985-87   5.122 net
agricultural sciences in the Commonwealth, List of research workers in the   1.113 uni
agriculture and food science, Information sources in   1.95 uni
agriculture guide, first edition, 1985, Saudi Arabian   3.57 uni
agriculture, second edition, Who's who in world   1.117 uni
air carriers, second edition, Inventory of world commercial   1.240 can
ALA yearbook of library and information services   7.2 uni
Alabama directory of mining and manufacturing 1985-86   7.154 usa
Alabama, Industrial   7.195 usa
Alaska, Oil directory of   7.145 usa
Alaska petroleum and industrial directory 1984   7.155 usa

aluminium survey, World   1.304 uni
American chemists and chemical engineers   7.92 usa
American Council of Independent Laboratories: directory   7.93 usa
American fisheries directory and reference book   7.71 usa
American library directory: a classified list of libraries in the United States and Canada with personnel and statistical data   7.3 uni
American malacologists: a national register of living professional and amateur conchologists   7.226 usa
American medical directory   7.227 usa
American men and women of science: physical and biological sciences, fifteenth edition   7.4 uni
American Speech and Hearing Association directory   7.230 usa
American universities and colleges, twelfth edition   7.5 gfr
American Veterinary Medical Association directory   7.72 usa
anaesthesiologists in Germany, Austria, and Switzerland, Index of   5.401 gfr.
ANEP 85: European petroleum yearbook   5.205 gfr
animal disease diagnostic laboratories, Directory of   7.77 usa
animal feed industry, Information sources on the   1.99 aut
animal health and disease data banks, International directory of   1.109 usa
Animal health international directory   1.310 uni
Annotated acronyms and abbreviations of marine sciences and related activities   7.73 usa
Annuaire CNRS   5.2 fra
Annuaire CNRS chimie   5.142 fra
Annuaire CNRS mathématiques, sciences physiques 1981-82   5.403 fra
Annuaire CNRS sciences de la terre, de l'océan, de l'atmosphère et de l'espace   5.175 fra
Annuaire dentaire   5.351 fra
Annuaire des stations hydro-minérales, climatiques balnéaires et établissements médicaux Français   5.352 fra
Annuario nazionale dell'energia   5.206 ita
Appropriate technology directory, volume 2   1.2 fra
Appropriate technology in situations: a directory   1.3 uni
Arab and Islamic International Organization Directory 1984-85   1.4 gfr
Arab oil and gas directory, tenth edition   1.203 uni
Arab oil and gas directory 1985   3.66 uni
Arab world, International who's who of the   1.46 uni
Arab world 1975-85, Major companies of the   3.88 uni
Arabian computer guide, 1985   3.69 uni
Arabian construction 1985   3.70 uni
Arabian government and public services 1985   3.1 uni
Arabian transport guide 1985   3.67 uni
architecture and planning 1985-86, Directory of official   5.225 uni
architecture, Information sources in   1.228 uni
Archive, bibliotheken und dokmumentationsstellen der Schweiz   5.3 swi
Archives, bibliothèques et centres de documentation en Suisse   5.3 swi
Archivi, biblioteche e centri di documentazione in Svizzera   5.3 swi
Arid lands research institution: a world directory 1977   1.162 usa
Arizona directory of manufacturers   7.156 usa
Arkansas directory of manufacturers   7.157 usa
ARPANET directory   7.118 usa

DIRECTORY TITLES INDEX

Art galleries and museums of New Zealand   4.1 nze
Ärzte-adressbuch, Nordbaden, verzeichnis der praktizierenden ärzte, krankenanstalten   5.353 gfr
Ärzte-adressbuch Nordwürttemberg, verzeichnis der praktizierenden ärzte, krankenanstalten   5.354 gfr
Ärzte-adressbuch Südwürttemberg: verzeichnis der praktizierenden ärzte, krankenanstalten   5.355 gfr
Asia and the Pacific, Directory of training institutions and resources in   1.20 pak
Asia and the Pacific engaged in distance education, Directory of institutions of higher education in   1.15 tha
Asia, Directory of national systems of technicians education in south and central   3.11 fra
Asia, Directory of social science research and training units -   3.30 fra
Asian agribusiness buyers' guide   3.52 uni
Asian and Australasian directories: a guide to directories published in or relating to all countries in Asia, Australia, and Oceania, Current   1.8 uni
Asian and Pacific electrical and electronics directory   1.185 sin
Asian computer yearbook   3.62 hng
Asian institutions of higher learning, Handbook: southeast   3.27 tha
Asian studies in Japan, 1981, Research institutes on   3.38 jap
Asian topics, Directory of current Hong Kong research on   3.7 hng
Asia's 7500 largest companies   3.71 uni
Asie, Inventaire descriptif des unités de recherche et de formation en sciences sociales -   3.30 fra
ASLIB directory of information services in the United Kingdom: volume 1 - information sources in science, technology and commerce   5.4 uni
ASLIB directory of information sources in the United Kingdom: volume 2 - social sciences, medicine and the humanities.   5.356 uni
Assam directory and tea areas handbook   3.53 ind
Association of Bronze and Brass Founders buyers guide   5.244 uni
Association of Consulting Engineers New Zealand: list of members   4.27 nze
Association of consulting scientists: members and services 1985-86   5.5 uni
Association of development research and training institutes of Asia and the Pacific (ADIPA): a directory of members   3.2 hng
Association of Hydraulic Equipment Manufacturers: directory of members   5.245 uni
Astronomy directory (western states)   7.107 usa
atomic energy, seventh edition, World nuclear directory: a guide to organizations and research activities in   1.357 uni
audio-visual sources: history of science, medicine, and technology, Directory of   7.11 usa
Australia: a directory, Map collections in   4.24 aus
Australia: current projects 1983, Water research in   4.25 aus
Australia, Kompass register: 4.29 aus
Australia, Medical directory of   4.32 aus
Australia, Scientific and technical research centres in   4.14 aus
Australian Academy of Science yearbook   4.2 aus
Australian companies, Database: 4.7 hng
Australian electronics directory   4.26 aus
Australian engineering directory   4.28 aus
Australian fishing industry directory 1985   4.16 aus

DIRECTORY TITLES INDEX

Australian libraries   4.3 uni
Australian maps 1984   4.22 aus
Australian reference books   4.4 uni
Australian sugar year book   4.17 aus
Automated manufacturing directory 1985   5.246 uni
automotive components, Who's who in West European   5.239 uni
Autotrade directory   5.247 uni
aviation directory, Japan   3.65 jap
Awards for Commonwealth university academic staff 1984-86   1.5 uni

Bangladesh trade directory   3.3 ind
Bee world - directory of the world's beekeeping museums issue   1.90 uni
beer and wine industry, Information sources on the   1.100 aut
Belgian ABC for commerce and industry   5.249 bel
Belgian Exports   5.248 bel
Belgisch ABC voor handel en industrie; ABC Belge pour le commerce et l'industrie   5.249 bel
Belgium research centres having library or documentation services, Directory of   5.67 bel
Berufsschulen und sonderberufsschulen in Bayern   5.6 gfr
beverage industry, Information sources on the non-alcoholic   1.106 aut
Biblioteksvejviser   5.7 den
Binsted's directory of food trade marks and brand names   5.250 uni
Bio-energy directory (biomass)   7.135 usa
bioconservation of agricultural wastes, Information sources on   1.96 aut
biological consulting practices, Directory of   5.364 uni
biological sciences (CABS), Current awareness in   1.313 uni
biological sciences, third edition, Information sources in the   1.326 uni
biologicals, Nature's directory of   1.343 uni
biologists, Organizations in the UK employing   5.394 uni
biomass directory, European   5.211 uni
biomass, Directory of industrial and technological research institutes: industrial conversion of   1.206 aut
biomedical engineers, Directory of   1.315 usa
biotechnology, Directory of research in   5.371 uni
biotechnology directory 1985, International   1.329 uni
Biotechnology engineers: biographical directory   7.6 usa
biotechnology firms worldwide directory, Genetic engineering and   1.323 usa
biotechnology, first edition, Directory of consultants in   7.236 usa
biotechnology funding sources, Federal   7.37 usa
biotechnology, Information sources in   1.325 uni
Biotechnology international   1.311 uni
biotechnology marketplace: Japan, New   3.37 usa
biotechnology marketplace: USA and Canada, New   7.50 usa
biotechnology sourcebook, Genetic engineering;   7.245 usa
biotechnology 1984, Directory of British   5.365 uni
Bird collections in the United States and Canada   7.74 usa
blind in the British Isles and overseas, Directory of agencies for the   5.362 uni
book and special collections in the United Kingdom and the Republic of Ireland, Directory of rare   5.39 uni
book trade, Directory of the German   5.1 gfr
Botanical gardens of the People's Republic of China   3.54 usa

Botanical Society, Directory of names and addresses and research specialisations for current members of the   7.239 usa
botany in the United States and Canada, Guide to graduate study in   7.247 usa
brachiopodologists, Doescher's directory of   1.171 usa
Brief directory of museums of India   3.4 ind
Brief guide to centres of international lending and photography   1.6 uni
British Aerosol Manufacturers' Association: 24th annual report 1984   5.251 uni
British archives: a guide to archive resources in the United Kingdom   5.8 uni
British associations and associations in Ireland, seventh edition, Directory of   5.24 uni
British calibration service: approved laboratories and their measurements   5.9 uni
British dental journal educational directory 1984   5.357 uni
British directories, tenth edition, Current   5.19 uni
British machine tools and equipment   5.252 uni
British official publications not published by HMSO 1985, Catalogue of   5.14 uni
British qualifications, fifteenth edition: a comprehensive guide to educational, technical, professional and academic qualifications in Britain   5.10 uni
British robot association members' handbook 1984-85   5.192 uni
British scientists 1980-81, third edition, Who's who of   5.118 uni
British universities' guide to graduate study 1985-86   5.11 uni
broadcasting, fifth edition, Directory of international   1.190 uni
Bronze and Brass Founders buyers guide, Association of   5.244 uni
Brown's directory of North America and international gas companies   1.256 usa
bryology: a world listing of herbaria, collectors, bryologists, current research, Compendium of   1.312 gfr
building boards from woods and other fibrous materials, Information sources on   1.135 aut
building research, information and development organisations, fifth edition, International directory of   1.234 uni
Bulgarian Academy of Sciences 1984 reference book   5.12 bul
Bulk carrier register, 17th edition   1.223 uni
Bulletin of special courses   5.13 uni
Bundes apotheken register: verzeichnis der apotheken in der bundes republik deutschland und in Berlin (west-sektoren)   5.358 gfr
bundes republik deutschland und in Berlin (west-sektoren), Bundes apotheken register: verzeichnis der apotheken in der   5.358 gfr
Buyers' guide to German industry   5.268 gfr
Buyers' guide to north east industry   5.253 uni
Buyers' guide to pumps 1985   5.254 uni

cacao, Directory of specialist workers on phytophthora palmivora with special reference to   1.92 uni
Calendar of the Pharmaceutical Society of Ireland   5.359 ire
California energy directory: a guide to organizations and information resources   7.136 usa
California manufacturers register   7.158 usa

California water resources directory: a guide to organizations and information resources   7.108 usa
Canada, Oil directory of   7.146 usa
Canada, Scientific and technical societies of   7.59 can
Canada's who's who of the poultry industry   7.75 can
Canadian conservation directory   7.7 can
Canadian forest industries   7.76 can
Canadian hospital directory   7.231 can
Canadian industry (1977), Directory of scientific and technological capabilities in   7.178 can
Canadian medical directory   7.232 can
Canadian trade index   7.159 can
cancer epidemiology, Directory of on-going research in   1.320 uni
cancer: professionals and facilities, Marquis who's who in   7.254 usa
cancer research and treatment establishments, International directory of specialized   1.334 swi
cancer research information resources, Directory of   1.316 usa
canning, freezing, preserving industries, Directory of the   7.180 usa
canning industry, Information sources on the   1.272 aut
cartography and in carto-bibliography, International directory of current research in the history of   1.174 uni
Cassava directory: directory of persons interested in tropical root crops (preliminary edition)   6.21 col
Catalogue of British official publications not published by HMSO 1985   5.14 uni
caving clubs 1984, Directory of British   5.176 uni
cement and concrete industry, Information sources on the   1.230 aut
ceramic directory, International   1.148 uni
ceramics industry, Information sources on the   1.139 aut
Chemaddressbook   1.122 gfr
Chemcyclopedia 85   7.94 usa
Chemfacts: Canada, first edition   7.96 uni
Chemfacts: France, first edition   5.145 uni
Chemfacts: Italy, first edition   5.146 uni
Chemfacts: Japan, first edition   3.58 uni
Chemfacts: Netherlands, second edition   5.147 uni
Chemfacts: Portugal, first edition   5.148 uni
Chemfacts: PVC   1.123 uni
Chemfacts: Scandinavia, second edition   5.149 uni
Chemfacts: Spain, second edition   5.150 uni
Chemfacts: United Kingdom, third edition   5.151 uni
chemical and allied industry and trade, Directory of the Dutch   5.154 net
Chemical buyers guide   7.95 can
Chemical company profiles: the Americas, second edition   1.124 uni
Chemical company profiles: Western Europe, third edition   5.152 uni
chemical directory, fourth edition, Worldwide   1.161 uni
chemical directory, Japan   3.61 jap
Chemical directory of Northern Europe: manufacturers and traders of chemicals and allied products in Scandinavia and Finland and Iceland   5.153 net
chemical engineering consultants, Directory of   7.99 usa
Chemical engineering faculties   7.97 usa
chemical engineering in Canadian universities, Directory of graduate programs in   7.102 can
chemical engineering research in Canadian universities, Twenty-first directory of   7.105 can

chemical industry and related products, Swiss   5.255 swi
Chemical industry directory and who's who   1.125 uni
Chemical industry yearbook, second edition   1.126 uni
chemical industry 1985-87, Federal Republic of Germany and West Berlin, Directory of the German   5.166 gfr
Chemical plant contractor profiles, fourth edition   1.127 uni
chemical producers, Directory of   7.106 usa
chemical producers, 1985-86, World directory of   1.158 usa
Chemical production index   1.129 gfr
Chemical research faculties: an international directory   1.128 usa
Chemical Suppliers Directory   1.130 gfr
chemical waste, ENREP directory of solid waste and   5.161 uni
Chemical yearbook, CNRS   5.142 fra
Chemical yearbook of the Royal Netherlands Chemical Society   5.156 net
Chemicalien adresboek   5.154 net
Chemicals address book   1.122 gfr
chemicals and petrochemicals, China's   3.60 hng
chemicals directory, Irish   5.170 ire
Chemicals 1985: chemicals on the UK market   5.155 uni
Chemische industrie der Schweiz und ihre nebenprodukte   5.255 swi
Chemische jaarboek der Koninklijke Nederlandse Chemische Vereniging   5.156 net
(chemistry), ACS directory of graduate research   7.91 usa
Chemistry and chemical engineering in the People's Republic of China, a trip report on the US delegation in pure and applied chemistry   3.59 usa
chemistry at private undergraduate colleges, Research in   7.104 usa
chemistry co-op, Directory of   7.100 usa
chemistry faculties, College   7.98 usa
chemists and chemical engineers, American   7.92 usa
chemists, Directory of German   5.141 gfr
Chemproductindex   1.129 gfr
Chemsuppliers directory   1.130 gfr
Chicago geographic edition 1985   7.160 usa
child guidance and school psychological services, Directory of   5.366 uni
Chile, Guide to special libraries and documentation centres in   6.10 chi
Chile: information guide, Scientific societies of   6.12 chi
Chilean universities, Graduate programmes offered by   6.18 chi
China, a trip report on the US delegation in pure and applied chemistry, Chemistry and chemical engineering in the People's Republic of   3.59 usa
China, Directory of the cultural organization of the Republic of   3.22 tai
China phone book and address directory   3.5 hng
China, Science and technology in   3.41 uni
China 1968-83, Directory of officials and organizations in   3.13 uni
China's chemicals and petrochemicals   3.60 hng
China's electronics and electrical products   3.63 hng
China's instruments and meters   3.64 hng
Chinese foreign trade 1985-86, Directory of   3.76 uni
Chinese manufacturers association of Hong Kong directory of members 1985   3.72 hng
Chirugenverzeichnis   5.360 gfr
Chung hua min-kuo hsueh-shu chi-ko lu   3.22 tai
CIRIA guide to sources of information in the construction industry   5.220 uni

DIRECTORY TITLES INDEX

Civic trust environmental directory, sixth edition   5.15 uni
climatology, Directory of UK research in   5.180 uni
Clinical and public health laboratory directory   7.233 usa
Clinical pharmacology: a guide to training programs   7.234 usa
clothing industry, Information sources on the   1.273 aut
CMERI directory of indigenous engineering products   3.73 ind
CNRS Chemical yearbook   5.142 fra
CNRS yearbook of earth, oceanic, atmospheric, and space sciences   5.175 fra
CNRS yearbook of mathematics and physical sciences   5.403 fra
coal industry manual 1985, Keystone   7.143 usa
coalfields, Guide to the   5.185 uni
Codata directory of data sources of science and technology   5.193 uni
coffee, cocoa, tea and spices industry, Information sources on the   1.101 aut
coil coating directory, European   5.270 uni
College chemistry faculties   7.98 usa
colleges of further and higher education: full time and sandwich courses in polytechnics and other colleges outside the university sector, Compendium of advanced courses in   5.16 uni
Colorado manufacturers 1985, Directory of   7.165 usa
commerce and industry, Netherlands ABC for   5.322 net
commerce, Norwegian directory of   5.323 nor
Commission of the European communities, Directory of the   5.41 lux
Commonwealth forestry handbook 1981, tenth edition   1.91 uni
Commonwealth universities yearbook 1985   1.7 uni
Commonwealth university academic staff 1984-86, Awards for   1.5 uni
community support agencies in the United Kingdom and Republic of Ireland, Someone to talk to directory 1985: a directory of self-help and   5.101 uni
Companies and suppliers in Africa, 34th edition   2.28 fra
company directories and summary of their contents, second edition, List of   1.283 aut
Compendium of advanced courses in colleges of further and higher education: full time and sandwich courses in polytechnics and other colleges outside the university sector   5.16 uni
Compendium of bryology: a world listing of herbaria, collectors, bryologists, current research   1.312 gfr
Compendium of rural research and development   4.18 aus
Compendium of university entrance requirements for first degree courses in the United Kingdom 1986-87   5.17 uni
Composite materials: a directory of European research   5.157 uni
Computer and telecommunications buyers' guide   7.119 usa
computer applications in United Kingdom libraries and information units, Directory of operational   5.37 uni
Computer books and serials in print 1985   1.186 uni
Computer companies in the UK, 1984, first edition   5.194 uni
computer executives, Directory of top   7.181 usa
computer graphics, Marquis who's who directory of   1.199 usa
computer graphics suppliers: hardware, software, systems and services, Klein, S., directory of   7.125 usa
computer guide, 1985, Arabian   3.69 uni

Computer publishers and publications 1985-86: an international directory and yearbook, second edition   1.187 usa
computer software, Directory of   7.15 usa
computer systems, third edition, Directory of consultants in   7.122 usa
computer training, Directory of   5.264 uni
Computer users' year book 1985   5.256 uni
computer yearbook, Asian   3.62 hng
computerized data files 1985, Directory of   7.16 usa
Computers and information processing world index   1.188 uni
Computing Services Association: directory of members and services   5.257 uni
Computing Services Association: members survey   5.258 uni
Computing Services Association: official reference book 1985   5.259 uni
confectionery and bakery products: manufacturers of bakery products, cocoa, chocolate, and their products, confectionery titbits, seasoned crackers, etc., European   5.271 net
Connecticut manufacturing directory   7.161 usa
conservation directory, Canadian   7.7 can
Conservation directory 1985   7.8 usa
construction directory, Irish   5.231 ire
construction industry, fourth edition, Guide to sources of information in the   5.228 uni
construction industry, Professional services to the   5.233 ire
Construction plant and equipment annual   5.260 uni
construction 1985, Arabian   3.70 uni
Consultants and consulting organizations directory, sixth edition   7.9 usa
Consulting and laboratory services   7.10 usa
Consulting engineers   5.221 uni
Consulting engineers who's who and yearbook   5.222 uni
consulting scientists, contract research organizations, and other scientific and technical services, Register of   5.89 uni
consulting scientists: members and services 1985-86, Association of   5.5 uni
Coral reef researchers: Pacific   4.5 ita
corporations, directors and executives, Standard and Poor's register of   1.295 bel
Corrosion: annual report, European Federation of   5.163 gfr
Corrosion prevention directory   5.158 uni
COSPAR directory of organization and members   1.163 fra
cotton and textile industries directory, twelfth edition, 1985-86, International federation of   1.279 swi
Councils, committees and boards: a handbook of advisory, consultative, executive and similar bodies in British public life   5.18 uni
Courses and training for careers in horticulture   5.123 uni
crystallographers, sixth edition, World directory of   1.159 net
CSIRO directory 1985   4.6 aus
CSIRO research programs 1985, Directory of   4.8 aus
Current Asian and Australasian directories: a guide to directories published in or relating to all countries in Asia, Australia, and Oceania   1.8 uni
Current awareness in biological sciences (CABS)   1.313 uni
Current bibliographic sources in education   1.9 swi
Current British directories, tenth edition   5.19 uni
Current European directories, second edition   5.20 uni
Current geological research in Texas: a directory of faculty and student research   7.109 usa

Current plant taxonomic research on Australian flora, 1980-81   4.19 aus
cytogenetics laboratory directory, International   1.330 usa
CZI register and buyer's guide   2.23 zim

dairy product manufacturing industry, Information sources on the   1.102 aut
Data base directory, 1984-85   7.120 usa
Data bases in Europe: a directory to machine-readable data bases and data banks in Europe   5.21 lux
Data processing professionals directory   7.121 usa
Data world international   1.189 usa
Database: Australian companies   4.7 hng
database guide, EUSIDIC   5.199 uni
databases and information systems and services, Directory of United Nations   1.21 swi
databases, Directory of online   1.18 usa
deaf, Reference issue: American annals of the   7.259 usa
defense directory, International   1.196 swi
Degree course guides   5.22 uni
Delaware directory of commerce and industry   7.162 usa
Denmark trade directory, Kingdom of   5.301 den
Dental directory   5.351 fra
dental journal educational directory 1984, British   5.357 uni
Dental schools of the world   1.314 usa
Dentists   5.378 fin
Dentists register, comprising the names and addresses of dental practitioners registered at 31 January 1985, together with the local list of names so registered and the list of bodies corporate carrying on the business of dentistry   5.361 uni
Deutsche branchen-fernsprechbuch   5.261 gfr
Deutsche firmen-alphabet: industrie, handel, verkehr, organisationen   5.262 gfr
Deutsches bundes-adressbuch   5.263 gfr
Diamond's Japan business directory   3.74 jap
Dictionary of Icelandic engineers and members of the Association of Chartered Engineers in Iceland   5.238 ice
Dictionary of Polish scientific societies: volume I: scientific societies at present active in Poland   5.99 pol
Dictionary of scientific biography   1.10 usa
Directoría de establecimientos de educación oficiales y privados: educación media y superior   6.1 col
Directorio de bases de·datos disponibles en el país (versión preliminar)   6.2 chi
Directorio de la educación superior en Colombia   6.3 col
Directorio de servicios de información y documentación en el Uruguay   6.4 uru
Directorio del sector eléctrico   6.24 uru
Directorio regional de unidades de información para el desarrollo   6.5 chi
Directory - development research and training institutes - Africa   2.13 fra
Directory: affiliates and offices of Japanese firms in the ASEAN countries   3.75 uni
Directory: affiliates and offices of Japanese firms in the USA   7.163 uni
Directory for scientific research organizations in South Africa   2.4 saf
Directory for the environment: organizations in Britain and Ireland 1984-85   5.23 uni
Directory: health systems agencies, state health planning and development agencies, statewide health coordinating councils   7.235 usa

DIRECTORY TITLES INDEX

Directory information service   1.11 usa
Directory of African experts   2.5 aut
Directory of African universities   2.6 gha
Directory of agencies for the blind in the British Isles and overseas   5.362 uni
Directory of agencies offering therapy counselling and support for psychosexual problems   5.363 uni
Directory of agricultural, horticultural and fishery co-operatives in the United Kingdom, sixth edition   5.124 uni
Directory of animal disease diagnostic laboratories   7.77 usa
Directory of Asian museums   3.6 fra
Directory of audio-visual sources: history of science, medicine, and technology   7.11 usa
Directory of Belgium research centres having library or documentation services   5.67 bel
Directory of biological consulting practices   5.364 uni
Directory of biomedical engineers   1.315 usa
Directory of British associations and associations in Ireland, seventh edition   5.24 uni
Directory of British biotechnology 1984   5.365 uni
Directory of British caving clubs 1984   5.176 uni
Directory of Canadian human services 1982-83   7.12 can
Directory of Canadian universities 1984-85   7.13 can
Directory of Canadian university resources for international development   7.14 can
Directory of cancer research information resources   1.316 usa
Directory of central Atlantic states - manufacturers   7.164 usa
Directory of centres for outdoor studies in England and Wales   5.25 uni
Directory of chemical engineering consultants   7.99 usa
Directory of chemical producers   7.106 usa
Directory of chemistry co-op   7.100 usa
Directory of child guidance and school psychological services   5.366 uni
Directory of Chinese foreign trade 1985-86   3.76 uni
Directory of Colorado manufacturers 1985   7.165 usa
Directory of computer software   7.15 usa
Directory of computer training   5.264 uni
Directory of computerized data files 1985   7.16 usa
Directory of consultants in biotechnology, first edition   7.236 usa
Directory of consultants in computer systems, third edition   7.122 usa
Directory of consultants in electronics, first edition   7.123 usa
Directory of consultants in energy technologies, first edition   7.137 usa
Directory of consultants in environmental science, first edition   7.78 usa
Directory of consultants in lasers and physics, first edition   7.263 usa
Directory of consultants in plastics and chemicals, first edition   7.101 usa
Directory of consultants in robotics and mechanics, first edition   7.150 usa
Directory of consulting practices in chemistry and related subjects   5.159 uni
Directory of courses of further and higher education in maintained colleges in the region   5.26 uni
Directory of CSIRO research programs 1985   4.8 aus
Directory of current Hong Kong research on Asian topics   3.7 hng

207

DIRECTORY TITLES INDEX

Directory of current UK r&d relevant to underwater and offshore instrumentation and measurement, 1976   5.207 uni
Directory of current United Kingdom research and development relating to offshore structures and pipelines   5.223 uni
Directory of current United Kingdom research and development relevant to underwater inspection and repair   5.224 uni
Directory of databases available in the country (preliminary version)   6.2 chi
Directory of development research and training institutes in Africa   2.7 fra
Directory of development research and training institutes in Latin America   6.6 fra
Directory of directories 1985, third edition   1.12 usa
Directory of directors   5.27 uni
Directory of East Flemish industry   5.242 bel
Directory of electric light and power companies   7.166 usa
Directory of electrical utilities, South America   6.24 uru
Directory of electronics, instruments and computers   5.195 uni
Directory of energy information centres in the world   1.204 fra
Directory of engineering societies and related organizations, eleventh edition, 1984   1.13 usa
Directory of environmental journals and media contacts   5.28 uni
Directory of environmental organizations in India   3.8 ind
Directory of establishments of private and official education: secondary and higher education   6.1 col
Directory of European agricultural organizations   5.125 uni
Directory of European associations, part 1: national industrial, trade and professional, third edition   5.29 uni
Directory of European associations, part 2: national learned, scientific and technical societies, third edition   5.30 uni
Directory of European Community trade and professional associations   5.265 bel
Directory of federal energy data sources: computer products and recurring publications   7.138 usa
Directory of federal technology resources, 1984   7.17 usa
Directory of federally supported information analysis centers 1979, fourth edition   7.18 usa
Directory of federally supported research in universities, 1983-84   7.19 can
Directory of fertilizer facilities: Africa   2.20 aut
Directory of first degree and diploma of higher education courses 1984-85   5.31 uni
Directory of Florida industries   7.167 usa
Directory of food and nutrition information services and resources   7.79 usa
Directory of forest tree nurseries in the United States   7.80 usa
Directory of further education   5.32 uni
Directory of gas utility companies   7.168 usa
Directory of geophysical research   1.164 net
Directory of geoscience departments - United States and Canada   7.110 usa
Directory of geoscience departments in universities in developing countries, third edition   1.165 tha
Directory of German chemists   5.141 gfr
Directory of graduate programs in chemical engineering in Canadian universities   7.102 can
Directory of graduate research   7.20 usa

Directory of health services 1985   5.367 uni
Directory of higher education in Colombia   6.3 col
Directory of Hong Kong industries 1985   3.77 hng
Directory of hydraulic research institutes and laboratories   1.224 net
Directory of industrial and technological research institutes in Africa   2.24 aut
Directory of industrial and technological research institutes: industrial conversion of biomass   1.206 aut
Directory of industrial information services and systems in developing countries   1.257 aut
Directory of industrial laboratories in Israel   3.78 isr
Directory of industrial research and development facilities in Canada 1985   7.21 can
Directory of industries   7.169 usa
Directory of information and documentation services in Uruguay, second edition   6.4 uru
Directory of information resources in the United States: geosciences and oceanography   7.111 usa
Directory of institutions and individuals active in environmentally-sound and appropriate technologies   5.33 uni
Directory of institutions and individuals active in the field of research in information science, librarianship and archival records management   1.14 net
Directory of institutions of higher education in Asia and the Pacific engaged in distance education   1.15 tha
Directory of inter-Arab organizations   3.79 uni
Directory of international and national medical related societies   1.317 uni
Directory of international and regional organizations conducting standards-related activities.   7.22 usa
Directory of international broadcasting, fifth edition   1.190 uni
Directory of international non-governmental organizations in consultative status with UNIDO   1.16 aut
Directory of international science organizations   3.9 jap
Directory of Japanese firms and representatives in Hawaii   7.170 usa
Directory of Kansas manufacturers and producers   7.171 usa
Directory of lectures in natural history and environmental issues   5.34 uni
Directory of libraries and special collections on Latin America and the West Indies   6.7 uni
Directory of libraries in Ireland   5.35 ire
Directory of libraries in Singapore   3.10 sin
Directory of major medical libraries worldwide   1.318 usa
Directory of manufacturers of vacuum plant, components and associated equipment in the UK, 1982   5.266 uni
Directory of manufacturers, State of Hawaii   7.172 usa
Directory of manufacturing establishments   2.25 sie
Directory of marine and freshwater scientists in Canada   7.81 can
Directory of marine science and fisheries related degree and diploma courses available at colleges and universities in the United Kingdom   5.126 uni
Directory of marine technology research   5.177 uni
Directory of Maryland manufacturers 1985-86   7.173 usa
Directory of medical libraries in New York state   7.237 usa
Directory of medical, public health and ayurvedic institutions in Himachal Pradesh   3.92 ind
Directory of medical schools worldwide, third edition 1983-84   1.319 usa
Directory of medical specialists   7.238 usa

DIRECTORY TITLES INDEX

Directory of metallurgical consultants and translators 1.131 usa
Directory of mines and quarries 5.178 uni
Directory of municipal wastewater treatment plants 5.160 uni
Directory of museums, second edition 1.17 uni
Directory of names and addresses and research specialisations for current members of the Botanical Society 7.239 usa
Directory of national systems of technicians education in south and central Asia 3.11 fra
Directory of natural history and related societies in Great Britain and Ireland 5.36 uni
Directory of natural science centers 7.82 usa
Directory of Nebraska manufacturers 1984-85 7.174 usa
Directory of neotropical protected areas 1.166 swi
Directory of New England manufacturers 7.175 usa
Directory of New Mexico manufacturing and mining 7.176 usa
Directory of non-governmental agricultural organization set up at European Community level 5.137 gfr
Directory of non-governmental organizations in environment 3.12 ind
Directory of North American geoscientists engaged in mathematics, statistics and computer applications 7.112 usa
Directory of North Carolina manufacturing firms 7.177 usa
Directory of North Wurttemberg, index of medical practitioners, hospitals 5.354 gfr
Directory of official architecture and planning 1985-86 5.225 uni
Directory of officials and organizations in China 1968-83 3.13 uni
Directory of on-going research in cancer epidemiology 1.320 uni
Directory of online databases 1.18 usa
Directory of operational computer applications in United Kingdom libraries and information units 5.37 uni
Directory of palaeontologists of the world, fourth edition 1.167 usa
Directory of personnel responsible for radiological health programs: 1985 7.240 usa
Directory of physics and astronomy staff members 7.264 usa
Directory of postgraduate and post-experience courses 1984-85 5.38 uni
Directory of private hospitals and health services 1985 5.370 uni
Directory of psychologists registered in the province of Ontario 7.241 can
Directory of rare book and special collections in the United Kingdom and the Republic of Ireland 5.39 uni
Directory of research and development activities in the United Kingdom in land survey and related fields 5.179 uni
Directory of research grants 1985 7.23 usa
Directory of research in biotechnology 5.371 uni
Directory of research institutes and industrial laboratories in Israel 3.14 isr
Directory of research institutes in Israel 3.15 isr
Directory of research lasers and expertise in universities, polytechnics and SRC establishments in the UK 5.196 uni

Directory of schools of medicine and nursing: British qualifications and training in medicine, dentistry, nursing and related professions 5.372 uni
Directory of sci-tech r&d institutions in ROC 3.18 tai
Directory of science resources for Maryland 1980-81 7.24 usa
Directory of scientific and technical societies in South Africa 2.8 saf
Directory of scientific and technological capabilities in Canadian industry (1977) 7.178 can
Directory of scientific directories 1.19 uni
Directory of scientific libraries in Thailand, third edition 3.16 tha
Directory of scientific research institutes in the People's Republic of China 3.17 usa
Directory of scientific resources in Georgia 1983-84 7.25 usa
Directory of sea-level research 1.168 swe
Directory of shipowners, shipbuilders and marine engineers 1.225 uni
Directory of social science research and training units - Asia 3.30 fra
Directory of southern African libraries, 1983 2.9 saf
Directory of Spanish industry, export and import 5.328 spa
Directory of special and research libraries in India 3.19 ind
Directory of special libraries and information centres 7.26 usa
directory of special libraries and information centres in New Zealand, DISLIC: 4.9 nze
Directory of special libraries and information sources in Indonesia 1981 3.20 ino
Directory of special libraries in Israel, fifth edition 3.21 isr
Directory of special libraries in Pittsburgh and vicinity 7.27 usa
Directory of special libraries in the German Federal Republic and West Berlin 5.112 gfr
Directory of special libraries in the Toronto area, tenth edition 7.28 can
Directory of specialist workers on phytophthora palmivora with special reference to cacao 1.92 uni
Directory of technical and further education 1986 5.40 uni
Directory of Texas manufacturers, 1985, 35th edition 7.179 usa
Directory of the canning, freezing, preserving industries 7.180 usa
Directory of the Commission of the European communities 5.41 lux
Directory of the cultural organization of the Republic of China 3.22 tai
Directory of the Dutch chemical and allied industry and trade 5.154 net
Directory of the forest products industry 7.83 usa
Directory of the German book trade 5.1 gfr
Directory of the German chemical industry 1985-87, Federal Republic of Germany and West Berlin 5.166 gfr
Directory of the learned societies in Japan 3.23 jap
Directory of the libraries of the University of Buenos Aires 6.14 arg
Directory of top computer executives 7.181 usa
Directory of training institutions and resources in Asia and the Pacific 1.20 pak
Directory of training research and information - producing centres in Iran 3.24 ira

DIRECTORY TITLES INDEX

Directory of transportation libraries in the United States and Canada   7.151 usa
Directory of UK renewable energy suppliers and services   5.208 uni
Directory of UK research in climatology   5.180 uni
Directory of United Nations databases and information systems   1.258 swi
Directory of United Nations databases and information systems and services   1.21 swi
Directory of university-industry liaison services   5.42 uni
Directory of US and Canadian marketing surveys and services 1985   7.182 usa
Directory of Utah manufacturers   7.183 usa
Directory of Western palearctic wetlands   1.169 swi
Directory of word processing systems   1.193 uni
Directory of world seismography stations   1.170 usa
Direktori perpustakaan khusus dan sumber informasi di Indonesia 1981   3.20 ino
DISLIC: directory of special libraries and information centres in New Zealand   4.9 nze
DOC Italia: annuario degli enti di studio, ricerca, cultura, e informazione   5.43 ita
DOC Italy: yearbook of research, cultural and information organizations   5.43 ita
Doctors   5.385 fin
Doescher's directory of brachiopodologists   1.171 usa
drugs and manufacturers, 45th edition, Italian directory of   5.384 ita
DSIR, Guide to   4.10 nze
Dutch companies with their UK agents, representatives   5.267 uni

Earth and astronomical sciences research centres: a world directory of organizations and programmes   1.172 uni
earth, oceanic, atmospheric, and space sciences, CNRS yearbook of   5.175 fra
earth-science agencies, Worldwide directory of national   1.183 usa
earth sciences, second edition, Information sources in the   1.173 uni
East-West business directory   1.259 uni
Eastern Europe: the state foreign trade organizations, the chambers of commerce and the state organizations in the field of transport, banking, tourism, trade fairs etc., Trade monopolies in   5.344 net
economic agencies: a world directory, State   1.62 uni
Economic and Social Research Council, Research supported by the   5.94 uni
Education and information science libraries   7.29 usa
education, Current bibliographic sources in   1.9 swi
education, Directory of further   5.32 uni
Education year book 1985   5.44 uni
education 1985: research, policy and practice, World yearbook of   1.85 uni
education 1986, Directory of technical and further   5.40 uni
educational and instructional technology 1984-85, International yearbook of   1.48 uni
Educational media yearbook, tenth edition 1984   1.22 usa
Egyptian directory of scientific centres and organizations: a directory of organizations in science, technology, agriculture, medicine and social sciences, second edition *   2.10 egy
Einkaufs der deutschen industrie   5.268 gfr
electric light and power companies, Directory of   7.166 usa

electrical and electronics directory, Asian and Pacific   1.185 sin
Electrical and electronics trades directory 1985   5.197 uni
electrical utilities, South America, Directory of   6.24 uru
Electricity supply handbook 1985   5.209 uni
electronics and electrical directory, Irish   5.200 ire
electronics and electrical products, China's   3.63 hng
electronics directory, Australian   4.26 aus
electronics, first edition, Directory of consultants in   7.123 usa
electronics industry, Information sources on the   1.195 aut
electronics, instruments and computers, Directory of   5.195 uni
electronics, Who's who in   7.132 usa
Encyclopedia of associations: regional, state, and local organizations - northeastern US   7.30 usa
Encyclopedia of associations: volume 1, national organizations of the US   7.31 usa
Encyclopedia of associations: volume 4, international organizations   1.23 usa
Encyclopedia of governmental advisory organizations   7.32 usa
Encyclopedia of information systems and services: volume 1 - international volume   1.24 usa
Encyclopedia of information systems and services, volume 2 - United States   7.33 usa
Encyclopedia of medical organizations and agencies   7.242 usa
Energy: a guide to organizations and information resources in the United States   7.139 usa
Energy: a register of research development and demonstration in the United Kingdom: part one - energy conservation; part three - renewable energy   5.210 uni
energy and nuclear sciences, International who's who in   1.353 uni
Energy balances of developing countries, 1971-82   1.207 usa
energy data and calculations, Handbook of   5.214 uni
energy data sources: computer products and recurring publications, Directory of federal   7.138 usa
energy directory: a guide to organizations and information resources, California   7.136 usa
energy directory: a guide to organizations and research activities in non-atomic energy, World   1.219 uni
Energy directory of qualified energy consultants, third edition, Institute of   5.215 uni
energy information centres in the world, Directory of   1.204 fra
Energy information directory   7.140 usa
energy information resources and research centres, International directory of new and renewable   1.212 usa
energy, Information sources on non-conventional sources of   1.211 aut
energy information - volume 1, World directory of   5.219 uni
Energy policies and programmes of IEA countries 1983 review   1.208 usa
energy professionals, AEE directory of   7.134 usa
Energy research programs directory   7.141 uni
energy suppliers and services, Directory of UK renewable   5.208 uni
energy supply companies of Western Europe, Major   5.217 uni
energy technologies, first edition, Directory of consultants in   7.137 usa
energy worldwide directory, Synthetic fuels and alternate   1.216 usa

DIRECTORY TITLES INDEX

energy yearbook, third edition, National 5.206 ita
ENEX directory 5.45 uni
engine digest, World 1.254 uni
Engineer buyers guide 5.269 uni
Engineering and industrial directory 1.226 uni
Engineering companies 1984-85 5.226 uni
engineering directory, Australian 4.28 aus
engineering directory 1985, Irish 5.232 ire
engineering education, 1985, Handbook of 3.68 ind
Engineering Guide, 20th edition 5.55 fra
engineering industry, Rylands directory of the 5.334 uni
engineering, International who's who in 1.239 uni
engineering products, CMERI directory of indigenous 3.73 ind
Engineering research centres: a world directory of organizations and programmes 1.227 uni
Engineering, science, and computer jobs 1986 7.34 usa
engineering, second edition, Information sources in 1.229 uni
engineering societies and related organizations, eleventh edition, 1984, Directory of 1.13 usa
engineering 1985, Who's who in 7.153 usa
engineers and members of the Association of Chartered Engineers in Iceland, Dictionary of Icelandic 5.238 ice
engineers, Consulting 5.221 uni
engineers, International directory of consulting 1.236 uni
engineers, International directory of consulting environmental and civil 1.237 usa
Engineers New Zealand: list of members, Association of Consulting 4.27 nze
engineers who's who and yearbook, Consulting 5.222 uni
ENREP directory 5.46 uni
ENREP directory of solid waste and chemical waste 5.161 uni
Entoma Europe 5.162 net
Entomology, Guide to New Zealand 4.31 nze
environment: organizations in Britain and Ireland 1984-85, Directory for the 5.23 uni
Environmental directory 7.35 usa
Environmental directory: national and regional organizations of interest to those concerned with amenity and environment 5.47 uni
environmental directory, sixth edition, Civic trust 5.15 uni
environmental directory, volumes 1 and 2, third edition, World 1.73 usa
Environmental education - list of institutions in the region of Asia and the Pacific 1.321 tha
Environmental education: Caribbean 6.8 ita
Environmental education: list of institutions in the region of Asia and the Pacific 3.25 tha
Environmental impact statement directory: the national network of EIS - related agencies and organizations 7.36 usa
environmental journals and media contacts, Directory of 5.28 uni
environmental organizations in India, Directory of 3.8 ind
environmental organizations, second edition, World directory of 1.69 usa
environmental pollution, Sources of information in 5.102 uni
environmental science, first edition, Directory of consultants in 7.78 usa
environmental sources, US directory of 7.65 usa
Environmental training programmes and policies in ASEAN: an overview 3.55 sin

environmentally-sound and appropriate technologies, Directory of institutions and individuals active in 5.33 uni
ESA membership directory 1984-85 7.243 usa
essential oils, Information sources on 1.97 aut
EURASIP directory: directory of European signal processing research institutions 5.198 net
Europa year book 1985: a world survey 1.25 uni
Europäische Föderation Korrosion: jahrsbericht 1978-84 5.163 gfr
European and North American scrap directory 1.260 uni
European associations, part 1: national industrial, trade and professional, third edition, Directory of 5.29 uni
European associations, part 2: national learned, scientific and technical societies, third edition, Directory of 5.30 uni
European biomass directory 5.211 uni
European coil coating directory 5.270 uni
European communities and other European organisations yearbook, sixth edition 1985 5.48 bel
European Community: the practical guide and directory for business, industry and trade, second edition, 1985 5.49 uni
European Community trade and professional associations, Directory of 5.265 bel
European companies: a guide to sources of information fourth edition 5.50 uni
European confectionery and bakery products: manufacturers of bakery products, cocoa, chocolate, and their products, confectionery titbits, seasoned crackers, etc. 5.271 net
European directories, second edition, Current 5.20 uni
European Federation of Corrosion: annual report 5.163 gfr
European glass directory and buyers' guide 5.272 uni
European offshore oil and gas directory 5.212 uni
European paint manufacturers: manufacturers of paints, varnishes, enamels, lacquers, printing inks, solvents and paint removers 5.164 net
European petroleum directory 1.209 usa
European plastics: manufacturers of raw material, semi-finished products and auxiliaries for the plastics industry 5.165 net
European research centres: a directory of organizations in science, technology, agriculture and medicine, sixth edition 5.51 uni
European sources of scientific and technical information, sixth edition 5.52 uni
Europe's 15 000 largest companies, tenth edition 1985 5.273 uni
EUSIDIC database guide 5.199 uni
Exploring science: a guide to contemporary science and technology museums 1.26 usa

Fab guide: a buyer's directory for welding fabrication engineers 5.274 uni
faculty directory, National 7.49 usa
Fairplay world shipping year book 1.261 uni
family planning addresses, World list of 1.349 uni
Far East and Australasia 1984-85 1.27 uni
farm surveys in the United Kingdom and Republic of Ireland 2, Register of 5.133 uni
fasteners, third edition, Handbook of industrial 5.284 uni
Federal biotechnology funding sources 7.37 usa
Federal register of pharmacies in Germany FR and West Berlin 5.358 gfr
Federal technology resources directory 7.38 usa

DIRECTORY TITLES INDEX

Ferro-alloy directory 1.262 uni
fertilizer facilities: Africa, Directory of 2.20 aut
fertilizer industry, Information sources on the 1.103 aut
fertilizer manufacturers, fifth edition, World directory of 1.305 uni
fertilizer plant equipment, World guide to 1.308 uni
fertilizer products, fifth edition, World directory of 1.160 uni
Field offices of the forest service 7.84 usa
Filters and filtration handbook, first edition 5.275 uni
finance and industry, Who's who in 7.223 usa
Financial aid for first degree study at Commonwealth universities 1984-86 1.28 uni
Financial Times industrial companies year book 1985 1.263 uni
Financial Times mining international year book 1986 1.264 uni
Financial Times oil and gas international year book 1985 1.265 uni
Financial Times who's who in world oil and gas 1982-83 1.266 uni
Fire directory 5.373 uni
fire protection and security, Handbook of industrial 5.379 uni
Firmenhandbuch chemische industrie 1985-87, Bundesrepublik Deutschland und Berlin (West) 5.166 gfr
First destinations of polytechnic students 5.53 uni
fish technology institutes, International directory of 1.110 ita
fisheries directory and reference book, American 7.71 usa
fishing industry directory 1985, Australian 4.16 aus
Flemish industry, Directory of East 5.242 bel
Florida industries, Directory of 7.167 usa
Florida medical directory 7.244 usa
flour milling and bakery products industries, Information sources on the 1.104 aut
food and nutrition information services and resources, Directory of 7.79 usa
food and nutrition, third edition, Sourcebook on 1.116 uni
food crisis: an international directory of organizations and information resources, World 1.119 usa
food directory, Made in Ireland 5.131 ire
food directory 1986-87, Middle East and Africa 1.114 leb
Food industry directory 1985-86 5.127 uni
food industry, Swiss beverage and 5.314 swi
Food manufacture: ingredient and machinery survey 5.276 uni
food trade marks and brand names, Binsted's directory of 5.250 uni
Food trade research register: volume 1 - food technology, tenth edition 5.277 uni
Food trade research register: volume 2 - food business research, tenth edition 5.278 uni
Food trades directory and food buyer's yearbook 1985-86 5.128 uni
Footwear, raw hides and skins and leather industry in OECD countries 1981-82 1.132 fra
Foreign medical school catalogue 1977 1.322 usa
Foreign scientific foundations : financing opportunities for the scientific community 6.9 chi
forest industries, Canadian 7.76 can
forest pathologists and entomologists, World directory of 1.118 aut
forest products industry, Directory of the 7.83 usa
forest service, Field offices of the 7.84 usa
forest tree nurseries in the United States, Directory of 7.80 usa
forestry and forest products libraries, International directory of 1.111 usa
forestry and wood science in Finland, Research in 5.134 fin
forestry handbook 1981, tenth edition, Commonwealth 1.91 uni
forestry schools, World list of 1.120 ita
Forestry vadecum for Latin America 6.23 chi
Forschungsstätten der landbauwissenschaften, ernährungswissenschaften, forstwissenschaften, holzwirtschaftswissenschaften, des naturschutzes, der landschaftspflege, der veterinärmedizin in der Bundesrepublik Deutschland 1984 5.129 gfr
Forthcoming international scientific and technical conferences 1.29 uni
foundation directory, International 1.38 uni
Foundry and metal industry of Switzerland 5.281 swi
Foundry directory and register of forges 5.279 uni
foundry industry, Information sources on the 1.274 aut
French Scientific Research, Yearbook of 5.2 fra
frontier science and technology, Who's who in 7.68 usa
Fundaciones científicas extranjeras: oportunidades de financiamiento para la comunidad científica 6.9 chi
furniture and joinery industry, Information sources on the 1.140 aut

gas companies, Brown's directory of North America and international 1.256 uni
Gas directory 1985 and who's who 5.280 uni
gas processing directory, 43rd edition, Worldwide refining and 1.222 usa
gas utility companies, Directory of 7.168 usa
Genetic engineering and biotechnology firms worldwide directory 1.323 uni
Genetic engineering; biotechnology sourcebook 7.245 uni
genetic services, International directory of 1.331 usa
geographic edition 1985, Chicago 7.160 usa
geography courses, Matter of degree: directory of 5.188 uni
geography of the United States, Schwendeman's directory of college 7.56 usa
Geological directory of the British Isles 5.181 uni
geological research in Texas: a directory of faculty and student research, Current 7.109 usa
geological sciences directory, United Kingdom research on the history of 5.190 uni
Geologists directory, third edition 1985 5.182 uni
Geologists yearbook 1977 5.183 uni
Geology in museums - a bibliography and index 5.184 uni
geophysical research, Directory of 1.164 net
Georgia manufacturing directory 7.184 usa
geoscience departments in universities in developing countries, third edition, Directory of 1.165 tha
geoscience departments - United States and Canada, Directory of 7.110 usa
geoscientists engaged in mathematics, statistics and computer applications, Directory of North American 7.112 usa
Geothermal world directory, twelfth edition 1.210 usa
German Federal Republic and West Berlin, Directory of special libraries in the 5.112 gfr
German industry, Buyers' guide to 5.268 gfr
German libraries, Special collections in 5.104 gfr
German scientific academies and societies handbook 5.59 gfr

DIRECTORY TITLES INDEX

German scientists directory 1983 5.69 gfr
German trade register of firms according to places 5.263 gfr
German trade register of firms classified according to trade 5.261 gfr
German trade register of firms from A-Z: industry, trade, commerce, and associations 5.262 gfr
Germany's industry and commerce, Kompass to West 5.297 gfr
Giesserei- und metallindustrie der Schweiz 5.281 swi
glass directory and buyers' guide, European 5.272 uni
glass industry, Information sources on the 1.141 aut
Good medicine guide 5.374 uni
government department and other libraries, 26th edition, Guide to 5.56 uni
Government programs and projects directory 7.39 usa
Government research directory 7.40 usa
Government research institutes in Japan 3.26 jap
governmental advisory organizations, Encyclopedia of 7.32 usa
Graduate programmes offered by Chilean universities 6.18 chi
Graduate programs in engineering and applied sciences 1986 7.41 usa
Graduate programs in physics, astronomy and related fields 1984-85 7.113 usa
Graduate programs in the biological, agricultural, and health sciences 1986 7.42 usa
Graduate programs in the physical sciences and mathematics 1986 7.43 usa
graduate research, Directory of 7.20 usa
Graduate studies 1985-86 5.54 uni
grain processing and storage, Information sources on 1.98 aut
Grants for study visits by university administrators and librarians 1985-87 1.30 uni
Green pages: directory of non-government environmental groups in Australia 4.20 aus
Guía de bibliotecas especializadas y centros de documentación de Chile 6.10 chi
Guía de escuelas y cursos de bibliotecología y documentación en America Latina 6.11 arg
Guía informativa: sociedades científicas de Chile 6.12 chi
Guía nacional de reuniones científicas 1982 6.13 chi
Guía preliminar de las bibliotecas de la universidad de Buenos Aires 6.14 arg
Guida di veterinaria e zootecnica 5.375 ita
Guida Monaci annuario sanitario 5.376 ita
Guide de l'Ingénierie 5.55 fra
Guide des services d'information en Tunisie, 1984-85 2.11 tun
Guide through museums and galleries of the Republic of Croatia 5.113 yug
Guide to agricultural information sources in Asia and Oceania 1.93 net
Guide to American directories, eleventh edition 7.44 usa
Guide to DSIR 4.10 nze
Guide to government department and other libraries, 26th edition 5.56 uni
Guide to graduate education in speech-language pathology and audiology, 1985 7.246 usa
Guide to graduate study in botany in the United States and Canada 7.247 usa
Guide to information services in marine technology 5.227 uni

Guide to Information Services in Tunisia, 1984-85 2.11 tun
Guide to library and documentation schools and courses in Latin America 6.11 arg
Guide to New Zealand Entomology 4.31 nze
Guide to parasite collections of the world 1.324 usa
Guide to postgraduate degrees, diplomas and courses in medicine 5.377 uni
Guide to professional services in speech pathology and audiology 7.248 usa
Guide to reference books for small and medium-sized libraries 1970-82 1.32 usa
Guide to reference books, ninth edition 1.31 uni
Guide to rubber and plastics test equipment 5.167 uni
Guide to scientific instruments 7.45 usa
Guide to sources of information in the construction industry, fourth edition 5.228 uni
Guide to special libraries and documentation centres in Chile 6.10 chi
Guide to specialist facilities and courses for handicapped people in post-school educational institutions in the region 5.57 uni
Guide to the coalfields 5.185 uni
Guide to the museums of Southern Africa 2.12 saf
Guide to the research and special libraries of Finland, sixth edition 5.107 fin
Guide to the Swiss museums 5.95 swi
Guide to universities 5.90 fra
Guide to US government directories volume 2, 1980-84 7.46 usa

Hammaslaakarit 5.378 fin
Handbok för nordisk träindustri 5.282 swe
Handbook of agricultural education 3.56 ind
Handbook of Austrian health organizations: address and reference book of public health departments, health organizations, and health establishments in Austria 5.381 aut
Handbook of energy data and calculations 5.214 uni
Handbook of engineering education, 1985 3.68 ind
Handbook of hose, pipes, couplings and fittings, first edition 5.283 uni
Handbook of indigenous manufacturers (engineering stores) 3.80 ind
Handbook of industrial fasteners, third edition 5.284 uni
Handbook of industrial fire protection and security 5.379 uni
Handbook of industrial materials, first edition 5.168 uni
Handbook of industrial safety and health 5.380 uni
Handbook of instruments and instrumentation, first edition 5.285 uni
Handbook of libraries: German Federal Republic, Austria, Switzerland 5.58 gfr
Handbook of mechanical power drives, third edition 5.286 uni
Handbook of medical education, 1985 3.93 ind
Handbook of noise and vibration control 5.287 uni
Handbook of Norwegian professional libraries 5.77 nor
Handbook of textile fibres, fifth edition 1.133 uni
Handbook of the northern wood industries, thirteenth edition 5.282 swe
Handbook of universities and faculties in the German Federal Republic, Austria, and Switzerland 5.60 gfr
Handbook: southeast Asian institutions of higher learning 3.27 tha

213

DIRECTORY TITLES INDEX

Handbuch der bibliotheken: Bundesrepublik Deutschland, Österreich, Schweiz   5.58 gfr
Handbuch der Deutschen wissenschaftlichen akademien und gesellschaften   5.59 gfr
Handbuch der Schweizerischen textil-, bekleidungs- und lederwirtschaft   5.288 swi
Handbuch der universitäten und fachhochschulen Bundesrepublik Deutschland, Österreich, Schweiz   5.60 gfr
Handbuch für die sanitätsberufe Österreichs. Adress- und nachschlagewerk über die sanitätsbehörden, sanitätsberufe und sanitäts-einrichtigung in Österreich   5.381 aut
handicapped people in post-school educational institutions in the region, Guide to specialist facilities and courses for   5.57 uni
Harris Illinois industrial directory   7.185 usa
Harris Indiana industrial directory   7.186 usa
Harris Michigan industrial directory   7.187 usa
Harris Pennsylvania industrial directory   7.188 usa
Hatrics directory of resources, eighth edition   5.61 uni
Hawaii agricultural export products directory   7.85 usa
Hawaii business directory   7.189 usa
Hawaii directory of manufacturers   7.190 usa
Health and welfare directory 13th edition   7.249 usa
Health care directory   5.382 uni
health care, Organization and institutions of the Finnish   5.400 fin
health directory, Monaco   5.376 ita
health information directory, third edition volume 1, Medical and   7.256 usa
health laboratory directory, Clinical and public   7.233 usa
health organizations: address and reference book of public health departments, health organizations, and health establishments in Austria, Handbook of Austrian   5.381 aut
Health organizations of the United States, Canada, and the world, fifth edition   7.250 usa
Health sciences information in Canada: associations   7.251 can
Health sciences information in Canada: libraries   7.252 can
Health sciences libraries: USA and Canada   7.253 usa
health services 1985, Directory of   5.367 uni
health systems agencies, state health planning and development agencies, statewide health coordinating councils, Directory: 7.235 usa
Heating, ventilating, refrigeration and air conditioning year book   5.229 uni
herbaria and working collections, List of California   7.88 usa
herbaria, Index of Argentinian   6.22 net
higher education courses 1984-85, Directory of first degree and diploma of   5.31 uni
higher education in Colombia, Directory of   6.3 col
Higher education in Malaysia - a bibliography   3.28 sin
Higher education in the United Kingdom, 1984-86   5.62 uni
Holland exports/commercial gardening and farming   5.289 net
Holland exports/consumer goods (food)   5.290 net
Holland exports/consumer goods (nonfood)   5.291 net
Holland exports/industrial products   5.292 net
Hong Kong directory of members 1985, Chinese manufacturers association of   3.72 hng
Hong Kong, fourth edition, 1978, Scientific directory of   3.45 hng
Hong Kong industries 1985, Directory of   3.77 hng

Horticultural research international: directory of horticultural research institutes and their activities in 61 countries   1.94 net
horticulture, Courses and training for careers in   5.123 uni
hose, pipes, couplings and fittings, first edition, Handbook of   5.283 uni
hospital directory, Canadian   7.231 can
Hospitals and health services year book 1985   5.383 uni
Hospitals and health services yearbook   4.33 aus
hospitals and health services 1985, Directory of private   5.370 uni
hospitals and house officer posts in the United Kingdom which are approved or recognised for pre-registration service, List of   5.386 uni
human services 1982-83, Directory of Canadian   7.12 can
HVAC redbook   5.230 uni
Hydraulic Equipment Manufacturers: directory of members, Association of   5.245 uni
Hydraulic handbook, eighth edition   5.293 uni
hydraulic research institutes and laboratories, Directory of   1.224 net
hydrographic organization yearbook, International   1.177 mon
Hydrological research in the United Kingdom 1965-70; 1970-75; 1975-80   5.186 uni

IBM (R) PC compatible computer directory: hardware, software and peripherals   1.194 uni
ICAR Inventory of Canadian agricultural research   7.86 can
Iceland - yearbook of trade and industry   5.294 ice
Idaho manufacturers directory   7.191 usa
IFLA directory 1984-85   1.33 net
Illinois industrial directory, Harris   7.185 usa
Illinois manufacturers directory   7.192 usa
Index of agricultural and food research   5.130 uni
Index of anaesthesiologists in Germany, Austria, and Switzerland   5.401 gfr
Index of Argentinian herbaria   6.22 net
Index of manufacturers of artificial sports surfaces   5.295 uni
India who's who   3.29 ind
Indian engineering and industry, Who's who in   3.50 ind
Indian pharmaceutical guide   3.94 ind
Indian science, Who's who in   3.51 ind
Indiana industrial directory   7.193 usa
Indiana industrial directory, Harris   7.186 usa
Indiana manufacturers' directory   7.194 usa
Indonesia, Kompass register: 3.83 ino
Indonesia 1981, Directory of special libraries and information sources in   3.20 ino
Industrial Alabama   7.195 usa
industrial design, World directory of institutions offering courses in   1.306 aut
Industrial development guide   5.296 uni
Industrial directory of the commonwealth of Pennsylvania, 1984-85   7.196 usa
industrial information services and systems in developing countries, Directory of   1.257 aut
industrial laboratories in Israel, Directory of   3.78 isr
Industrial Laboratory Directory, Irish   5.298 ire
industrial maintenance and repair, Information sources on   1.269 aut
industrial materials, first edition, Handbook of   5.168 uni
Industrial minerals directory   1.267 uni

214

DIRECTORY TITLES INDEX

industrial quality control, Information sources on   1.270 aut
industrial research and development facilities in Canada 1985, Directory of   7.21 can
Industrial research in the United Kingdom: a guide to organizations and programmes, eleventh edition   5.63 uni
Industrial research laboratories of the United States, nineteenth edition   7.47 uni
Industrial technology: a guide to sources of information in the United Kingdom available to developing countries   5.64 uni
industrial training, Information sources on   1.271 aut
Industrie- Kompass Deutschland, thirteenth edition   5.297 gfr
industry, Buyers' guide to north east   5.253 uni
Informatics, Studies and Research in   5.204 gfr
information analysis centers 1979, fourth edition, Directory of federally supported   7.18 usa
Information business 1984: a guide to companies and individuals: the annual directory of the Information Industry Association   1.268 net
Information consultants, freelancers and brokers directory 1985   4.11 aus
Information professionals directory: who's who in librarianship and information science   5.65 uni
information science, librarianship and archival records management, Directory of institutions and individuals active in the field of research in   1.14 net
Information sources in agriculture and food science   1.95 uni
Information sources in architecture   1.228 uni
Information sources in biotechnology   1.325 uni
Information sources in engineering, second edition   1.229 uni
Information sources in physics, second edition   1.351 uni
Information sources in science and technology, second edition   1.34 uni
Information sources in the biological sciences, third edition   1.326 uni
Information sources in the earth sciences, second edition   1.173 uni
Information sources in the medical sciences, third edition   1.327 uni
Information sources of the vegetable oil processing industry   1.107 aut
Information sources on bioconservation of agricultural wastes   1.96 aut
Information sources on building boards from woods and other fibrous materials   1.135 aut
Information sources on essential oils   1.97 aut
Information sources on grain processing and storage   1.98 aut
Information sources on industrial maintenance and repair   1.269 aut
Information sources on industrial quality control   1.270 aut
Information sources on industrial training   1.271 aut
Information sources on leather and leather products   1.136 aut
Information sources on natural and synthetic rubber   1.137 aut
Information sources on non-conventional sources of energy   1.211 aut
Information sources on the animal feed industry   1.99 aut
Information sources on the beer and wine industry   1.100 aut
Information sources on the canning industry   1.272 aut
Information sources on the cement and concrete industry   1.230 aut
Information sources on the ceramics industry   1.139 aut
Information sources on the clothing industry   1.273 aut
Information sources on the coffee, cocoa, tea and spices industry   1.101 aut
Information sources on the dairy product manufacturing industry   1.102 aut
Information sources on the electronics industry   1.195 aut
Information sources on the fertilizer industry   1.103 aut
Information sources on the flour milling and bakery products industries   1.104 aut
Information sources on the foundry industry   1.274 aut
Information sources on the furniture and joinery industry   1.140 aut
Information sources on the glass industry   1.141 aut
Information sources on the iron and steel industry   1.275 aut
Information sources on the machine tool industry   1.231 aut
Information sources on the meat processing industry   1.105 aut
Information sources on the non-alcoholic beverage industry   1.106 aut
Information sources on the packaging industry   1.276 aut
Information sources on the paint and varnish industry   1.142 aut
Information sources on the pesticides industry   1.143 aut
Information sources on the petrochemical industry   1.144 aut
Information sources on the pharmaceutical industry   1.328 aut
Information sources on the printing and graphics industry   1.145 aut
Information sources on the pulp and paper industry   1.146 aut
Information sources on the soap and detergent industry   1.147 aut
Information sources on utilization of agricultural residues for the production of panels, pulp and paper   1.138 aut
Information sources on woodworking industry machinery   1.232 aut
information systems and services: volume 1 - international volume, Encyclopedia of   1.24 usa
information systems and services, volume 2 - United States, Encyclopedia of   7.33 usa
Information technology in the UK   5.66 uni
Information trade directory 1983   1.277 uni
Institute of Energy directory of qualified energy consultants, third edition   5.215 uni
Institute of Water Pollution Control yearbook   5.169 uni
Institutes, programmes, projects: developing research   6.15 arg
Institutos, programas, proyectos: temas de investigación en desarrollo   6.15 arg
instrumentation, volume 6, ISA directory of   7.124 usa
instruments and instrumentation, first edition, Handbook of   5.285 uni
instruments and meters, China's   3.64 hng
instruments, Guide to scientific   7.45 usa
Interavia aerospace directory   1.233 swi
Intergovernmental organization directory 1984-85   1.35 gfr
International biotechnology directory 1985   1.329 uni
International ceramic directory   1.148 uni
International congress calendar, 25th edition 1985   1.36 gfr
International Council of Scientific Unions, Yearbook of the   1.87 fra

DIRECTORY TITLES INDEX

International cytogenetics laboratory directory 1.330 usa
International defense directory 1.196 swi
International directory of acid deposition researchers, North American and European edition 1.149 usa
International directory of agricultural engineering institutions 1.108 ita
International directory of animal health and disease data banks 1.109 usa
International directory of building research, information and development organisations, fifth edition 1.234 uni
International directory of consultants and technical sources, third edition 1985 1.235 uni
International directory of consulting engineers 1.236 uni
International directory of consulting environmental and civil engineers 1.237 usa
International directory of current research in the history of cartography and in carto-bibliography 1.174 uni
International directory of fish technology institutes 1.110 ita
International directory of forestry and forest products libraries 1.111 usa
International directory of genetic services 1.331 usa
International directory of higher education research institutes 1982 1.37 swi
International directory of marine scientists, third edition 1.175 ita
International directory of market research organizations 1985 1.278 uni
International directory of mining 1.176 usa
International directory of new and renewable energy information resources and research centres 1.212 usa
International directory of organizations concerned with aging 1.332 aut
International directory of psychologists: exclusive of the USA, fourth edition 1.333 net
International directory of specialized cancer research and treatment establishments 1.334 swi
International directory of telecommunications: market trends, companies, statistics, products, and personnel 1.197 uni
International federation of cotton and textile industries directory, twelfth edition, 1985-86 1.279 swi
International foundation directory 1.38 uni
International guide to library archival and information science associations, second edition 1.39 uni
International handbook of universities and other institutions of higher education 1.40 uni
International hydrographic organization yearbook 1.177 mon
International index of laboratory animals 1.335 uni
International index on training on conservation of cultural property 1.41 ita
International ISBN publishers' directory including publishers' addresses from some countries outside the ISBN system 1.49 gfr
International medical who's who, second edition 1.337 uni
International mycological directory 1.336 uni
International nuclear energy guide, thirteenth edition 1985 1.352 fra
International organization abbreviations and addresses 1984-85 1.42 gfr
International organizations: a dictionary and directory 1.43 uni
International pulp and paper directory-IPPD 1.150 usa
International register of specialists and current research in plant systematics 1.338 usa

International research centers directory, second edition 1.44 usa
International robotics yearbook 1.238 uni
international science organizations, Directory of 3.9 jap
International union of independent laboratories register of members 1983 1.66 uni
International who's who in energy and nuclear sciences 1.353 uni
International who's who in engineering 1.239 uni
International who's who of the Arab world 1.46 uni
International who's who 1984-85 1.45 uni
International yearbook and statesmen's who's who 1.47 uni
International yearbook of educational and instructional technology 1984-85 1.48 uni
International zinc and galvanizing directory, third edition 1.280 uni
International zoo yearbook 1.339 uni
internationales bibliothekshandbuch, sixth edition 1.75 gfr
internationales handbuch der spezialbibliotheken 1.77 gfr
Internationales ISBN-verlagsverzeichnis einschliesslich verlagsadressen aus ländern ohne ISBN-system 1.49 gfr
internationales verzeichnis der wirtschaftsverbände 1.78 gfr
internationales verzeichnis wissenschaftlicher verbände und gesellschaften 1.76 gfr
Introductory guide to information sources in physics 5.404 uni
Inventaire - instituts de recherche et de formation en matière de développement - Afrique 2.13 fra
Inventaire des centres Belges de recherche disposant d'une bibliotheque ou d'un service de documentation 5.67 bel
Inventaire des instituts de recherche et de formation en matière de développement en Amérique Latine 6.16 fra
Inventaire descriptif des unités de recherche et de formation en sciences sociales - Asie 3.30 fra
Inventory of agricultural research 7.87 usa
Inventory of data sources in science and technology 1.50 fra
Inventory of power plants in the US 7.142 usa
Inventory of water resources research in Australia 4.23 aus
Inventory of world commercial air carriers, second edition 1.240 can
Iran, Directory of training research and information - producing centres in 3.24 ira
Irish chemicals directory 5.170 ire
Irish construction directory 5.231 ire
Irish electronics and electrical directory 5.200 ire
Irish engineering directory 1985 5.232 ire
Irish Industrial Laboratory Directory 5.298 ire
Irish plastics and rubber directory 5.171 ire
iron and steel industry, Information sources on the 1.275 aut
iron and steel industry, Japan's 3.82 jap
Iron and steel works of the world 1.281 uni
ISA directory of instrumentation, volume 6 7.124 usa
Israel, Directory of research institutes and industrial laboratories in 3.14 isr
Israel, Directory of research institutes in 3.15 isr
Italian directory of drugs and manufacturers, 45th edition 5.384 ita

Jahrbuch der Deutschen bibliotheken 5.68 gfr
Jahrbuch fur bergbau, energie, mineralöl und chemie 5.187 gfr
Japan aviation directory 3.65 jap
Japan business directory, Diamond's 3.74 jap

DIRECTORY TITLES INDEX

Japan chemical directory 3.61 jap
Japan company handbook 3.81 jap
Japan, Directory of the learned societies in 3.23 jap
Japan directory of professional associations 3.31 jap
Japan, Government research institutes in 3.26 jap
Japan, Science and technology in 3.42 uni
Japan, 1983-84, Pharmaceutical manufacturers of 3.91 jap
Japan, 29th edition, 1985-86, Standard trade index of 3.46 uni
Japanese firms and representatives in Hawaii, Directory of 7.170 usa
Japanese firms in the ASEAN countries, Directory: affiliates and offices of 3.75 uni
Japanese firms in the USA, Directory: affiliates and offices of 7.163 uni
Japanese research institutes funded by ministries other than education 3.32 usa
Japanese research institutes funded by the ministry of education 3.33 usa
Japan's iron and steel industry 3.82 jap

Kansas manufacturers and producers, Directory of 7.171 usa
Kelly's manufacturers and merchants directory 5.299 uni
Kelly's UK Exports 5.300 uni
Kentucky directory of manufacturers 7.199 usa
Keystone coal industry manual 1985 7.143 usa
Kingdom of Denmark trade directory 5.301 den
Klein, S., directory of computer graphics suppliers: hardware, software, systems and services 7.125 usa
Kokusai gakujutsu dantai soran 3.9 jap
Kompass: Belgium 5.302 bel
Kompass buku merah 3.86 may
Kompass: Denmark 5.303 den
Kompass Deutschland, thirteenth edition, Industrie- 5.297 gfr
Kompass: France 5.304 fra
Kompass: Holland 5.305 net
Kompass: Italy 5.306 ita
Kompass: Norway 5.307 nor
Kompass register: Australia 4.29 aus
Kompass register: Indonesia 3.83 ino
Kompass register: Malaysia 3.84 may
Kompass register: Morocco 2.26 mor
Kompass register: Singapore 3.85 sin
Kompass: Spain 5.308 spa
Kompass: Sweden 5.309 swe
Kompass: Switzerland 5.310 swi
Kompass to West Germany's industry and commerce 5.297 gfr
Kompass trade information book 3.86 may
Kompass: United Kingdom 5.311 uni
Korea directory 3.34 kor
Korea, 1982-83, Research institutes on social sciences and humanities in the Republic of 3.39 jap
Kunststoff-industrie der Schweiz 5.312 swi
Kurschners deutscher gelehrten-kalender 1983 5.69 gfr

Lääkärit 5.385 fin
laboratory animals, International index of 1.335 uni
Laboratory equipment directory 1984 5.313 uni
Land drilling and oil well servicing contractors directory, eleventh edition 1.213 usa
land survey and related fields, Directory of research and development activities in the United Kingdom in 5.179 uni
lasers and expertise in universities, polytechnics and SRC establishments in the UK, Directory of research 5.196 uni
lasers and physics, first edition, Directory of consultants in 7.263 usa
Latin America, Science and technology in 6.19 uni
Latin American petroleum directory 6.25 usa
Leading consultants in technology, second edition 7.48 usa
leather and leather products, Information sources on 1.136 aut
Leather guide, sixteenth edition 1986 1.151 uni
leather industry in OECD countries 1981-82, Footwear, raw hides and skins and 1.132 fra
Lebensmittel und- getranke-industrie der Schweiz 5.314 swi
lending and photography, Brief guide to centres of international 1.6 uni
librarianship and information science, Information professionals directory: who's who in 5.65 uni
libraries and information centers, Subject directory of special 7.62 usa
libraries and information centres, Directory of special 7.26 usa
libraries and special collections on Africa, in the United Kingdom and Western Europe, fourth edition, SCOLMA directory of 2.17 uni
libraries and special collections on Latin America and the West Indies, Directory of 6.7 uni
libraries, Australian 4.3 uni
libraries, Education and information science 7.29 usa
libraries: German Federal Republic, Austria, Switzerland, Handbook of 5.58 gfr
libraries in India, Directory of special and research 3.19 ind
libraries in Ireland, Directory of 5.35 ire
libraries in Israel, fifth edition, Directory of special 3.21 isr
libraries in Pittsburgh and vicinity, Directory of special 7.27 usa
libraries in Singapore, Directory of 3.10 sin
libraries in the Toronto area, tenth edition, Directory of special 7.28 can
Libraries in West Africa: a bibliography 2.14 uni
libraries of Finland, sixth edition, Guide to the research and special 5.107 fin
libraries of the University of Buenos Aires, Directory of the 6.14 arg
libraries of the world: a selective guide, Major 1.52 uni
libraries, Subject collections in European 5.106 uni
libraries: USA and Canada, Health sciences 7.253 usa
libraries: USA and Canada, Science and technology 7.57 usa
libraries, 1983, Directory of southern African 2.9 saf
library and documentation schools and courses in Latin America, Guide to 6.11 arg
Library and information networks in the United Kingdom 5.70 uni
Library and information networks in Western Europe 5.71 uni
library and information services, ALA yearbook of 7.2 uni
library archival and information science associations, second edition, International guide to 1.39 uni
Library directory 5.7 den

DIRECTORY TITLES INDEX

library directory: a classified list of libraries in the United States and Canada with personnel and statistical data, American 7.3 uni

Licht's, F.O., international sugar economic yearbook and directory 1.282 gfr

Licht's, F.O., internationales zuckerwirtschaftliches jahr- und adressbuch 1.282 gfr

L'informatore farmaceutico 5.384 ita

Liquid gas carrier register, 20th edition 1.241 uni

List of California herbaria and working collections 7.88 usa

List of company directories and summary of their contents, second edition 1.283 aut

List of company directories and summary of their contents, second edition 1.51 swi

List of hospitals and house officer posts in the United Kingdom which are approved or recognised for pre-registration service 5.386 uni

List of registered medical and surgical practitioners 3.95 hng

List of research workers in the agricultural sciences in the Commonwealth 1.113 uni

Loadstar bulk handling directory, 1985 1.242 uni

Local area networks: a European directory of suppliers and systems 5.216 uni

London directory of industry and commerce 1985 5.315 uni

machine tool industry, Information sources on the 1.231 aut

machine tools and equipment, British 5.252 uni

machine tools and related production engineering, Register of research on 5.235 uni

Machinery buyers' guide 5.316 uni

Made in Ireland food directory 5.131 ire

Maine marketing directory 7.200 usa

Major companies in the Netherlands 5.317 uni

Major companies of Argentina, Brazil, Mexico and Venezuela 1982 6.26

Major companies of Europe, fifth edition 1985 5.318 uni

Major companies of Europe 1979-85 5.319 uni

Major companies of the Far East 1985, second edition 3.90 uni

Major companies of Nigeria, fourth edition, 1983 2.27 uni

Major companies of Saudi Arabia 1985 3.87 uni

Major companies of Scandinavia 1985 5.320 uni

Major companies of the Arab world 1985, ninth edition 3.89 uni

Major companies of the Arab world 1975-85 3.88 uni

Major energy supply companies of Western Europe 5.217 uni

Major libraries of the world: a selective guide 1.52 uni

malacologists: a national register of living professional and amateur conchologists, American 7.226 usa

Malaysia - a bibliography, Higher education in 3.28 sin

Malaysia, Kompass register: 3.84 may

Malaysian universities, Science and technology education in 3.40 sin

manufacturers and merchants directory, Kelly's 5.299 uni

manufacturing establishments, Directory of 2.25 sie

Map collections in Australia: a directory 4.24 aus

marine and freshwater scientists in Canada, Directory of 7.81 can

Marine environmental centres: Caribbean 6.17 ita

Marine environmental centres: East Asian Seas 3.35 ita

Marine environmental centres: Indian Ocean and Antarctic 3.36 ita

Marine environmental centres: Mediterranean 1.53 ita

Marine environmental centres: South Pacific 4.12 ita

Marine research centres: Africa 2.15 ita

marine science and fisheries related degree and diploma courses available at colleges and universities in the United Kingdom, Directory of 5.126 uni

marine sciences and related activities, Annotated acronyms and abbreviations of 7.73 usa

marine scientists, third edition, International directory of 1.175 ita

marine scientists, US directory of 7.117 usa

marine technology, Guide to information services in 5.227 uni

marine technology research, Directory of 5.177 uni

market research organizations 1985, International directory of 1.278 uni

Marquis international who's who directory of optical science and engineering 1.198 usa

Marquis who's who directory of computer graphics 1.199 usa

Marquis who's who directory of online professionals 1.285 usa

Marquis who's who in cancer: professionals and facilities 7.254 usa

Marquis who's who in rehabilitation: professionals and facilities 7.255 usa

Maryland manufacturers 1985-86, Directory of 7.173 usa

Massachusetts directory of manufacturers 7.201 usa

Materials research centres: a world directory of organizations and programmes in materials science 1.152 uni

mathematics and physical sciences, CNRS yearbook of 5.403 fra

mathematics, World directory of 1.356 usa

Matter of degree: directory of geography courses 5.188 uni

McGraw-Hill modern scientists and engineers 1.54 uni

meat processing industry, Information sources on the 1.105 aut

Medical and health information directory, third edition volume 1 7.256 usa

Medical and healthcare books and serials in print 1985: an index to literature in the health sciences 1.340 uni

medical and surgical practitioners, List of registered 3.95 hng

medical appliances: sources of supply reference for practice and hospital, Seibt 5.398 gfr

medical directory, American 7.227 usa

medical directory, Canadian 7.232 can

medical directory, Florida 7.244 usa

Medical directory, North Baden, index of medical practitioners, hospitals 5.353 gfr

Medical directory of Australia 4.32 aus

Medical directory of South Wurttemberg: index of medical practitioners, hospitals 5.355 gfr

medical directory, sixth edition 1983-84, US 7.262 usa

Medical directory 1985, 141st edition 5.387 uni

medical education, 1985, Handbook of 3.93 ind

medical establishments, Yearbook of French hydro-mineral, climatic and balneological resorts and 5.352 fra

Medical libraries and information services in the German Democratic Republic 5.390 gdr

medical libraries in New York state, Directory of 7.237 usa

medical libraries worldwide, Directory of major 1.318 usa

medical organizations and agencies, Encyclopedia of 7.242 usa

DIRECTORY TITLES INDEX

medical practitioners, hospitals, Directory of North Wurttemberg, index of 5.354 gfr
medical, public health and ayurvedic institutions in Himachal Pradesh, Directory of 3.92 ind
Medical register 5.388 uni
medical related societies, Directory of international and national 1.317 uni
Medical research centres: a world directory of organizations and programmes, sixth edition 1.341 uni
Medical research directory 5.389 uni
Medical research institutes in the Soviet Union 5.391 gfr
medical schools, fifth edition, World directory of 1.346 swi
medical schools, New directory of 7.257 usa
medical schools worldwide, third edition 1983-84, Directory of 1.319 usa
medical sciences, Research programs in the 7.260 uni
medical sciences, third edition, Information sources in the 1.327 uni
medical specialists, Directory of 7.238 usa
medical who's who, second edition, International 1.337 uni
medical yearbook, Swiss 5.397 swi
medicine and nursing: British qualifications and training in medicine, dentistry, nursing and related professions, Directory of schools of 5.372 uni
medicine, fifth edition, Who's who in 5.402 gfr
medicine guide, Good 5.374 uni
medicine, Guide to postgraduate degrees, diplomas and courses in 5.377 uni
medicine, World meetings: 1.350 usa
Medizinische bibliotheken und informationsstellen in der Deutschen Demokratischen Republik 5.390 gdr
Medizinische forschungsinstitute in der Sowjetunion 5.391 gfr
metal industry of Switzerland, Foundry and 5.281 swi
Metal traders of the world 1.286 uni
metal works of the world, Non-ferrous 1.290 uni
metallurgical consultants and translators, Directory of 1.131 usa
Metallurgical plant makers of the world 1.287 uni
metallurgy, Source journals in 1.156 usa
metals survey and directory, World precious 1.309 uni
metals survey, Minor 1.289 uni
Metalworking directory - a Dun's industrial guide 7.103 uni
Meteorological services of the world 1.178 swi
Michigan industrial directory, Harris 7.187 usa
Michigan manufacturers directory 7.203 usa
Michigan manufacturers directory 7.202 usa
Microcomputer companies in the UK 5.201 uni
Microcomputer users year book 5.321 uni
microorganisms, World directory of collections of cultures of 1.345 aus
Microwaves and rf product data directory 1.200 usa
Middle East and Africa construction directory 1986-87 1.243 leb
Middle East and Africa food directory 1986-87 1.114 leb
Middle East and North Africa 1984-85 1.55 uni
Middle East and world water directory, second edition 1983-84 1.153 leb
Middle East, Science and technology in the 3.43 uni
mineralogists, World directory of 1.181 fra
minerals directory, Industrial 1.267 uni
Mines and mining equipment and service companies worldwide, third edition 1.288 uni
mines and quarries, Directory of 5.178 uni

mining and manufacturing 1985-86, Alabama directory of 7.154 usa
Mining, energy, mineral oil and chemistry yearbook 5.187 gfr
mining, International directory of 1.176 usa
mining international year book 1986, Financial Times 1.264 uni
Minnesota Manufacturers Register 1985 7.204 usa
Minor metals survey 1.289 uni
MIRDAB: microbiological resource databank catalogue 5.392 net
Mississippi manufacturers directory 7.205 usa
Missouri directory of manufacturing and mining industrial services 7.206 usa
Modern pharmaceuticals of Japan VII 1985 3.96 jap
Monaci health directory 5.376 ita
Montana manufacturers and products directory 7.207 usa
Morocco, Kompass register: 2.26 mor
Museji Jugoslavije 5.72 yug
museum gids, Nederlandse 5.75 net
museums and related institutions, Official directory of Canadian 7.51 can
museums, Directory of Asian 3.6 fra
museums guide, Netherlands 5.75 net
museums, Guide to the Swiss 5.95 swi
Museums in Great Britain with scientific and technological collections 5.73 uni
museums of India, Brief directory of 3.4 ind
museums of Southern Africa, Guide to the 2.12 saf
Museums of the world, third edition 1.56 gfr
museums, second edition, Directory of 1.17 uni
Museums yearbook 5.74 uni
mycological directory, International 1.336 uni

National energy yearbook, third edition 5.206 ita
National faculty directory 7.49 usa
National guide to scientific events 1982 6.13 chi
National register of research projects - part I: natural sciences: biological, medical and related sciences 2.31 saf
National register of research projects - part II: natural sciences: physical, engineering, and related sciences 2.16 saf
natural history and environmental issues, Directory of lectures in 5.34 uni
natural history and related societies in Great Britain and Ireland, Directory of 5.36 uni
natural science centers, Directory of 7.82 usa
Natural Science Society, Swiss Academy of Sciences: translations, administration volume, Swiss 5.96 swi
Naturalists' directory and almanac (international) 1.342 usa
Nature's directory of biologicals 1.343 uni
Nebraska manufacturers 1984-85, Directory of 7.174 usa
Nederlands ABC voor handel en industrie 5.322 net
Nederlandse museum gids 5.75 net
neotropical protected areas, Directory of 1.166 swi
Netherlands ABC for commerce and industry 5.322 net
Netherlands, Major companies in the 5.317 uni
Netherlands museums guide 5.75 net
networks: a European directory of suppliers and systems, Local area 5.216 uni
Nevada industrial directory 7.208 usa
New biotechnology marketplace: Japan 3.37 usa
New biotechnology marketplace: USA and Canada 7.50 usa
New directory of medical schools 7.257 usa

DIRECTORY TITLES INDEX

New England manufacturers, Directory of 7.175 usa
New Hampshire marketing directory 7.209 usa
New Jersey directory of manufacturers 7.210 usa
New Mexico manufacturing and mining, Directory of 7.176 usa
New trade names in the rubber and plastics industry 1.154 uni
New York manufacturers directory 7.211 usa
New Zealand, Art galleries and museums of 4.1 nze
New Zealand manufacturers directory 4.30 nze
Nigeria, fourth edition, 1983, Major companies of 2.27 uni
noise and vibration control, Handbook of 5.287 uni
Non-ferrous metal works of the world 1.290 uni
Norges Handels-Kalender 5.323 nor
Norges Teknisk-Naturskapelige Forskningsråd årsberetning 5.76 nor
Norske vitenskapelige og faglige biblioteker: en håndbok 5.77 nor
North American on-line directory 7.126 uni
North Carolina manufacturing firms, Directory of 7.177 usa
North Sea oil and gas directory, thirteenth edition 1985 5.218 uni
Norwegian directory of commerce 5.323 nor
Norwegian professional libraries, Handbook of 5.77 nor
NSCA reference book 5.78 uni
nuclear energy guide, thirteenth edition 1985, International 1.352 fra
Nuclear engineering international buyers guide 1.354 uni
Nuclear industry almanac: volume 1, Western Europe 5.405 uni
Nuclear industry almanac - volume 2 Asia 1.355 uni

Official directory of Canadian museums and related institutions 7.51 can
Offshore contractors and equipment directory, 17th edition 1.214 usa
Offshore drilling register 1983 1.244 uni
offshore industry, United Kingdom government departments and other agencies concerned with the 5.236 uni
offshore industry, Vendor profiles of suppliers to the 5.237 uni
Offshore service vessel register, eighth edition 1.245 uni
offshore structures and pipelines, Directory of current United Kingdom research and development relating to 5.223 uni
Ohio industrial directory 7.212 usa
Ohio research and development facilities directory 7.52 usa
oil and gas directory, European offshore 5.212 uni
oil and gas directory, tenth edition, Arab 1.203 uni
oil and gas directory, thirteenth edition 1985, North Sea 5.218 uni
oil and gas international year book 1985, Financial Times 1.265 uni
oil and gas 1982-83, Financial Times who's who in world 1.266 uni
Oil directories of companies outside the USA and Canada 1.215 usa
Oil directories: producing and drilling in the USA 7.144 usa
Oil directory of Alaska 7.145 usa
Oil directory of Canada 7.146 usa
Oil directory of Houston: Texas 7.147 usa
oil industry directory, USA 7.148 usa
oilfield service, supply and manufacturers directory 1985, third edition, USA 7.149 usa
Oklahoma directory of manufacturers and products 7.213 usa
Oncology directory 7.258 uni
Online bibliographic databases 1.57 uni
On-line database search services directory 7.127 usa
on-line directory, North American 7.126 uni
On-line information retrieval 1965-76 5.80 uni
online professionals, Marquis who's who directory of 1.285 usa
Optical industry and systems purchasing directory 7.128 usa
optical science and engineering, Marquis international who's who directory of 1.198 usa
Optical sensor component directory 5.202 uni
Opticians register, comprising the opticians act 1958, rules made by the council: register of, ophthalmic opticians and dispensing opticians, and lists of bodies corporate carrying on the business as ophthalmic or dispensing opticians 5.393 uni
organic organisations in the UK and other relevant bodies, WWOOF directory of 5.139 uni
Organization and functions of the Eastern Regional Research Center federal research 7.89 usa
Organization and institutions of the Finnish health care 5.400 fin
Organizations in the UK employing biologists 5.394 uni
outdoor studies in England and Wales, Directory of centres for 5.25 uni

Pacific research centres: a directory of organizations in science, technology, agriculture and medicine 4.13 uni
packaging industry, Information sources on the 1.276 aut
Paint and resin directory 5.172 uni
paint and varnish industry, Information sources on the 1.142 aut
paint manufacturers: manufacturers of paints, varnishes, enamels, lacquers, printing inks, solvents and paint removers, European 5.164 net
palaeontological collections, World 1.182 usa
palaeontologists of the world, fourth edition, Directory of 1.167 usa
Pan European associations 5.81 uni
paper trade directory 1985: mills of the world, Phillips 1.291 uni
parasite collections of the world, Guide to 1.324 usa
Patent information and documentation in Western Europe 5.82 gfr
peace research institutions, World directory of 1.71 fra
Pennsylvania industrial directory, Harris 7.188 usa
Pennsylvania, 1984-85, Industrial directory of the commonwealth of 7.196 usa
Personnel and training databook, second edition 5.324 uni
pesticides industry, Information sources on the 1.143 aut
petrochemical, and natural gas processing plants of the world, Refining, construction, 1.155 usa
petrochemical directory, Worldwide 1.221 usa
petrochemical industry, Information sources on the 1.144 aut
petroleum and industrial directory 1984, Alaska 7.155 usa
petroleum directory, European 1.209 usa
petroleum directory, Latin American 6.25 usa
petroleum yearbook, ANEP 85: European 5.205 gfr
pharmaceutical chemists 1985, Register of 5.396 uni
pharmaceutical engineering: buyer's guide for pharmaceutical engineering and for laboratory and pharmaceutical equipment, Seibt 5.399 gfr

DIRECTORY TITLES INDEX

pharmaceutical guide, Indian 3.94 ind
pharmaceutical industry, Information sources on the 1.328 aut
Pharmaceutical manufacturers of Japan, 1983-84 3.91 jap
pharmaceutical manufacturers, World directory of 1.347 uni
Pharmaceutical Society of Ireland, Calendar of the 5.359 ire
pharmacies in Germany FR and West Berlin, Federal register of 5.358 gfr
Pharmacology and pharmacologists: an international directory 1.344 uni
Phillips paper trade directory 1985: mills of the world 1.291 uni
physical and biological sciences, fifteenth edition, American men and women of science: 7.4 uni
physics and astronomy staff members, Directory of 7.264 usa
physics, astronomy and related fields 1984-85, Graduate programs in 7.113 usa
physics, at United Kingdom universities and polytechnics, seventh edition, Research fields in 5.406 uni
physics, Introductory guide to information sources in 5.404 uni
physics, second edition, Information sources in 1.351 uni
pipelines and international directory, World 1.220 uni
plant systematics, International register of specialists and current research in 1.338 usa
plant taxonomic research on Australian flora, 1980-81, Current 4.19 aus
plastics and chemicals, first edition, Directory of consultants in 7.101 usa
plastics and rubber directory, Irish 5.171 ire
Plastics directory and buyers' guide 1984 7.214 usa
Plastics in building: index of applications and suppliers 5.173 uni
plastics industry, Swiss 5.312 swi
plastics: manufacturers of raw material, semi-finished products and auxiliaries for the plastics industry, European 5.165 net
Pneumatic handbook, sixth edition 5.325 uni
Polish Academy of Sciences: directory 5.83 pol
Pollution research and the research councils 5.84 uni
Pollution research index: a guide to world research in environmental pollution, second edition 1.179 uni
polymer science and technology, sixth edition, Review of research activities in 5.174 uni
Polymers, paint and colour year book 1985 5.326 uni
Polytechnic courses handbook 1985-86 5.85 uni
Polytechnics directory 5.86 uni
Population and related organizations: international address list 1.58 usa
postgraduate and post-experience courses 1984-85, Directory of 5.38 uni
Post's pulp and paper directory 7.215 usa
potato research in the UK 1983, Survey of 5.136 uni
poultry industry, Canada's who's who of the 7.75 can
Poultry world disease directory 5.132 uni
Poultry world international directory 1985 1.115 uni
power drives, third edition, Handbook of mechanical 5.286 uni
power generation markets, second edition, Who's who in world 1.217 uni
power plants in the US, Inventory of 7.142 usa
printing and graphics industry, Information sources on the 1.145 aut

Process engineering directory, 1985 5.327 uni
PRODEI 5.328 spa
Product directory 1984-85 5.203 uni
Professional services to the construction industry 5.233 ire
Programas de post-grado ofrecidos por las universidades Chilenas 6.18 chi
psychologists: exclusive of the USA, fourth edition, International directory of 1.333 net
psychologists registered in the province of Ontario, Directory of 7.241 can
psychosexual problems, Directory of agencies offering therapy counselling and support for 5.363 uni
public health and postgraduate training in public health, World directory of schools of 1.348 swi
Public health laboratory service directory 5.395 uni
Public reference services in the UK: a directory of information and specialist staff, second edition 5.87 uni
Publishers directory 7.53 usa
pulp and paper directory-IPPD, International 1.150 usa
pulp and paper directory, Post's 7.215 usa
pulp and paper industry, Information sources on the 1.146 aut
Pump selection systems and applications, second edition 5.329 uni
Pump users' handbook, second edition 5.330 uni
Pumping manual, seventh edition 5.331 uni
pumps 1985, Buyers' guide to 5.254 uni
PVC, Chemfacts: 1.123 uni
Pyttersen's Nederlandse almanak: handbook van personen en instellingen in Nederland de Nederlandse Antillen en Suriname 5.88 net
Pyttersen's Netherlands almanac: handbook of persons and organizations in the Netherlands, the Netherlands Antilles and Surinam 5.88 net

Quality technology handbook 5.332 uni

radiological health programs: 1985, Directory of personnel responsible for 7.240 usa
Railway directory and year book 1.246 uni
Reference book for world traders 1.292 usa
Reference issue: American annals of the deaf 7.259 usa
Refining, construction, petrochemical, and natural gas processing plants of the world 1.155 usa
Refrigeration and air conditioning year book 5.234 uni
Regional directory of development information units 6.5 chi
Register of consulting scientists, contract research organizations, and other scientific and technical services 5.89 uni
Register of farm surveys in the United Kingdom and Republic of Ireland 2 5.133 uni
Register of pharmaceutical chemists 1985 5.396 uni
Register of research on machine tools and related production engineering 5.235 uni
rehabilitation: professionals and facilities, Marquis who's who in 7.255 usa
Remote sensing of earth resources: list of UK groups and individuals engaged in remote sensing with a brief account of their activities and facilities, fifth edition 5.189 uni
Repertoire des universités 5.90 fra
Research centers directory, ninth edition 7.54 usa
research, cultural and information organizations, DOC Italy: yearbook of 5.43 ita
Research establishments 5.91 uni

DIRECTORY TITLES INDEX

Research establishments in agriculture, nutrition, veterinary medicine, forestry and timber in the German Federal Republic, 1984   5.129 gfr
Research fields in physics, at United Kingdom universities and polytechnics, seventh edition   5.406 uni
research grants 1985, Directory of   7.23 usa
Research in British universities, polytechnics and colleges   5.92 uni
Research in chemistry at private undergraduate colleges   7.104 usa
Research in forestry and wood science in Finland   5.134 fin
Research institutes on Asian studies in Japan, 1981   3.38 jap
Research institutes on social sciences and humanities in the Republic of Korea, 1982-83   3.39 jap
Research libraries and collections in the UK: a selective inventory and guide   5.93 uni
Research opportunities in Commonwealth developing countries   1.59 uni
Research programs in the medical sciences   7.260 uni
Research services directory, second edition   7.55 usa
Research supported by the Economic and Social Research Council   5.94 uni
Research, training, test and production reactor directory   7.265 usa
Review of research activities in polymer science and technology, sixth edition   5.174 uni
Rhode Island directory of manufacturers   7.216 usa
robot association members' handbook 1984-85, British   5.192 uni
Robotics: a worldwide guide to information sources   1.247 uni
robotics and mechanics, first edition, Directory of consultants in   7.150 usa
Robotics, CAD   1.248 uni
Robotics industry directory   7.129 usa
robotics research and development, World yearbook of   1.255 uni
robotics yearbook, International   1.238 uni
Royal College of Veterinary Surgeons registers and directory   5.135 uni
Royal Norwegian Council for Scientific and Industrial Research, annual report 1984   5.76 nor
rubber and plastics industry, New trade names in the   1.154 uni
rubber and plastics test equipment, Guide to   5.167 uni
rubber industry, Rubbicana Europe: a directory of suppliers to the   5.333 uni
rubber, Information sources on natural and synthetic   1.137 aut
Rubbicana Europe: a directory of suppliers to the rubber industry   5.333 uni
Rural industry directory 1983   4.21 aus
rural research and development, Compendium of   4.18 aus
Rylands directory of the engineering industry   5.334 uni

Satellite directory (telecommunications)   1.202 usa
Saudi Arabia 1985, Major companies of   3.87 uni
Saudi Arabian agriculture guide, first edition, 1985   3.57 uni
Scandinavia's 5000 largest companies   5.335 uni
Scholarships guide for Commonwealth postgraduate students 1985-87   1.60 uni
Schweizer museumsfuhrer   5.95 swi
Schweizerische Naturforschende Gesellschaft: verhandlungen, administrativer teil   5.96 swi
Schweizerisches medizinisches jahrbuch   5.397 swi
Schwendeman's directory of college geography of the United States   7.56 usa
Science and technology education in Malaysian universities   3.40 sin
Science and technology in China   3.41 uni
Science and technology in Japan   3.42 uni
Science and technology in Latin America   6.19 uni
Science and technology in the Middle East   3.43 uni
Science and technology libraries: USA and Canada   7.57 usa
Science and technology report   5.97 uni
Science research institutes under the jurisdiction of the Ministry of Education, Science and Culture   3.44 jap
science resources for Maryland 1980-81, Directory of   7.24 usa
scientific and technical information in Ireland, Sources of   5.103 ire
Scientific and technical organizations and agencies directory   7.58 usa
Scientific and technical research centres in Australia   4.14 aus
Scientific and technical societies of Canada   7.59 can
Scientific directory and annual bibliography   7.261 usa
Scientific directory of Hong Kong, fourth edition, 1978   3.45 hng
Scientific institutions of Latin America   6.20 usa
scientific resources in Georgia 1983-84, Directory of   7.25 usa
Scientific societies of Chile: information guide   6.12 chi
SCOLMA directory of libraries and special collections on Africa, in the United Kingdom and Western Europe, fourth edition   2.17 uni
Scottish conservation directory   5.98 uni
scrap directory, European and North American   1.260 uni
sea-level research, Directory of   1.168 swe
Sea technology buyers guide   7.115 usa
Seals and sealing handbook, first edition   5.336 uni
secondary and higher education, Directory of establishments of private and official education: 6.1 col
Securitech, the annual international guide to security equipment and services   1.293 uni
security equipment and services, Securitech, the annual international guide to   1.293 uni
Seibt medical appliances: sources of supply reference for practice and hospital   5.398 gfr
Seibt medizinische technik: bezugsquellennachweis für artz- und krankenhaus und zahnärztliche und zahntechnische arbeitsmittel und werkstoffe   5.398 gfr
Seibt pharma-technik: bezugsquellennachweis für die pharmazeutische technik und für labor- und apothekenausstatung   5.399 gfr
Seibt pharmaceutical engineering: buyer's guide for pharmaceutical engineering and for laboratory and pharmaceutical equipment   5.399 gfr
seismography stations, Directory of world   1.170 usa
Selected federal computer-based information systems   7.60 usa
Sheet metal industries yearbook 1985   5.337 uni
Ship and boat international guide to the small ship and workboat market   1.250 uni
shipowners, shipbuilders and marine engineers, Directory of   1.225 uni
shipping year book, Fairplay world   1.261 uni
signal processing research institutions, EURASIP directory: directory of European   5.198 net

DIRECTORY TITLES INDEX

Singapore, Kompass register: 3.85 sin
Słownik Polskich towarzystw naukowych: volume I towarzystwa naukowe działajace obecnie w Polsce   5.99 pol
soap and detergent industry, Information sources on the   1.147 aut
social and behavioral sciences, human services and management, World meetings: 1.81 usa
Social services year book 1985-86   5.100 uni
Sociétés et fournisseurs d'Afrique noire 1984-85   2.28 fra
Software users year book   5.338 uni
Someone to talk to directory 1985: a directory of self-help and community support agencies in the United Kingdom and Republic of Ireland   5.101 uni
Source journals in metallurgy   1.156 usa
Sourcebook of global statistics   1.61 uni
Sourcebook on food and nutrition, third edition   1.116 uni
Sources of information in environmental pollution   5.102 uni
Sources of scientific and technical information in Ireland   5.103 ire
South America, Central America and the Caribbean, 1986   6.27 uni
South Pacific research register   4.15 fij
Space activities and resources: review of United Nations international and national programmes   1.180 usa
Special collections in German libraries   5.104 gfr
Speech and Hearing Association directory, American   7.230 usa
speech-language pathology and audiology, 1985, Guide to graduate education in   7.246 usa
speech pathology and audiology, Guide to professional services in   7.248 usa
sports surfaces, Index of manufacturers of artificial   5.295 uni
Stainless steel directory 1985-86   5.339 uni
Stainless steel international survey and directory   1.294 uni
Standard and Poor's register of corporations, directors and executives   1.295 bel
Standard trade index of Japan, 29th edition, 1985-86   3.46 uni
standards-related activities., Directory of international and regional organizations conducting   7.22 usa
State economic agencies: a world directory   1.62 uni
Statistics Africa: sources for social, economic and market research   2.18 uni
Statistics America: sources for social, economic and market research (North, Central and South America), second edition   1.63 uni
Statistics Asia and Australasia: sources for social, economic and market research   1.64 uni
Statistics Europe: sources for social, economic, and market research, fourth edition   5.105 uni
statistics, Sourcebook of global   1.61 uni
Steel traders of the world   1.296 uni
Studien- und Forschungsführer Informatik   5.204 gfr
Studies and Research in Informatics   5.204 gfr
Subject collections: a guide to special book collections and subject emphases as reported by university, college, public and special libraries and museums in the United States and Canada   7.61 uni
Subject collections in European libraries   5.106 uni
Subject directory of special libraries and information centers   7.62 usa

sugar economic yearbook and directory, Licht's, F.O., international   1.282 gfr
Sugar industry buyers' guide   1.297 uni
Sugar year book   1.298 uni
sugar year book, Australian   4.17 aus
Suomen tieteellisten kirjastojen opas vetenskapliga bibliotek i Finland   5.107 fin
Surgeons' register   5.360 gfr
Survey of potato research in the UK 1983   5.136 uni
Swiss archives, libraries, and documentation centres   5.3 swi
Swiss beverage and food industry   5.314 swi
Swiss chemical industry and related products   5.255 swi
Swiss export directory: export products and services of Switzerland   5.340 swi
Swiss medical yearbook   5.397 swi
Swiss Natural Science Society, Swiss Academy of Sciences: translations, administration volume   5.96 swi
Swiss plastics industry   5.312 swi
Swiss textile, clothing and leather directory   5.288 swi
Synthetic fuels and alternate energy worldwide directory   1.216 usa

Taiwan buyers' guide   3.47 tai
Tanker register 1985, 25th edition   1.251 uni
tea areas handbook, Assam directory and   3.53 ind
Technical and scientific writers' register   5.108 uni
technology resources directory, Federal   7.38 usa
technology resources, 1984, Directory of federal   7.17 usa
technology, second edition, Leading consultants in   7.48 usa
technology, second edition, Who's who in   5.117 gfr
technology today, fourth edition, Who's who in   7.69 usa
Technology 1984   7.90 usa
telecommunications: market trends, companies, statistics, products, and personnel, International directory of   1.197 uni
Telecommunications systems and services directory   7.130 usa
Terveydenhuollon organisaatio ja laitokset   5.400 fin
Texas manufacturers, 1985, 35th edition, Directory of   7.179 usa
Texas trade and professional associations and other selected organizations 1985   7.217 usa
textile, clothing and leather directory, Swiss   5.288 swi
textile fibres, fifth edition, Handbook of   1.133 uni
Thailand, third edition, Directory of scientific libraries in   3.16 tha
Timber trades directory, 25th edition   5.341 uni
TNO: a key to research facilities   5.109 net
TNO: research applied   5.110 net
Top 3000 directories and annuals 1985-86   1.65 uni
Trade associations and professional bodies of the United Kingdom, seventh edition   5.342 uni
Trade directories of the world, 31st edition   1.299 usa
Trade directory information in journals, fifth edition   5.343 uni
Trade monopolies in Eastern Europe: the state foreign trade organizations, the chambers of commerce and the state organizations in the field of transport, banking, tourism, trade fairs etc.   5.344 net
Trade names dictionary company index   1.301 usa
Trade shows and professional exhibits directory   1.302 usa
traders, Reference book for world   1.292 usa
Transport engineers handbook, second edition   1.252 uni
transport guide 1985, Arabian   3.67 uni

DIRECTORY TITLES INDEX

transportation libraries in the United States and Canada, Directory of 7.151 usa
Tunisia, 1984-85, Guide to Information Services in 2.11 tun
Tunnelling Directory 1985 1.253 uni
Twenty-first directory of chemical engineering research in Canadian universities 7.105 can

UK on-line search services 5.111 uni
UK's 7500 largest companies 1985-86 5.345 uni
underwater and offshore instrumentation and measurement, 1976, Directory of current UK r&d relevant to 5.207 uni
underwater inspection and repair, Directory of current United Kingdom research and development relevant to 5.224 uni
Union internationale des laboratoires independants: register of members 1.66 uni
United Kingdom government departments and other agencies concerned with the offshore industry 5.236 uni
United Kingdom research on the history of geological sciences directory 5.190 uni
United Nations databases and information systems, Directory of 1.258 swi
United Nations international and national programmes, Space activities and resources: review of 1.180 usa
United States and the global environment: a guide to American organizations concerned with the international environmental issues 7.63 usa
universities and colleges, twelfth edition, American 7.5 gfr
universities and faculties in the German Federal Republic, Austria, and Switzerland, Handbook of 5.60 gfr
universities and other institutions of higher education, International handbook of 1.40 uni
universities, Guide to 5.90 fra
Universities handbook 1983-84 3.48 ind
Universities of Pakistan year book 3.49 pak
universities, polytechnics and colleges, Research in British 5.92 uni
universities, World list of 1.79 uni
universities, 1983-84, Directory of federally supported research in 7.19 can
universities 1984-85, Directory of Canadian 7.13 can
university entrance requirements for first degree courses in the United Kingdom 1986-87, Compendium of 5.17 uni
university-industry liaison services, Directory of 5.42 uni
University postgraduate degrees by course and research in applied mathematics and pure mathematics in the United Kingdom 1985-86 5.407 uni
university resources for international development, Directory of Canadian 7.14 can
Urban mass transit: a guide to organizations and information sources 7.152 usa
Uruguay, second edition, Directory of information and documentation services in 6.4 uru
US code name directory 7.64 usa
US directory of environmental sources 7.65 usa
US directory of marine scientists 7.117 usa
US government directories volume 2, 1980-84, Guide to 7.46 usa
US industrial directory 7.218 usa
US medical directory, sixth edition 1983-84 7.262 usa
USA oil industry directory 7.148 usa
USA oilfield service, supply and manufacturers directory 1985, third edition 7.149 usa
Utah manufacturers, Directory of 7.183 usa

vacuum plant, components and associated equipment in the UK, 1982, Directory of manufacturers of 5.266 uni
Vadecum forestal para America Latina 6.23 chi
Valves, piping and pipelines, first edition 5.346 uni
vegetable oil processing industry, Information sources of the 1.107 aut
Vendor profiles of suppliers to the offshore industry 5.237 uni
Verkfraedingtal: aeviagrip islenzkra verkfraedinga og annarra felagsmanna verkfraedinafélags Islands 5.238 ice
Vermont business phone book 1984-85 7.219 usa
Verzeichnis der ärzte für anaesthesiologie in der Bundesrepublik Deutschland, Österreich und der Schweiz 5.401 gfr
Verzeichnis der land-und ernährungswissenschaflichen verbände zusammengeschlossen im Rahmen der EG 5.137 gfr
Verzeichnis der spezialbibliotheken in der Bundesrepublik Deutschland einschliesslich West-Berlin 5.112 gfr
Veterinary drugs: food additives and manufacturers guide, fourth edition 5.375 ita
Veterinary Medical Association directory, American 7.72 usa
Veterinary register of Ireland 5.138 ire
Veterinary surgeons directory of the Federal Republic of Germany 5.119 gfr
Veterinary Surgeons registers and directory, Royal College of 5.135 uni
Video register 1985-86, eighth edition 7.131 usa
Virginia industrial directory 1984-85 7.220 usa
Vocational schools in Bavaria 5.6 gfr
Vodič kroz muzeje, galerije i zbirke u SR Hrvatskoj 5.113 yug
Volcanological research in the United Kingdom 1971-75: a survey of research activities 5.191 uni
Walker's manuals of western US corporations 1985 7.221 uni
Washington information directory 1985-86 7.66 usa
Washington manufacturers register 7.222 usa
Waste management research 1978 5.114 uni
wastewater treatment plants, Directory of municipal 5.160 uni
water directory, second edition 1983-84, Middle East and world 1.153 leb
water industry, Who's who in the 5.350 uni
Water Pollution Control yearbook, Institute of 5.169 uni
Water research in Australia: current projects 1983 4.25 aus
water resources directory: a guide to organizations and information resources, California 7.108 usa
water resources research in Australia, Inventory of 4.23 aus
Water services yearbook 1985 5.115 uni
welding fabrication engineers, Fab guide: a buyer's directory for 5.274 uni
Welding 85 5.347 uni
Wer liefert was? 5.348 gfr
Who owns whom 1985, United Kingdom and Republic of Ireland 5.349 uni
Who supplies what? 5.348 gfr
Who's who in consulting: a reference guide to professional personnel engaged in consultation for business; industry, and government 7.67 usa
Who's who in electronics 7.132 usa
Who's who in engineering 1985 7.153 usa
Who's who in finance and industry 7.223 usa

DIRECTORY TITLES INDEX

Who's who in frontier science and technology   7.68 usa
Who's who in Indian engineering and industry   3.50 ind
Who's who in Indian science   3.51 ind
Who's who in medicine, fifth edition   5.402 gfr
Who's who in microcomputing   7.133 uni
Who's who in science in Europe: a biographical guide in science, technology, agriculture and medicine, fourth edition   5.116 uni
Who's who in technology, second edition   5.117 gfr
Who's who in technology today   1.67 usa
Who's who in technology today, fourth edition   7.69 usa
Who's who in the water industry   5.350 uni
Who's who in West European automotive components   5.239 uni
Who's who in the world   1.68 usa
Who's who in world agriculture, second edition   1.117 uni
Who's who in world power generation markets, second edition   1.217 uni
Who's who of British scientists 1980-81, third edition   5.118 uni
wines, alcohol and spirits of the Common Market, Yearbook of   5.140 bel
Wire industry yearbook 1985   1.303 uni
Wisconsin manufacturers register   7.224 usa
wood-based panel producers, World directory of   1.307 usa
wood industries, thirteenth edition, Handbook of the northern   5.282 swe
woodworking industry machinery, Information sources on   1.232 uni
word processing systems, Directory of   1.193 uni
World aluminium survey   1.304 uni
World calendar of forthcoming meetings, metallurgical and related fields   1.157 uni
World directory of chemical producers, 1985-86   1.158 usa
World directory of collections of cultures of microorganisms   1.345 aus
World directory of crystallographers, sixth edition   1.159 net
World directory of energy information - volume 1   5.219 uni
World directory of environmental organizations, second edition   1.69 usa
World directory of fertilizer manufacturers, fifth edition   1.305 uni
World directory of fertilizer products, fifth edition   1.160 uni
World directory of forest pathologists and entomologists   1.118 aut
World directory of institutions offering courses in industrial design   1.306 aut
World directory of mathematics   1.356 usa
World directory of medical schools, fifth edition   1.346 swi
World directory of mineralogists   1.181 fra
World directory of national science and technology policy making bodies   1.70 fra
World directory of peace research institutions   1.71 fra
World directory of pharmaceutical manufacturers   1.347 uni
World directory of research projects, studies and courses in science and technology policy   1.72 fra
World directory of schools of public health and postgraduate training in public health   1.348 swi
World directory of wood-based panel producers   1.307 usa
World energy directory: a guide to organizations and research activities in non-atomic energy   1.219 uni
World engine digest   1.254 uni
World environmental directory, volumes 1 and 2, third edition   1.73 usa

World food crisis: an international directory of organizations and information resources   1.119 usa
World guide to abbreviations of organizations, seventh edition   1.74 uni
World guide to fertilizer plant equipment   1.308 uni
World guide to libraries   1.75 gfr
World guide to scientific associations and learned societies   1.76 gfr
World guide to special libraries   1.77 gfr
World guide to trade associations   1.78 gfr
World list of family planning addresses   1.349 uni
World list of forestry schools   1.120 ita
World list of universities   1.79 uni
World meetings: medicine   1.350 usa
World meetings: outside United States and Canada   1.80 usa
World meetings: social and behavioral sciences, human services and management   1.81 usa
World meetings: United States and Canada   7.70 usa
World museums publications   1.82 uni
World nuclear directory: a guide to organizations and research activities in atomic energy, seventh edition   1.357 uni
World of learning 1984-85, 35th edition   1.83 uni
World palaeontological collections   1.182 usa
World pipelines and international directory   1.220 uni
World precious metals survey and directory   1.309 uni
World problems and human potential, second edition 1985-86   1.84 gfr
World yearbook of education 1985: research, policy and practice   1.85 uni
World yearbook of robotics research and development   1.255 uni
Worldwide chemical directory, fourth edition   1.161 uni
Worldwide directory of national earth-science agencies   1.183 usa
Worldwide petrochemical directory   1.221 usa
Worldwide refining and gas processing directory, 43rd edition   1.222 usa
WWOOF directory of organic organisations in the UK and other relevant bodies   5.139 uni
Wyoming directory of manufacturing and mining   7.225 usa

Yearbook of French hydro-mineral, climatic and balneological resorts and medical establishments   5.352 fra
Yearbook of French Scientific Research   5.2 fra
Yearbook of German Libraries, 51st edition   5.68 gfr
Yearbook of international organizations 1984-85, 21st edition   1.86 gfr
Yearbook of the International Council of Scientific Unions   1.87 fra
Yearbook of wines, alcohol and spirits of the Common Market   5.140 bel
Yugoslavian museums   5.72 yug

Zambia directory 1985   2.29 zam
Zambia industrial and commercial directory 1984-85   2.30 zam
Zenkoku gakkyokai soran   3.23 jap
Zimbabwe research index   2.19 zim
zinc and galvanizing directory, third edition, International   1.280 uni
zoo yearbook, International   1.339 uni

225

PUBLISHER INDEX

AB Svensk Trävarutidning   5.282 swe
ABC Belge pour le Commerce et l'Industrie BV   5.248 bel; 5.249 bel
ABC voor Handel en Industrie CV   5.291 net; 5.292 net; 5.322 net
Academy of Scientific Research and Technology   2.10 egy
Acarology Laboratory, Ohio State University   1.88 usa
Agricultural and Food Research Council   5.121 uni; 5.130 uni
Alabama Development Office   7.154 usa; 7.195 usa
Alliance for Engineering in Medicine and Biology   1.315 usa
American Association for the Advancement of Science   7.45 usa
American Association of Engineering Societies   1.13 usa; 7.153 usa
American Chemical Society   1.128 usa; 3.59 usa; 7.20 usa; 7.91 usa; 7.92 usa; 7.94 usa; 7.98 usa; 7.100 usa
American Geological Institute   7.110 usa
American Institute of Chemical Engineers   7.97 usa; 7.99 usa
American Institute of Physics   7.113 usa; 7.264 usa
American Iron and Steel Institute   7.198 usa
American Library Association   1.31 uni; 7.2 uni
American Malacologists   7.226 usa
American Mathematical Society   1.356 usa
American Nuclear Society   7.265 usa
American Psychological Association   7.229 usa
American Society of Agricultural Engineers   7.90 usa
American Society of Parasitologists   1.324 usa
American Speech-Language-Hearing Association   7.230 usa; 7.246 usa; 7.248 usa
American Veterinary Medical Association   7.72 usa
Amt für Wissenschaft und Forschung/Office de la Science et de la Recherche   5.3 swi
Applied Computer Research Incorporated   7.181 usa
Arab Construction World   1.243 leb
Arab Petroleum Research Centre   1.203 uni
Arab Water World   1.153 leb
Armstrong, Alan, and Associates Limited   1.65 uni
ARPANET Network Information Center   7.118 usa
A/S Forlaget Kompass   5.303 den
Asian and Pacific Skill Development Programme   1.20 pak
Aslib   1.29 uni; 1.57 uni; 5.4 uni; 5.37 uni; 5.70 uni; 5.71 uni; 5.79 uni; 5.80 uni; 5.111 uni; 5.356 uni
Assam Review Publishing Company   3.53 ind
Association for Child Psychology and Psychiatry   5.366 uni
Association for Population/Family Planning Libraries and Information Centers - International (APLIC)   1.58 usa
Association of African Universities   2.6 gha
Association of Bronze and Brass Founders   5.244 uni
Association of Commonwealth Universities   1.5 uni; 1.7 uni; 1.28 uni; 1.30 uni; 1.59 uni; 1.60 uni; 5.11 uni; 5.17 uni
Association of Consulting Scientists   5.5 uni
Association of Cytogenetic Technologists   1.330 usa
Association of Geoscientists for International Development   1.165 tha
Association of Hydraulic Equipment Manufacturers   5.245 uni
Association of Indian Universities   3.48 ind; 3.56 ind; 3.68 ind; 3.93 ind
Association of Science-Technology Centers   1.26 usa
Association of Southeast Asian Institutions of Higher Learning   3.27 tha
Association of Universities and Colleges of Canada   7.13 can; 7.14 can
Astronomy Society of the Pacific   7.107 usa
ATE Information Services   7.151 usa
Athlone Press   6.7 uni
Aurea Publications   7.257 usa
Australasian Medical Publishing Company   4.32 aus
Australian Academy of Science   4.2 aus
Australian Conservation Foundation   4.20 aus
Australian Government Publishing Service   4.21 aus; 4.23 aus
Australian Scientific Industries Association   4.14 aus

Badgemore Park Enterprises Limited   5.264 uni
Bayerisches Landesamt für Statistik und Datenverarbeitung, München   5.6 gfr
Beacon Publications   3.1 uni; 3.57 uni; 3.67 uni; 3.69 uni; 3.70 uni
Benn Business Information Services Limited   1.125 uni; 1.151 uni; 1.291 uni; 5.213 uni; 5.280 uni; 5.341 uni
Berita Kompass sdn Bhd   3.84 may
Bezugsquellennachweis für den Einkauf   5.348 gfr
Biblioteca de la Universidad de Buenos Aires   6.14 arg
Biblioteca Nacional   6.4 uru
Bio-Energy Council   7.135 usa
Blackie Publishing Group   1.74 uni
Blackwell Scientific Publications   5.66 uni
Blandford Press Limited   5.183 uni

PUBLISHER INDEX

Botanical Society of America   7.239 usa; 7.247 usa
Bowker Publishing Company   1.39 uni; 1.52 uni; 1.82 uni; 1.186 uni; 1.247 uni; 1.248 uni; 1.340 uni; 5.106 uni; 5.315 uni; 7.3 uni; 7.4 uni; 7.47 uni; 7.61 uni; 7.126 uni; 7.141 uni; 7.260 uni
Braynart Group Limited   4.30 nze
British Aerosol Manufacturers' Association Limited   5.251 uni
British Association for Counselling   5.363 uni
British Dental Journal   5.357 uni
British Geological Survey   5.178 uni
British Library   5.92 uni
British Library, Science Reference Library   5.56 uni
British Museum (Natural History)   5.36 uni
British Overseas Trade Board   1.278 uni
British Plastics Federation   5.173 uni
British Pump Manufacturers' Association   5.254 uni
British Robot Association   5.192 uni
British Sulphur Corporation   1.160 uni; 1.305 uni; 1.308 uni
Brunel University Industrial Services Bureau   5.42 uni
Bryce Francis Limited   4.27 nze
BSO Publications Limited   1.190 uni
Buchhändler-Vereinigung GmbH   1.49 gfr; 5.1 gfr
Bureau de Recherches Géologiques et Minières   1.181 fra
Bureau of Business Research, University of Texas at Austin   7.179 usa; 7.217 usa
Bureau of Economic Geology   7.109 usa
Business Press International Limited   1.225 uni; 1.246 uni
Business Publishers Incorporated   1.73 usa
Butterworth and Company Limited   1.34 uni; 1.95 uni; 1.173 uni; 1.228 uni; 1.229 uni; 1.326 uni; 1.327 uni; 1.351 uni; 5.157 uni
Butterworths Scientific Limited   5.214 uni; 5.240 uni; 5.332 uni

Cahners Publishing Company   7.218 usa
California Institute of International Studies   6.20 usa
California Institute of Public Affairs   1.119 usa; 7.63 usa; 7.108 usa; 7.136 usa; 7.139 usa; 7.152 usa;
Canada Institute for Scientific and Technical Information   7.19 can; 7.59 can; 7.251 can; 7.252 can
Canadian Agricultural Research Council   7.86 can
Canadian Hospital Association   7.231 can
Canadian Manufacturers' Association   7.159 can
Canadian Museums Association   7.51 can
Canadian Nature Federation   7.7 can
Canadian Society for Chemical Engineering   7.102 can; 7.105 can
Capel Editorial Distribuidora SA   5.328 spa
Cartermill Publishing Limited   5.365 uni
CBD Research Limited   1.8 uni; 1.63 uni; 1.64 uni; 2.18 uni; 2.22 uni; 5.18 uni; 5.19 uni; 5.20 uni; 5.24 uni; 5.29 uni; 5.30 uni; 5.50 uni; 5.81 uni; 5.105 uni;
Central Mechanical Engineering Research Institute   3.73 ind
Central Statistics Office   2.25 sie
Centre de Documentation Nationale   2.11 tun
Centre for East African Cultural Studies   3.38 jap
Centre for East Asian Cultural Studies   3.39 jap
Centre National de Documentation Scientifique et Technique   5.67 bel
Centre National de la Recherche Scientifique   5.2 fra; 5.142 fra; 5.175 fra; 5.403 fra
Chadwyck-Healey Limited   5.14 uni
Charles Scribner's Sons   1.10 usa
Chemical Daily Company Limited   3.61 jap

Chemical Data Services   1.123 uni; 1.124 uni; 1.127 uni; 1.161 uni; 3.58 uni; 5.143 uni; 5.144 uni; 5.145 uni; 5.146 uni; 5.147 uni; 5.148 uni; 5.149 uni; 5.150 uni; 5.151 uni; 5.152 uni; 7.96 uni
Chemical Industries Association   5.155 uni
Chemical Information Services Limited   1.158 usa
China Phone Book Company Limited   3.5 hng; 3.60 hng; 3.63 hng; 3.64 hng
China Productivity Centre   3.47 tai
Chinese Manufacturers Association of Hong Kong   3.72 hng
Civic Trust   5.15 uni; 5.47 uni
Clarkson, H., and Company Limited   1.223 uni; 1.241 uni; 1.244 uni; 1.245 uni; 1.251 uni
Clive Bingley Limited   4.3 uni; 5.93 uni
Cocoa, Chocolate and Confectionery Alliance   1.92 uni
Codata Bulletin, 24   5.193 uni
Collier Macmillan Limited   7.1 uni
Colliery Guardian   5.185 uni
Comisión de Integración Eléctrica Regional   6.24 uru
Comisión Nacional de Investigación Científica y Tecnológica, Centro de Información y Orientación de Estudios en el Extranjero   6.9 chi; 6.18 chi
Comisión Nacional de Investigación Científica y Tecnológica, Centro Nacional de Información y Documentación   6.10 chi
Comisión Nacional de Investigación Científica y Tecnológica, Departamento de Fomento   6.12 chi
Comisión Nacional de Investigación Científica y Tecnológica, Dirección de Información y Documentación   6.2 chi; 6.13 chi
Commission of the European Communities   5.21 lux
Commission on Economic Development   7.208 usa
Committee for Middle East Trade   3.79 uni
Committee for Scientific Co-ordination Hong Kong, Hong Kong Government Printer   3.45 hng
Committee of 100   7.169 usa
Committee of Directors of Polytechnics   5.53 uni
Committee on Space Research (COSPAR)   1.163 fra
Commonwealth Agricultural Bureaux   1.113 uni
Commonwealth Forestry Association   1.91 uni
Commonwealth Mycological Institute   1.336 uni
Commonwealth Scientific and Industrial Research Organization, Australia   4.6 aus; 4.8 aus; 4.14 aus
Communications Trends Incorporated   1.187 usa
Compass Publications Incorporated   1.249 usa; 7.115 usa
Computer Publications Limited   3.62 hng
Computing Services Association   5.257 uni; 5.258 uni; 5.259 uni
Conference of Professors of Applied Mathematics   5.407 uni
Conference of Radiation Control Program Directors Incorporated   7.240 usa
Congressional Information Service Incorporated   7.87 usa
Congressional Quarterly Incorporated   7.66 usa
Connecticut Labor Department, Office of Research and Information   7.161 usa
Consejo Nacional de Investigaciones Científicas y Técnicas (PRODAT)   6.15 arg
Construction Industry Research and Information Association   5.207 uni; 5.220 uni; 5.223 uni; 5.224 uni; 5.228 uni; 5.236 uni
Convention of American Instructors of the Deaf and Conference of Educational Administrators Serving the Deaf   7.259 usa
Council for Environmental Conservation   5.28 uni; 5.34 uni
Council for Environmental Education   5.25 uni

PUBLISHER INDEX

Council for National Academic Awards   5.31 uni; 5.38 uni
Council for Postgraduate Medical Education in England and Wales   5.377 uni
Council for Scientific and Industrial Research   2.4 saf; 2.8 saf
Council of International Relations   1.314 usa
Council of Polytechnic Librarians   5.86 uni
Council on Undergraduate Research   7.104 usa
Crain Communications Limited   5.333 uni
Cramer, J.   1.312 gfr
Croner Publications Incorporated   1.292 usa; 1.299 usa
Cuadra Associates Incorporated   1.18 usa

D. Reidel Publishing Company   5.198 net
Danish Library Association   5.7 den
Data Base Asia   4.7 hng
Data Processing Professionals Directory Incorporated   7.121 usa
Data Research Group   5.91 uni; 5.221 uni; 5.226 uni
DECHEMA   1.121 gfr; 5.163 gfr
Departamento Administrativo Nacional de Estadística - DANE   6.1 col
Department of Agriculture, State of Hawaii   7.85 usa
Department of Economic and Community Development   7.52 usa
Department of Food and Agriculture   7.88 usa
Department of Primary Industry   4.16 aus; 4.18 aus
Department of Resources and Energy   4.25 aus
Department of the Environment and Department of Transport Library Services   5.102 uni; 5.114 uni
Department of the Environment, Environmental Information System   3.12 ind
Department of Trade and Industry   5.189 uni
Deutscher Adressbuch-Verlag GmbH   5.262 gfr; 5.263 gfr; 5.268 gfr
Deutscher Apotheker Verlag   5.358 gfr
Diamond Lead Company   3.74 jap
Dick, J., Publishing (Research Publications)   1.67 usa; 7.48 usa; 7.69 usa; 7.78 usa; 7.101 usa; 7.122 usa; 7.123 usa; 7.137 usa; 7.150 usa; 7.236 usa; 7.263 usa
Directory of Health Services   3.92 ind
Directory Publishers of Zambia   2.29 zam
DIT   5.55 fra
DMS Incorporated   7.64 usa
Doescher, Rex A.   1.171 usa
Droit, Le   7.12 can
Dun and Bradstreet   4.4 uni; 5.349 uni; 7.103 uni
Duncan Publications   1.259 uni

East Anglian Regional Advisory Council for Further Education   5.26 uni
EC Publications Office   5.41 lux
Econ Verlag GmbH   5.166 gfr
Economic and Social Research Council   5.94 uni
Economic Development Department, State of Oklahoma   7.213 usa
EDIAFRIC-la documentation Africaine   2.28 fra
Editions de Chabassol   5.351 fra
Editions Delta   5.48 bel; 5.140 bel; 5.265 bel
Editoriale Italiana   5.43 ita
EIC Intelligence   3.37 usa; 7.50 usa
ELC International   3.71 uni; 5.273 uni; 5.335 uni; 5.345 uni
Electrical - Electronic Press   1.354 uni
Electronic Engineering Association   5.203 uni
Elm Publications   5.65 uni

Elsevier Science Publishers   1.333 net; 1.164 net
Elsevier Science Publishers, Biomedical Division   5.392 net
Enercom   1.352 fra
Energy Information Agency   7.140 usa; 7.142 usa
Energy Publications   5.215 uni
Energy Publications Incorporated   1.256 usa
Entomological Society of America   7.243 usa
Entomological Society of New Zealand   4.31 nze
Environmental Protection Agency   7.65 usa; 1.149 usa
ERA Technology Limited   5.202 uni
Etas Kompass   5.306 ita
Eurolec/David Raynor   5.194 uni; 5.201 uni
Europa Publications   1.25 uni; 1.27 uni; 1.38 uni; 1.45 uni; 1.55 uni; 1.83 uni; 2.3 uni; 6.27 uni
Expansion Scientifique Française   5.352 fra

Fairmont Press Incorporated   7.134 usa
Fairplay Publications Limited   1.261 uni
Farm Papers Limited   7.75 can
Farmers Publishing Group   1.115 uni; 5.132 uni
Federal Energy Administration   7.138 usa
Federation for Community Planning   7.249 usa
Fédération Internationale de Documentation (FID)   1.14 net; 1.93 net
Findlay Publications Limited   5.316 uni
Finnish Research Library Association   5.107 fin
Fishery Data Centre   1.112 ita
Flora and Fauna Publications   1.342 usa
Florida Medical Association Incorporated   7.244 usa
Food and Agriculture Organization of the United Nations   1.53 ita; 1.108 ita; 1.110 ita; 1.120 ita; 1.175 ita; 2.15 ita; 3.35 ita; 3.36 ita; 4.12 ita; 6.8 ita; 6.17 ita; 6.23 chi
Food Trade Press Limited   5.250 uni
Foreign Medical School Information   1.322 usa
Franz Steiner Verlag Wiesbaden GmbH   5.59 gfr
Freie Universität Berlin   5.391 gfr
Fuel and Metallurgical Journals Limited   5.270 uni; 5.272 uni; 5.274 uni; 5.326 uni; 5.337 uni

Gale Research Company   1.1 usa; 1.11 usa; 1.12 usa; 1.23 usa; 1.24 usa; 1.44 usa; 1.301 usa; 7.9 usa; 7.26 usa; 7.29 usa; 7.30 usa; 7.31 usa; 7.32 usa; 7.33 usa; 7.39 usa; 7.40 usa; 7.49 usa; 7.53 usa; 7.54 usa; 7.55 usa; 7.57 usa; 7.58 usa; 7.62 usa; 7.67 usa; 7.127 usa; 7.130 usa; 7.242 usa; 7.250 usa; 7.253 usa; 7.256 usa
General Dental Council   5.361 uni
General Medical Council   5.386 uni; 5.388 uni
General Optical Council   5.393 uni
General Trade Directories Private Limited   1.185 sin
Geo Books   1.174 uni; 5.188 uni
Geographical Studies and Research Center   7.56 usa
Geological Society   5.181 uni
Georgia Tech Research Institute   7.25 usa
Geothermal World   1.210 usa
Government of India Department of Publications, Delhi   3.80 ind
Gower Publishing Company Limited   1.188 uni; 1.218 uni; 5.219 uni
Graham and Trotman Limited   1.284 uni; 2.27 uni; 3.66 uni; 3.87 uni; 3.88 uni; 3.89 uni; 3.90 uni; 5.217 uni; 5.318 uni; 5.319 uni; 5.320 uni
Guardian Communications Limited   5.334 uni
Guida Monaci SpA   5.376 ita

PUBLISHER INDEX

Hampshire Technical Research Industrial Commercial Service   5.61 uni
Harris Publishing Company Incorporated   7.132 usa; 7.185 usa; 7.186 usa; 7.187 usa; 7.188 usa; 7.193 usa; 7.196 usa; 7.212 usa
Hawaii Business Directory Incorporated   7.189 usa
Hawaii State Department of Planning and Economic Development   7.170 usa
Hayden Publishing Company Incorporated   1.200 usa
Health and Human Services Department   7.235 usa
Heating and Ventilating Contractors' Association   5.229 uni
Heating and Ventilating Publications (Developments) Limited   5.230 uni
Her Majesty's Stationery Office   5.9 uni; 5.210 uni
Hilger, Adam, Limited   5.89 uni
Hobsons Limited   5.22 uni; 5.32 uni; 5.54 uni
Hong Kong Productivity Centre   3.77 hng
Horticultural Education Association   5.123 uni
Hunt Institute for Botanical Documentation   1.338 usa

ICCROM (International Centre for Conservation)   1.41 ita
Iceland Review   5.294 ice
IFLA Office for International Lending   1.6 uni
IMS World Publications   1.310 uni; 1.311 uni; 1.347 uni
Indian Association of Special Libraries and Information Centres   3.19 ind
Indonesian National Scientific Documentation Centre   3.20 ino
Industrial Aids Limited   5.241 uni
Industrial and Marine Publications Limited   5.115 uni
Industrial Press   5.237 uni
INFA Publications   3.29 ind
Information Management and Consulting Association   4.11 aus
Information Resources Press   7.60 usa
Institut Français de l'Énergie   1.204 fra
Institut für Wissenschaftsinformation in der Medizin   5.390 gdr
Institute for Industrial Research and Standards   5.103 ire; 5.171 ire; 5.200 ire; 5.231 ire; 5.232 ire; 5.233 ire; 5.298 ire
Institute for Industrial Research and Standards, Industrial Research Centre   5.170 ire
Institute of Biology   5.364 uni; 5.394 uni
Institute of Geologists   5.182 uni
Institute of Grocery Distribution   5.277 uni; 5.278 uni
Institute of Health Services Management   5.383 uni
Institute of Hydrology   5.186 uni
Institute of Marine Engineers   1.235 uni
Institute of Offshore Engineering   5.227 uni
Institute of Physics   5.404 uni; 5.406 uni
Institute of Water Pollution Control   5.160 uni
Institute of Water Pollution Control   5.169 uni
Instituto Bibliotecnológico, Universidad de Buenos Aires   6.11 arg
Instituto Colombiano para el Fomento de la Educación Superior   6.3 col
Instrument Society of America   7.124 usa
Interavia SA   1.196 swi; 1.233 swi
Intercontinental Marketing Group   3.31 jap
Inter-ed srl   5.206 ita
Intermediate Technology Publications Limited   1.3 uni
International Agricultural Centre   5.122 net
International Association for Hydraulic Research   1.224 net
International Bee Research Association   1.90 uni

International Civil Aviation Organization   1.240 can
International Council of Scientific Unions   1.87 fra
International Energy Agency   1.207 usa; 1.208 usa
International Federation of Library Associations and Institutions   1.33 net
International Hydrographic Bureau   1.177 mon
International Mineralogical Association   1.181 fra
International Palaeontological Association   1.167 usa
International Planned Parenthood Federation   1.349 uni
International Research Services Incorporated   1.237 usa
International Sugar Journal Limited   1.297 uni
International Sugar Organization   1.298 uni
International Textile Manufacturers Federation   1.279 swi
International Trade Publications   3.52 uni
International Union against Cancer   1.334 swi
International Union for the Conservation of Nature and Natural Resources   1.166 swi; 1.169 swi
International Union of Crystallography   1.159 net
International Union of Forestry Research Organizations   1.118 aut
International Union of Independent Laboratories   1.66 uni
International Who's Who of the Arab World Limited   1.46 uni
Intertrade Publications Private Limited   3.3 ind
Iranian Documentation Centre   3.24 ira

Japan Chamber of Commerce and Industry   3.46 uni
Japan Society for the Promotion of Science   3.44 jap
Japan Pharmaceutical, Medical and Dental Supply Exporters Association   3.96 jap
JETRO   3.75 uni; 7.163 uni
John Martin Publishing   5.239 uni
Journal: Bulletin American Association of Botanical Gardens, Arbor 13, 33-44   3.54 usa
Journal: Canadian Special Publication of Fisheries and Aquatic Sciences, 73   7.81 can
Journal: Taxon. (29/5-6), 731-41   6.22 net
Judge, James J., Incorporated   7.180 usa

Kawata Publicity Incorporated   3.82 jap
Kelly's Directories   5.299 uni; 5.300 uni
Kentucky Department of Economic Development   7.199 usa
Keystone   7.143 usa
Klein, B. Publications   7.44 usa
Knowledge Industry Publications Incorporated   7.120 usa; 7.131 usa
Kogan Page Limited   1.48 uni; 1.85 uni; 1.238 uni; 1.252 uni; 1.255 uni; 5.10 uni; 5.125 uni; 5.324 uni
Kompass Belgium SA   5.302 bel
Kompass Deutschland Verlags- und Vertriebsges mbH   5.297 gfr
Kompass Embassy Information Proprietary Limited   3.85 sin
Kompass España SA   5.308 spa
Kompass-Indonesia   3.83 ino
Kompass Maroc   2.26 mor
Kompass Nederland BV   5.305 net
Kompass Norge A/S   5.307 nor
Kompass Publishers Limited   5.311 uni
Kompass Schweiz Verlag   5.310 swi
Kompass Sverige AB   5.309 swe
Kon Danmarks Handels-Kalendar   5.301 den
Koninklijke Nederlandse Chemische Vereniging   5.156 net
Korea Directory Company   5.34 kor
Kothari Publications   3.50 ind; 3.51 ind

PUBLISHER INDEX

Learned Information Limited   1.277 uni; 5.199 uni
Libarary Association - Reference, Special and Information Section   5.87 uni
Libraries Unlimited   1.22 usa; 1.32 usa
Library Association of Ireland and Library Association (Northern Ireland Branch)   5.35 ire
Library Association of Singapore   3.10 sin
Library Association Publishing   5.39 uni; 5.369 uni
Library of Congress   7.111 usa
Licht, F.O., GmbH   1.282 gfr
Loadstar Publications   1.242 uni
London and Sheffield Publishing Company Limited   1.148 uni
London and South Eastern Regional Advisory Council for Further Education   5.13 uni; 5.16 uni; 5.57 uni
Longman Group Limited   1.19 uni; 1.61 uni; 1.62 uni; 1.89 uni; 1.117 uni; 1.152 uni; 1.172 uni; 1.179 uni; 1.184 uni; 1.197 uni; 1.219 uni; 1.227 uni; 1.263 uni; 1.264 uni; 1.265 uni; 1.266 uni; 1.337 uni; 1.341 uni; 1.353 uni; 1.357 uni; 3.41 uni; 3.42 uni; 3.43 uni; 4.13 uni; 5.40 uni; 5.44 uni; 5.51 uni; 5.52 uni; 5.62 uni; 5.63 uni; 5.100 uni; 5.116 uni; 5.225 uni; 5.296 uni; 5.367 uni; 5.370 uni; 5.387 uni; 6.19 uni
Longman Professional/China Council for the Promotion of International Trade   3.76 uni
Lundy, J.W., Enterprises   7.10 usa

Machine Tool Industry Research Association   5.235 uni
Machine Tool Trades Association   5.252 uni
Maclaren Publishers Limited   5.234 uni
Macmillan Press   1.40 uni; 1.43 uni; 1.79 uni; 1.325 uni; 1.329 uni; 5.8 uni; 5.49 uni; 5.211 uni
Macmillan Publishing Company   1.80 usa; 1.81 usa; 1.343 uni; 1.350 usa; 7.70 usa; 7.245 uni
Magnum Publications Limited   1.303 uni
Mansell Publishing Limited   1.182 usa
Manufacturers' News Incorporated   7.155 usa; 7.156 usa; 7.157 usa; 7.158 usa; 7.160 usa; 7.162 usa; 7.164 usa; 7.167 usa; 7.171 usa; 7.175 usa; 7.176 usa; 7.177 usa; 7.183 usa; 7.184 usa; 7.190 usa; 7.191 usa; 7.192 usa; 7.194 usa; 7.200 usa; 7.201 usa; 7.202 usa; 7.204 usa; 7.205 usa; 7.206 usa; 7.207 usa; 7.209 usa; 7.210 usa; 7.211 usa; 7.216 usa; 7.219 usa; 7.222 usa; 7.224 usa; 7.225 usa
March of Dimes Birth Defects Foundation   1.331 usa
Market Research Society   1.278 uni
Marquis Who's Who Incorporated   1.68 usa; 1.116 usa; 1.198 usa; 1.199 usa; 1.285 usa; 7.68 usa; 7.223 usa; 7.238 usa; 7.254 usa; 7.255 usa; 7.258 usa
Martin, John, Publishers   1.217 uni; 1.254 uni
Maryland Department of Economic and Community Development   7.24 usa; 7.173 usa
McGraw Hill Publications   1.176 usa
McGraw-Hill Book Company (UK) Limited   1.54 uni; 7.133 uni
Medical and Health Department, Hong Kong   3.95 hng
Medical Research Council   1.335 uni
Melrose Press Limited   1.239 uni
Mental Health Foundation   5.101 uni
Merrow Publishing Company Limited   1.133 uni
Metal Bulletin Books Limited   1.260 uni; 1.262 uni; 1.267 uni; 1.281 uni; 1.286 uni; 1.287 uni; 1.289 uni; 1.290 uni; 1.294 uni; 1.296 uni; 1.304 uni; 1.309 uni; 5.279 uni
Metal Bulletin Journals Limited   1.250 uni
Metals Information   1.131 usa; 1.156 usa; 1.157 usa
Middle East Food   1.114 leb

Midwest Oil Register Incorporated   1.155 usa; 1.205 usa; 1.215 usa; 7.144 usa; 7.145 usa; 7.146 usa; 7.147 usa; 7.166 usa; 7.168 usa
Miller Freeman Publications Incorporated   1.150 usa; 1.307 usa; 7.83 usa; 7.215 usa
Ministry of Agriculture, Fisheries and Food, Directorate of Fisheries Research   5.126 uni
Modern Metals Publications Limited   5.339 uni
Morgan-Grampian Book Publishing Company Limited   1.253 uni; 5.195 uni; 5.246 uni; 5.247 uni; 5.260 uni; 5.269 uni; 5.313 uni; 5.327 uni
Morgan-Grampian (Process Press) Limited   5.243 uni; 5.276 uni
Municipal Publications   5.222 uni
Museums Association   5.74 uni
Museums Association of India   3.4 ind
Muzejski Dokumentacioni Centar   5.72 yug; 5.113 yug

National Academy of Sciences   7.117 usa
National Agricultural Library   1.109 usa
National Board of Health   5.378 fin; 5.385 fin; 5.400 fin
National Bureau of Standards   7.22 usa
National Central Library, Bureau of International Exchange of Publications   3.22 tai
National Centre of Scientific and Technological Information   3.14 isr; 3.15 isr; 3.21 isr; 3.78 isr
National Consultative Committee for Agricultural Education   5.120 uni
National Corrosion Service   5.158 uni
National Council for US-China Trade   3.17 usa
National Environmental Engineering Research Institute   3.8 ind
National Institutes of Health   7.261 usa
National Library of Australia   4.22 aus; 4.24 aus
National Museum of Wales   5.184 uni
National Oceanic and Atmospheric Administration   1.170 usa
National Oceanographic Data Centre   7.73 usa
National Referral Center   7.18 usa
National Science Council, Science and Technology Information Centre   3.18 tai
National Society for Clean Air   5.78 uni
National Technical Information Service   7.17 usa; 7.38 usa
National Wildlife Federation   7.8 usa
Natural Environment Research Council   5.84 uni
Natural Science for Youth Foundation   7.82 usa
Neale Watson Academic Publications Incorporated   7.11 usa
Nebraska Department of Economic Development   7.174 usa
Netherlands British Chamber of Commerce   5.267 uni; 5.317 uni
New Zealand Library Association   4.9 nze
Newman Books Limited   5.127 uni; 5.128 uni
Nik Ibrahim Kamil   3.86 may
Nils-Axel Mörner   1.168 swe
North of England Development Council   5.253 uni
Nuclear Energy Intelligence   1.355 uni; 5.405 uni

Office National d'Information sur les Enseignements et les Professions   5.90 fra
Office of Environmental Quality Control   7.35 usa
Office of Naval Research   3.32 usa; 3.33 usa
Oldham, Anne   5.176 uni
OMEC Publishing Company   7.6 usa; 7.37 usa
Online Publications   5.216 uni
Ontario Board of Examiners in Psychology   7.241 can

PUBLISHER INDEX

Optical Publishing Company Incorporated   1.201 usa; 7.128 usa
Organization for Economic Cooperation and Development   1.2 fra; 1.132 fra; 2.7 fra; 2.13 fra; 3.30 fra; 6.6 fra; 6.16 fra
Organizzazione Editoriale Medico Farmaceutica SRL   5.375 ita; 5.384 ita
Oriental Economist   3.81 jap
Oryx Press   7.23 usa; 7.46 usa; 7.79 usa
Ossolineum, Publishing House of the Polish Academy of Sciences   5.83 pol; 5.99 pol
Otto Vieth Verlag   5.205 gfr
Oxford University Press   1.320 uni; 1.344 uni

Pacific Information Centre, in association with University of the South Pacific Library   4.15 fij
Pacific Southwest Forest and Range Experiment Station, US Forest Service   1.111 usa
Pamposh Publications   3.94 ind
Paul Haupt Berne Publishers (Switzerland)   5.95 swi
PennWell Publishing Company   1.209 usa; 1.213 usa; 1.214 usa; 1.216 usa; 1.221 usa; 1.222 usa; 6.25 usa; 7.116 usa; 7.148 usa; 7.149 usa
Pergamon Press Limited   1.154 uni; 1.313 uni; 1.317 uni; 5.33 uni; 5.266 uni; 5.342 uni
Peter Isaacson Publications Proprietary Limited   4.29 aus; 4.33 aus
Peter Peregrinus Limited   5.45 uni; 5.46 uni; 5.161 uni; 5.197 uni
Peterson's Guides Incorporated   7.34 usa; 7.41 usa; 7.42 usa; 7.43 usa; 7.234 usa
Pharmaceutical Society of Great Britain   5.396 uni
Pharmaceutical Society of Ireland   5.359 ire
Phillips Publications Incorporated   1.202 usa; 7.114 usa
Pick Publications Incorporated   7.203 usa
Pitman Publishing Limited   5.85 uni
Plastics and Rubber Institute   5.174 uni
Plenum Publishing Corporation   7.36 usa
Plunkett Foundation for Co-operative Studies   5.124 uni
Pluto Press   5.97 uni
Pretoria State Library   2.9 saf
PSG Publishing Company Incorporated   7.227 usa
Public Affairs Clearing House   1.69 usa
Public Health Laboratory Service   5.395 uni
Publishing House of the Bulgarian Academy of Sciences   5.12 bul
PUDOC, Centre for Agricultural Publishing and Documentation   1.94 net

Queensland Department of Primary Industries   4.19 aus

Rapra Technology Limited   1.134 uni; 5.167 uni; 5.295 uni
Rauch Associates Incorporated   7.182 usa
Reed, A.H. and A.W., Limited   4.1 nze
Regional Development Authority for East-Flanders (GOMOV)   5.242 bel
Regional Institute of Higher Education and Development   3.28 sin; 3.40 sin; 3.55 sin
Repro Incorporated   7.93 usa
Rhys Jones Marketing   1.236 uni
Riksbibliotektjenesten   5.77 nor
Routledge and Kegan Paul   5.23 uni
Royal College of Veterinary Surgeons   5.135 uni
Royal National Institute for the Blind   5.362 uni

Royal Norwegian Council for Scientific and Industrial Research   5.76 nor
Royal Society   5.190 uni; 5.191 uni
Royal Society of Chemistry   5.159 uni
Rutherford Appleton Laboratory   5.196 uni

Saur Verlag, K.G.   1.4 gfr; 1.35 gfr; 1.36 gfr; 1.42 gfr; 1.56 gfr; 1.75 gfr; 1.76 gfr; 1.77 gfr; 1.78 gfr; 1.84 gfr; 1.86 gfr; 2.1 uni; 2.2 gfr; 2.14 uni; 2.17 uni; 5.58 gfr; 5.60 gfr; 5.137 gfr
Savory Milln and Company   1.226 uni; 5.382 uni
Schlütersche Verlagsanstalt und Druckerei GmbH und Co   5.119 gfr
Schwabe and Company Limited   5.397 swi
Science and Engineering Research Council   5.177 uni; 5.180 uni; 5.371 uni
Science and Planning Directorate, Department of Constitutional Planning   2.16 saf
Science and Technology Agency   3.26 jap
Science Council of Japan   3.9 jap; 3.23 jap
Science Information Publishing Centre, Department of Scientific and Industrial Research.   4.10 nze
Science Museum   5.73 uni
Science Planning Directorate, Department of Constitutional Development and Planning   2.31 saf
Science Reference Library   1.300 uni; 5.343 uni
Scientific Liaison Office   2.19 zim
Scientific Surveys Limited   1.220 uni
Scottish Development Agency   5.98 uni
Seibt Verlag GmbH   5.398 gfr; 5.399 gfr
Sharpe, M.E.   3.13 uni
Simon Books Limited   5.118 uni
Sittig and Noyes   1.323 usa
Skinner, Thomas, Directories   5.27 uni
S.M. Bryde A/S   5.323 nor
Society of Authors   5.108 uni
Society of Forestry in Finland   5.134 fin
Solar Energy Research Institute   1.212 usa
Solar Energy Unit, University College   5.208 uni
Southam Communications Limited   7.95 can; 7.214 usa; 7.232 can
Southern African Museums Association   2.12 saf
Spearhead Publications Limited   5.218 uni
Special Libraries Association, Pittsburgh Chapter   7.27 usa
Special Libraries Association, Toronto Chapter   7.28 can
Spon, E. & F.N.   1.234 uni; 1.288 uni
Springer-Verlag   5.204 gfr; 5.360 gfr; 5.401 gfr
SRI International   7.106 usa
Staatsuitgeverij (J. Nijland)   5.75 net
Stamex bv   5.153 net; 5.154 net; 5.162 net; 5.164 net; 5.165 net; 5.271 net; 5.344 net
Standard and Poor's Corporation   1.295 bel
State of Hawaii Department of Planning and Economic Development   7.172 usa
Statistics Canada, Department of Supply and Services Canada   7.21 can; 7.178 can
Strand Publishing Proprietary Limited   4.17 aus
Surveyors Publications   5.179 uni
Swiss Academy of Sciences   5.96 swi
Swiss Office for the Development of Trade   5.340 swi

Technical Database Corporation   7.129 usa
Technical Indexes Proprietary Limited   4.26 aus; 4.28 aus
Technology and Business Communications Incorporated   7.125 usa

PUBLISHER INDEX

Telford, Thomas, Limited   5.212 uni
Thai National Documentation Centre   3.16 tha
Thames and Hudson   5.374 uni
Thebal-Verlag   5.353 gfr; 5.354 gfr; 5.355 gfr
Thomas Skinner Directories   1.47 uni
Thompson Publications   2.23 zim
TNO Corporate Communication Department   5.109 net; 5.110 net; 5.168 uni; 5.275 uni; 5.283 uni; 5.284 uni; 5.285 uni; 5.286 uni; 5.287 uni; 5.293 uni; 5.325 uni; 5.329 uni; 5.330 uni; 5.331 uni; 5.336 uni; 5.346 uni; 5.379 uni; 5.380 uni
Treasure Island Publishing Incorporated   7.119 usa
Tropical Development and Research Institute   5.64 uni
Turret-Wheatland Limited   5.172 uni; 5.350 uni

UNESCO   1.70 fra; 1.71 fra; 1.72 fra; 1.175 ita; 3.6 fra;
Unesco Regional Office for Education in Asia and the Pacific   1.15 tha; 1.321 tha; 3.25 tha
Unesco, Regional Office of Science and Technology for South and Central Asia   3.11 fra
UNISAF Publications Limited   1.293 uni; 5.373 uni
United Nations Economic Commission for Latin America   6.5 chi
United Nations Education, Scientific, and Cultural Organization   1.9 swi; 1.37 swi; 1.50 fra
United Nations Industrial Development Organization   1.16 aut; 1.96 aut; 1.97 aut; 1.98 aut; 1.99 aut; 1.100 aut; 1.101 aut; 1.102 aut; 1.103 aut; 1.104 aut; 1.105 aut; 1.106 aut; 1.107 aut; 1.135 aut; 1.136 aut; 1.137 aut; 1.138 aut; 1.139 aut; 1.140 aut; 1.141 aut; 1.142 aut; 1.143 aut; 1.144 aut; 1.145 aut; 1.146 aut; 1.147 aut; 1.195 aut; 1.206 aut; 1.211 aut; 1.230 aut; 1.231 aut; 1.232 aut; 1.257 aut; 1.269 aut; 1.270 aut; 1.271 aut; 1.272 aut; 1.273 aut; 1.274 aut; 1.275 aut; 1.276 aut; 1.283 aut; 1.306 aut; 1.328 aut; 1.332 aut; 2.5 aut; 2.20 aut; 2.24 aut
United Nations Organization   1.21 swi; 1.51 swi; 1.191 swi; 1.192 swi
United Nations Publications   1.180 usa
United States Department of Agriculture, Eastern Regional Research Center   7.89 usa
United States Department of Health, Education and Welfare   1.316 usa
University Grants Commission   3.49 pak
University of Arizona, Mathematics Department   7.112 usa
University of Arizona Press   1.162 usa

University of Colorado, Bureau of Business Research   7.165 usa
University of Exeter, Agricultural Economics Unit   5.133 uni
University of Hong Kong Centre of Asian Studies   3.2 hng; 3.7 hng
University of the State of New York, State Education Department   7.237 usa
US Department of Agriculture, Forest Service   7.80 usa; 7.84 usa
US Department of Commerce   7.15 usa; 7.16 usa; 1.318 usa; 1.319 usa; 7.233 usa; 7.262 usa
US Geological Survey   1.183 usa
US Government Printing Office, Superintendent of Documents   7.77 usa

Van Loghum Slaterus   5.88 net
Verkfraeoingafélag Islands   5.238 ice
Verlag Chemie GmbH   5.141 gfr
Verlag Dieter Göschl   5.381 aut
Verlag für Wirtschaftsliteratur GmbH   5.255 swi; 5.281 swi; 5.288 swi; 5.312 swi; 5.314 swi
Verlag Glückauf GmbH   5.187 gfr
Verlag Otto Harrassowitz   5.68 gfr
Veterinary Council   5.138 ire
Vieweg, Friedr., und Sohn Verlagsgesellschaft mbH   5.112 gfr
Virginia Chamber of Commerce   7.220 usa
VNU Business Publications bv   5.256 uni; 5.321 uni; 5.338 uni

Walker's Manuals   7.221 uni
Walter de Gruyter   1.122 gfr; 1.129 gfr; 1.130 gfr; 5.69 gfr; 5.104 gfr; 7.5 gfr
Welding Institute   5.347 uni
Who's Who International Red Series   5.117 gfr; 5.402 gfr
Wiley, John, and Sons Limited   1.193 uni; 1.194 uni; 5.389 uni
Wing Aviation Press   3.65 jap
Working Weekends on Organic Farms   5.139 uni
World Data Center, University of Queensland   1.345 aus
World Health Organization   1.346 swi
World Meteorological Organization   1.178 swi
Zambian Industrial and Commercial Association   2.30 zam
Zentralstelle für Agrardokumentation und -information   5.129 gfr
Zoological Society of London   1.339 uni